"CLEAN" ENERGY EXPLOITATIONS

"CLEAN" ENERGY EXPLOITATIONS

Helping Citizens Understand the Environmental and Humanity Abuses that Support "Clean" Energy

RONALD STEIN / TODD ROYAL

Archway Publishing books may be ordered through booksellers or by contacting:

Archway Publishing
1663 Liberty Drive
Bloomington, IN 47403
www.archwaypublishing.com
844-669-3957

ISBN: 978-1-6657-0497-7 (sc)
ISBN: 978-1-6657-0496-0 (hc)
ISBN: 978-1-6657-0495-3 (e)

Library of Congress Control Number: 2021906299

Print information available on the last page.

Archway Publishing rev. date: 05/28/2021

CONTENTS

- *We get very excited when we "see" EV's powered by batteries and clean intermittent electricity being generated from wind and solar. What we do not see is the environmental degradations from the origins of the minerals and metals around the world required to support and manufacture all that clean electricity.*

CHAPTER 5 *HUMANITY ATROCITIES* 167

- *Humanity abuses mining for the minerals and metals to support solar, wind, and EV batteries are occurring worldwide.*
- *We get very excited when we "see" EV's powered by batteries and clean intermittent electricity being generated from wind and solar. What we do not see is the environmental degradation from the origins of the minerals and metals required to support and manufacture all that clean intermittent electricity.*

CHAPTER 6 *CHINA AND INDIA* 217

- *The two largest countries in the world desperately need affordable, reliable, scalable, abundant, and flexible sources for electricity.*
- *With more than 35 percent of the world's 8 billion population, they have amassed the most coal fired power plants and are building more at a record pace.*

CHAPTER 7 *FINANCIAL AND ENVIRONMENTAL RACIAL BIASING* 253

- *Energy and environmental policies impose economic stresses for energy on billions of low-income and minority communities that can least afford it.*
- *The financial racial biasing of climate and environmental policies against the poorest residents of the world is unconscionable.*

CHAPTER 8 *SUMMARY OF THE 2020 ELECTIONS IMPACT ON THE ECONOMY AND ENVIRONMENT* 287

Endnotes 293

INTRODUCTION

Here is the most important fact about today's environmental movement, and the clean energy exploitations this book will explore:

- The United States of America, the largest economy in the history of mankind, representing 4 percent of the world's population (330 million vs 7.8 billion) could literally shut down, and cease to exist, and the opposite of what you have been told and believe will take place. Simply put, in the United States, every person, animal, or anything that causes emissions to harmfully rise could vanish off the face of the earth; or even die off, and global emissions will still explode in the coming years and decades ahead over the population and economic growth of China, India, and Africa.[1]

That is right - the U.S. could cease to exist, and global emissions from the world's approximately 7.8 billion population will still rise. It is why the American or European version of the Green New Deal will never work or lower emissions.[2] The entire thing is based on intermittent electricity from solar panels and wind turbines, which also do not come close to working as advertised.[3] But still the green energy exploiters continue their quest. The question is why?

With the election of former Vice President Joe Biden to the Presidency of the United States in 2021, more than 150 world leaders, including Amazon boss Jeff Bezos, Salesforce CEO Marc Benioff, and Ford Motor Company executive chairman Bill Ford, signed an open letter to President Joe Biden to pledge their support to the new administration's goals of combatting climate change. In their Open Letter titled "President Biden: You Can be The Climate President" those wealthy leaders mentioned many of the issues with their own conclusions. [4 5]

A few points made in the Open Letter that are discussed at length in this book are:

1. The letter refers to "*the devastating global health and economic crises*"
    - **Chapter One** in this book introduces everyone to the environmental degradation and humanity abuses that support "clean" energy (wind turbines and solar farms defined as "renewables").

2. The letter then invokes the mantra of "*transforming the world's energy and electrical grids into platforms of decarbonization*": meaning all fossil fuels (oil, petroleum, diesel, jet fuel, natural gas, and coal) will be eliminated.
    - **Chapters One, Two, and Three** of our book discusses at length how California, Germany, and Australia have taken the lead to go "green", and now vie for having the highest electricity costs in the world. Please understand that the world starts and stops with electricity. Without it, you cannot have modern life, health, wealth, or basic security.

3. Mr. Bezos, Benioff, and Ford rightfully implore how "*reducing harmful pollution and tackling economic, racial,*

and health inequality" within the umbrella of climate change is the duty of mankind.

- A wonderful mantra except **Chapters Four and Five** show the global environmental degradation and human atrocities committed for solar panels, wind turbines, electric vehicle car batteries, and battery energy storage systems for electrical grids need rare earth metals and minerals that have nothing to do with human equality or the betterment of mankind.

4. All three billionaires, Bezos, Benioff, and Ford then applaud President Biden for "*rejoining the Paris Climate Agreement on day one*" of the new administration without understanding.
    - Our **Chapter Six** tackles the vexing issue of China and India. How neither country has financial accountability or enforcement mechanisms to keep them from being the largest emitters in the world and how both are destroying their nations over pollution and environmental degradations.

5. The theme of justice rears its rightful head in the letter over "*restoring justice and building an equitable, inclusive, and just future for all*".
    - **Chapter Seven** adjudicates justice in clean energy within the framework of financial and environmental racial biasing that energy and environmental polices promoting clean energy imposes on the billions without reliable electricity and the high costs of renewables on the low-income and minority communities that can least afford energy racism and inequality.

6. Each billionaire then ends with "*the legacy of environmental injustice*" in the letter that demands action for the sake of science and future demands on a global economy.
    - **Chapter Eight** rightfully discusses the election impact of the U.S. moving from a pro-fossil fuel administration to the Biden's administration's all-in for climate change policy movement and what that means for the U.S., Europe, and western-aligned nation's economy and overall environmental health.

But more importantly, the 150 signatories of this letter and President Biden need to comprehend is that energy is more than intermittent electricity from wind, solar, and EV batteries. Ever since the discovery of the versatility of products available from petroleum derivatives, and the beginning of manufacturing and assembly of cars, truck, airplanes, and military equipment, the world has had almost 200 years to develop clones or generics to replace the crude oil derivatives that are the basis of more than 6,000 products we use such as: medications, electronics, communications, tires, asphalt, and fertilizers.

The social needs of our materialistic societies are most likely going to remain for continuous, uninterruptable, and reliable electricity from coal, natural gas, or nuclear energy to electricity. Otherwise, the chemicals derivatives and over 6,000 products that get manufactured out of crude oil, which is the crux of our daily lifestyles and economies will cease to exist.

Green advocates often say they support sustainable and ethical coffee, sneakers, handbags, and diamonds. No child labor sweat shops or unsafe conditions tolerated. But it is a different story with green energy, battery energy storage systems for electrical grids and electric vehicles (EVs).

Renewables have a role in our energy usage, but we need to consider what they can do, and what they cannot. Science shows that wind and solar can generate electricity, but science also reveals wind and solar cannot manufacture the oil derivatives that are the basis of more than 6,000 products that we did not have before 1900. Indeed, virtually every aspect of our daily lifestyles and modern economics on every continent and region of the world, are based on the oil derivatives for those thousands of products we now take for granted.

Let us believe in the undisputable science. Renewables can only generate electricity, and intermittent electricity at best. The undisputable science is that renewables CANNOT manufacture any of the oil derivatives that is the basis of a hydrocarbon-based world. Without oil as an example, the coronavirus would ravage the world unchecked: the COVID-19 vaccine would not work without oil and fossil fuels.[6]

Imagine how life was without the oil industry that did not exist before 1900 when we had, NO medications and medical equipment, NO vaccines, NO water filtration systems, NO sanitation systems, NO fertilizers to help feed billions, NO pesticides to control locusts and other pests, NO communications systems, including cell phones, computers, and iPads, NO vehicles, NO airlines that now move 4 billion people around the world, NO cruise ships that now move 25 million passengers around the world, NO merchant ships that are now moving billions of dollars of products monthly throughout the world, NO tires for vehicles, and NO asphalt for roads, and NO space program. Looking back just a few short centuries, we have come a long way since the pioneer days. This is only scratching the surface of what the world did not have before 1900.

Our second book released in 2020: **Just GREEN Electricity – *Helping Citizens Understand a World without Fossil Fuel*** explains that no one has the answer at this time as to how we can continue to make products and move things without fossil fuels and the derivatives from petroleum.[7] Most of all, the "aha moment" has arrived – electricity alone from wind and solar cannot replace the fuels and products that are manufactured from petroleum.

People across the world are repeatedly told from the media, movie and television stars, media influencers, most universities, think tanks, non-profits, non-governmental organizations movies, television, streaming services, legal systems, and many multinational corporations to stop every human activity imaginable over anthropogenic (man-made) global warming/climate change and only use the wind and sun for intermittent electricity. What should be understood is the era of the "deadly green hypocrite is in play," since "huge amounts of CO2 are emitted at every stage of the Green energy manufacturing process."[8]

The deadliest part to understand about environmental injustices taking place from the U.S. to China – the power of green political interests – organizations like Greenpeace, Friends of the Earth, Extinct Rebellion, and Antifa – want to eliminate all exploration and production drilling (E&P) for oil, petroleum, natural gas, and coal using hydraulic fracturing (fracking). Add pipelines, hydroelectric plants using dammed water, nuclear generation of zero emission electricity are all slated for eradication – replaced by mined rare earth minerals and metals needed for transitioning to EVs, solar panels, wind turbines, and battery energy storage systems for electrical grids.[9]

There are 62,500 power plants around the world operating today, all types, generating electricity for the world's inhabitants.[10] Of that total, more than 2,449 are coal-fired power plants. Over half (1,363) of the world's coal power plants (2,449) are in China and India whose populations of mostly poor peoples is roughly 2.7 billion. Together they are in the process of building 284 new ones of the 546 total.[11] [12]

Insidiously, the work is done in parts of the world less regulated, and where loss of life does not mean the same thing the way it does in the U.S., European Union (EU), or western-aligned capitals. Vulnerable populations are then exploited to import the exotic minerals and metals to build wind turbines, solar panels, and batteries for EVs, without the benefit of profiting from the over 6,000 products that come from a barrel of crude oil.[13]

Consider most of the world's rare earth minerals needed for solar panels, wind turbines, electric vehicle batteries and battery energy storage systems for electrical grids are mined near Baotou, Inner Mongolia. This process involves pumping toxic acid into the earth while processing it further with additional chemicals and acids. One ton of rare earth metals and minerals releases upwards of 420,000 cubic feet of toxic gases, 2,600 cubic feet of acidic wastewater into water reserves and over a ton of radioactive waste. What Mongolia is left with is black sludge, lifeless lakes and local Mongolians and workers suffering from severe skin and respiratory disease – additionally children have been shown to have soft bones with soaring cancer rates.[14]

Lithium is the main rare earth mineral needed for the clean energy transition; however, lithium is mainly extracted from Tibet, and the highlands of Argentina-Bolivia-Chile known as the "lithium triangle." Far from Tibet being known for the Dali Lama now this majestic country is producing abundant amounts of lithium coupled with dead, toxic fish, and carcasses of cows and yaks within the Liqi River. The Ganzizhou Rongda mine has poisoned this Tibetan river. Unsophisticated, native peoples in the lithium triangle now complain of contaminated streams needed for human consumption, livestock watering, irrigation systems with mountains left desolate over discarded salt from the lithium brining process.[15]

Cobalt is second in line when it comes to importance of rare earth minerals needed for solar panels, wind turbines, and all types of batteries to exist; but the world's top producer, the Democratic Republic of Congo has at least 40,000 children – some as young as 4-years old – working with their parents for less than $2 a day. Cave-in's, constant exposure to toxic, radioactive water, dust, and dangerous air loaded with cobalt, lead, and uranium with other heavy metals breathed into lungs day-after-day so western citizens can feel good about their Tesla or wind turbine. Cobalt ore is sent to China since one of the larger mines in the Congo is Chinese-owned Congo Dongfang International Mining Company.[16]

The above three paragraphs describing these injustices is just to meet current raw material requirements for wind and solar intermittently generated electricity, and for battery energy storage systems for electrical grids, and EV batteries. Try to picture the raw material demands, Third World mining and child labor conditions, and ecological destruction, under the humongous demands of the Green New Deal.[17]

Global advertising touts the virtue of cleaner and greener western nations. They are nowhere near clean or green. Essential products for so-called carbon-free societies, decarbonization of the electrical sector, or a just, green world is coming off the backs of some of the most oppressed people on earth.[18] Current mining practices will cause grotesque human rights abuses to proliferate during the global clean energy transition.

The least of the problems caused by Green zealots are electrical grid blackouts, which further chapters will examine.[19] Physics and economics instead of prudent policy decisions dictate more blackouts, higher electric rates (hurting the poor and vulnerable), hundreds of billions in new battery energy storage systems for electrical grids to meet Renewable Portfolio Standards (RPS), and excessive land-use requirements, which will wreck global biodiversity.[20] Moreover, the inefficient ways coal and gas-fired power plants, and nuclear energy plants sit at a simmering standstill to back up intermittent solar and wind electricity farms is a disaster in the electricity-for-life-mix. However, should the public have an expectation for reliable electricity when so many are suffering from green energy?

Yes, because continuous uninterruptible electricity is essential for any type of economic activity, and the ability to counter the Green movement killing off entire countries.[21] Most aspects of increasingly sophisticated electrical grids use of information technology, computerized controls, and sensitive electronics ensures the need for on-demand electricity. For electrical grids to function the entire system needs the ability to meet load demand. Deficiencies or shortfalls affects every person on the planet – the documentary film, *Juice: How Electricity Explains the World* does a visually stunning job highlighting how life begins and ends with electricity from stable grids.[22]

If current and future investments are not made towards upgraded generation stations, storm-resilient transmission lines, and effective distribution systems, then costs will skyrocket for consumers and businesses. Ultimately, power unreliability and expensive industrial processes are the norm. Then competitive reductions in a post-COVID world flourish unabated, household incomes plummet and the clarion call for more renewables (solar panels and wind turbines), battery energy storage systems for electrical grids, and EVs that destroy the environment and kill humanity are the model for green havoc spreading globally.[23]

When public expectations are considered then what does that mean for daily electricity and the push for a destructive green transition? Vulnerable people's and nations should groan. Nearly all infrastructures rely on electricity, and modern commerce from advanced to emerging economies depend on functional grid networks. Allowing climate change to overtake these realities is gross negligence on the part of voters and elected officials.[24]

Underperforming continents like Africa to the failed state status of Venezuela extracts costs in these industry sectors: manufacturing, health care, data centers, finance, insurance, real estate, construction, retail trade, transportation services, mining, utilities, agriculture, education, and entertainment. What takes place is the UN blames climate change and calls for more rare earth mineral extraction in China.[25] It is a failed strategy that plummets countries like the Congo and Haiti into further despair. They need new electrical grids, powered by coal-fired power plants; not a wind turbine or solar panel farm that is unreliable and intermittent.

It is also on the backs of African lives these minerals and metals will be mined. Shouldn't African lives matter too and allow extensive exploration and production (E&P) – the same way America and Europe did over a 150 years ago during their Industrial Revolutions?[26] The same could be said for poor Chinese, rural Indians in India, and South Americans, over the drudgery of lithium mining so wealthy westerners can have solar panels, wind turbines, battery energy storage systems for electrical grids, and EVs to feel good about themselves environmentally.[27] Africans, Chinese, Indians, South Americans, et al., all deserve the life-giving opportunity presented from fossil fuels and nuclear over deadly minerals and metals that support renewables, and the utter failure they present towards building a modern society.

Using the U.S. as an example, and the absurdity to imagine Americans relying on renewables for their electrical grid. Why – because the American grid is in shambles, and does any sane person believe U.S. citizens will allow darkness to cover their days and a return to a world without fossil fuels?[28] Most electric transmission and distribution (T&D) lines in the U.S. were constructed in the 1950's and 1960's with a 50-year life expectancy. Fifty years later most of them, 450,000 miles of transmission lines connecting 5,800 major power plants across the country, have not been replaced.

Permitting for new power lines typically needs 31 permits from 22 different Federal Agencies with this process taking up to ten years to build a line that can be built in one year. Our second book: ***Just Green Electricity: Helping Citizens Understand a World Without Fossil Fuels*** deeply examines the intricacies of electricity, and the tens of trillions of dollars it will take to construct new electrical grids in the U.S. and across the world.[29]

Just in California an aging electrical grid that consistently causes out-of-control wildfires have transmission lines over 50-100 years old, and an attempt by the California Democratic Party to shift all electrical generation to intermittent, industrial solar and wind farms is failing.[30] Does the public expect reliable electricity under these circumstances and a permitting process that will take at least a decade or more? An all-electric economy will require a new grid wherever these policies are attempted.

Additionally, Federal permits (before acquiring State, County, City, Town, Native American tribal land permits, or Villages) the builder of said electrical grid T&D's will need private property easement from owner(s), permission to cut down dangerous trees impeding newly constructed lines, railroad permits, highway permits, joint and common use pole permits/agreements, and wire crossing permission of other utilities, gas companies, and fiber optics for cell phones and high-speed internet buy-in.[31]

These are just some of the reasons why misguided environmentalists, conservationists, nature-lovers, or people who do not want anything new built near them are generally opposed to fracking. It is unlikely fracking will ever be fully banned unless there is a suitable replacement for oil, petroleum, and natural gas that fracking exploration and production (E&P) unlocks from shale rock formations. More fossil fuels allow more electricity to be used and then will involve new grids being built to handle additional generation.

Moreover, Fracking provides trillions of dollars in economic activity without being nearly as invasive as an updated, or newly built electrical grid.[32] But western environmentalists are doing everything in their power to slow the permitting process while touting an all-electric society; whereas China makes a big show they are all-in for renewables.

The exact opposite is taking place in China as they are massively pushing and building coal-fired and nuclear power plants. China, with thousands of coal-fired power plants, already has

46 nuclear reactors in operation and 11 more currently under construction to provide continuous uninterruptible zero-emission electricity.[33] Evidence also points to the fact the Chinese Communist government is reducing western influence (the U.S. and NATO) in the Middle East to secure Middle Eastern oil supplies "on a scale dwarfing current and probable future wind energy output."[34] A Green New Deal or total electrification of society while punishing fossil fuel exploration and production (E&P) means oil, natural gas, and possibly coal will come from the Middle East, Russia, China, Mexico and Canada.

The U.S. Democratic Party and European Union have pushed the Green New Deal based on the premise mankind is uncontrollably warming the earth, and the solution are renewables backed up by battery energy storage systems for electrical grids. Say that is the case, how do they propose to overcome 81-85 percent of current electrical generation globally comes from fossil fuels, another 5-9 percent from nuclear, and the remainder from wind, solar, hydroelectric, and biomass?[35]

We are talking about a staggering human cost to mine, process, transport the exotic minerals and metals to support wind turbines, solar panels, battery energy storage systems for electrical grids, and EV batteries that are trying to displace an entire global economy based on fossil fuels.[36] Still, the environmentalists of the U.S. Democratic Party, and its allies in capitals from Paris to Brussels persist without examining the billions of lives suffocated for environmental purity that will never exist.[37] Even globally famous environmentalists like Bjorn Lomberg who embrace carbon taxes, and mankind is heating the earth uncontrollably says there are bigger issues to tackle other than climate change like malaria that kills millions of Africans every year.

Bjorn Lomberg's new book, *False Alarm: How Climate Change Panic Costs Us Trillions, Hurts the Poor, and Fails to Fix the Planet* contends the way to build prosperous, healthy nations and continents is with prodigious economic growth.[38]

God Bless him for saying CO2 will rush upwards no matter how many solar panels or wind turbines are deployed for electricity.

Time Magazine's Environmentalist of the Year, Michael Shellenberger has a new book, *Apocalypse Never: Why Environmental Alarmism Hurts Us All* essentially drives home the same point as Bjorn Lomberg's book.[39] That climate change is real but does not come close to our most serious environmental problems. Lack of continuous, reliable electricity, access to education, birth control, food security and overall health continue as issues that wreck the environment more than undefined anthropogenic (man-made) global warming/climate change. But the western environmentalists backed by their governments and the UN marches forward without examining the facts of the environment. Or, how to care for the earth while nurturing economic growth and the 6,000 products that come from a barrel of crude oil. Not to mention "Big Oil" is curing the coronavirus pandemic.[40]

Besides the human rights crisis caused by the mining for rare earth minerals that support decarbonization, climate risk assessments have been categorically wrong for decades. As an example, take the book about New Zealand (a country the size of the U.S. state of Colorado) titled, *First National Climate Change Risk Assessment for New Zealand*.[41] The authors admit this report is not solely for New Zealand, but the world. Many problems persist throughout this climate report, which does not assess climate, but makes political assertions left for elected officials. The breathtaking scope of the report asserts it can predict weather in 2050 out to 2100 while identifying all significant risks of human caused anthropogenic (man-made) global warming/climate change under the guise of computer models.

Never mind that most westerners suffer internet outages when too many people are on their cell phone network, but a computer model can predict precipitation, temperatures decades ahead, and the thousands of variables that cause climate.[42]

Now imagine you are a starving African or South American undergoing environmental persecution in a collapsing mine so a faux-virtuous westerner can drive a Tesla in Paris or tout their green credentials by having solar panels on the roof of their home in Los Angeles. The fallacy of this report from a small country is phenomenal in its environmental arrogance and racism towards people of color. This book will go into greater detail in coming chapters over this very subject – the subjugation, racism, and anthropogenic (man-made) global warming/climate change -slavery billions are under to satisfy westerners misguided virtue and environmental guilt.

This anthropogenic (man-made) global warming/climate change -slavery, western-based environmental lobby stretches into other areas of public and private policy. No one of sound mind or logic doubts how the coronavirus has altered most parts of human existence. Here are a few examples to consider. For the environmentalist they have always wanted people out of emission-belching cars in favor of public transit systems. U.S. transit numbers have significantly declined since COVID-19, because people are instructed to keep 6-feet apart.[43] Buses, subways, light rail, and all forms of public transportation systems favors density over spacing.

Washington D.C. is now leading the charge globally to rid the U.S. and world of suburbs.[44] Favoring high-density apartment buildings over single-family homes. Bill Gates who has an enormous mansion, other large homes, private jets, and all the trappings of being a successful billionaire has now embraced public transportation and suburb-ridding fallacy, because of anthropogenic (man-made) global warming/climate change.[45] It is important to understand these people want you to change – from your choice of vehicle to where and how you live. There is a climate fraud of epic proportions taking place on a scale greater than the 2007-09 global recession.[46] When JP Morgan in their *Annual Energy Paper* documented that making climate change predictions, and advocating for renewables, energy policies that benefit every human on the planet is broken, and in need of repair.[47]

The wretched absurdity behind a computer software mogul and bankers commenting on energy policy and climate change is they have absolutely zero idea – or don't care – because of personal, or corporate investments that renewables are the worse form of intermittent electricity possible.[48] Every single wind turbine and solar panel is a waste of money at this time.[49] Every new solar panel or wind turbine added to any electrical grid results in higher costs, blackouts (see technologically advanced California for verification), and less reliability since coal-fired, gas-fired, or nuclear power plants, which always run 24/7/365 are generating continuous uninterruptible electricity over intermittent electricity from renewables.[50]

Do Bill Gates and JP Morgan care, or understand, that sunny climates have bigger problems when using solar panels for electricity. A new study in February 2020th edition of the *Journal of Renewable and Sustainable Energy* conclusively found solar panels degrade faster than previously thought?[51]

Now that Bill Gates is an environmentalist and believer in anthropogenic (man-made) global warming/climate change what is his answer to this dilemma? Likely he, like other billionaire environmentalists such as Tom Steyer of California never will answer the question for how to overcome the technologically inferior nature of wind turbines and solar panels. The bright spot is the U.S. is rediscovering the role of lower cost, carbon-free, continuous, uninterruptable electricity that nuclear power plants provide.[52]

In a mad-mad post COVID-19 world, any environmentalist, non-governmental organization, non-profit, or billionaire claiming they care for the poor and victimized over anthropogenic (man-made) global warming/climate change while railing against carbon dioxide gas and not promoting carbon-free nuclear power sounds unhinged or self-interested in taxpayer subsidies renewables consistently receive. The energy transition, which overwhelmingly benefits the wealthy and well-educated has disadvantages for anyone without the means to qualify for taxpayer subsidized renewables. Increased costs from this energy transition results in clean energy exploitation, environmental injustice, and labor use atrocities that are sweeping the globe through energy poverty, income inequality, and household economic stagnation.[53]

Thus, it is radical, foolish, and dangerous to miners and children in poor countries to secure energy reliance on renewable electricity. Particularly, when discarded solar panels dumped in landfills are full of toxic "gallium arsenide, tellurium, silver, crystalline silicon, lead, cadmium, and other heavy metals."[54] Wind turbines spanning 45-70 meters apiece and weighing between 10-25 tonnes are either being ground up into concrete and used in the base-anchoring system of other turbines (a process that exerts tremendous amount of energy and CO2 emissions), or simply dumped in landfills.[55] The turbine blades are highly toxic since they contain Bisphenol A; the EU and Canada have banned this type of blade from entering landfills.[56]

CONCLUSION OF THE INTRODUCTION

Why is this type of environmental crusade so destructive? Why has it gone from great intentions to utter destruction? They are now the worst form of environmental nihilism morphed into ideological fervor the world has not witnessed since the rise of National Socialism in Germany in the early 1930s. Adolf Hitler appeared on the cover of Time Magazine on multiple occasions, and most famously, on January 2, 1939, when he was named their Man of the Year. The mainstream media endorsed National Socialism, and it is now promoting globally destructive green transition on all seven continents.

If this movement and its leaders like Al Gore, Tom Steyer, and the entire western environmental-political-cabal is not exposed, (what this book is attempting), and stopped we will witness death and destruction on a global scale not seen since World War II over their all-out attempt to eliminate all forms of fossil fuels and emission free nuclear-generated electricity.

But environmentalists keep blindly chugging ahead witnessed by the possibility of World War III breaking out between India and China. A conflict that has been brewing for over 50 years and is now playing out along their long-disputed border. Environmentalists for their part are now preoccupied with COVID-19, anthropogenic (man-made) global warming/climate change, renewables and shaming the west over anything that made it the beacon of hope for hundreds of years.

Whereas China and India are engaged in an "unprecedented military build-up."[57] Strategic ambiguity on China's part has now forced India to assertively seek closer ties with the U.S. and an anti-China alliance under the guise of the Quadrilateral Security Dialogue that includes the U.S., Japan, and Australia.[58] India assertiveness from New Delhi now signals a willingness to decouple economically from China. No serious foreign analyst expects China to allow economic decoupling or military decisiveness to materialize unchecked. Chinese aggression will have to be countered.[59]

Let a war break out between billions of Indians and Chinese and global emissions will darken the skies worse than what the environmentalists of California have done to the State's forests, public lands, and hillside communities.[60] CO2 will not destroy the planet the way Extinction Rebellion, teenage, climate-sensation, Greta Thunberg, and groups like the Sierra Club adamantly imply, but war will.

If western environmentalists were truly concerned about fossil fuels, they would follow Michael Shellenberger's advice, and have zero-carbon nuclear power plants on every continent for continuous, uninterruptable electricity.[61] If there is no time to wait, then switch to no-emitting nuclear. These climate alarmist's rail against modern society, but their dysfunctional message never includes nuclear, instead believing intermittent electricity from wind and solar alone can power modern societies like the U.S., Germany, China, and all the militaries around the world.

Why should nuclear be praised and used – because it is the only stand-alone power generation source for electricity that does not emit carbon dioxide gas during the process the way coal and natural gas does. As a result of the safety of nuclear power reactors and ability to provide continuous uninterruptable zero emission electricity, today there are about 440 nuclear reactors operating in 30 countries around the world with 50 more under construction. [62] Significant further capacity is being created by plant upgrading of existing reactors. [63] Additionally, there are 140 nuclear powered ships that have accumulated 12,000 reactor years of "safe" marine operation.[64]

Every part in the manufacturing process of solar panels and wind turbines would not exist without crude oil, all the parts are made from derivatives from petroleum.[65] Both nuclear, coal, and increasingly natural gas, are hamstrung by well-organized and unprincipled political opposition from environmental organizations by scaring people with imaginary dangers.[66] The nuclear industry has been on life-support for decades now over a well-funded, and coordinated effort by environmentalists to destroy them.[67]

The Global Warming Hysteria is one of the most successful junk science campaigns ever launched for money, votes, and power when there are so many questions anthropogenic (man-made) global warming/climate change needs to answer.[68] Many respectable academics, government employees, and scientists such as Dr. Judith Curry are attacked if they do not wholeheartedly embrace anthropogenic (man-made) global warming/climate change. Money trumps ethics every time.

Environmental interest groups need looming catastrophes at every turn for the science establishment to maintain the anthropogenic (man-made) global warming/climate change narrative.

The tragedy this book will uncover and use heavily sourced material to bring to light is our legislators and policymakers in the U.S. and countries who incorporate the Paris Climate Agreement into their energy policies are the lies swallowed, and trillions wasted on renewable boondoggles.[69]

Ironically, increasing CO2 in the atmosphere has a bountiful effect on plant growth, greening the Earth, and abundant agricultural production.[70] Rather than a threat, CO2 is a blessing towards higher food production on less land. The days of screaming CLIMATE CHANGE when epic policy failures are the reason why electricity prices skyrocket when renewables are employed, or forests turn into tinderboxes over mismanaged fire policies needs to stop. Having an environmentalist telling anyone to turn up their thermostat or use their appliances in the middle of the night to set a "climate example" are the true deniers.[71]

U.S. and other western environmentalists can scream from the mountaintops "climate change," while knowing full well China, India, and Africa are not going to stop using fossil fuels for billions desperate for reliable and abundant electricity. Indisputable data based on facts, "that if everyone in the world, per capita, used half as much energy as Americans do, global energy production would have to double."[72]

No renewable portfolio standard will overcome for decades ahead that 85 percent of global energy production comes from oil, petroleum, natural gas, and coal (fossil fuels). Hydroelectric power (dams) and nuclear make up approximately 11 percent, and the other 4 percent is from intermittent electricity (solar panels, wind turbines, biomass). It is not changing anytime soon according to the *British Petroleum Statistical Review of World Energy 2020*, but environmentalists want to change this reality for you, but never for them.[73]

If you doubt us, what is written in this sourced-introduction, our other two books – ***"Energy Made Easy"*** and ***"Just Green Electricity"*** confirm these facts. Energy and electricity reality points towards dumping toxic and poisonous solar panels, wind turbines, and battery energy storage systems for electrical grids in favor of natural gas-fired power plants, and nuclear energy to electricity plants. As we said in our first two books, have repeated numerous times in this introduction, and will continue writing about in this book, please remember that over 6,000 daily products come from a barrel of crude oil.

Without those thousands of petroleum products, modern life will revert backwards toward the caveman era. The global environmental degradations and human atrocities associated with the mining of exotic, rare earth minerals and metals to support renewables, battery energy storage systems for electrical grids and EVs should have you question why you have been told the world is uncontrollably warming and only a deadly green transition will save you.

Keep reading and learn why this is the case.

RON & TODD

CHAPTER ONE

CALIFORNIA

- ***The 5th largest economy in the world has the highest costs for electricity and fuels in America.***
- ***With only 0.5 percent of the world's 8 billion population, the state continues to make climate change its number one energy priority leading to higher costs for energy.***
- ***A minor player in the world's population and emissions arena, uses deceptive techniques to displace emissions, or "leak" them, to other states and countries, to help achieve its low emission goals.***

  - Former Governor Brown's secret war on clean energy
  - Leakage or Emission Displacement to others
  - California: A Minor Player on the World's Emissions Stage
  - Automobile manufacturing "leaked" to other countries.
  - Energy Imports
  - Forestry Mismanagement
  - Disparate cost effects on poorer, less educated residents of all races and ethnicities
  - Cost of Living Burdens

FORMER GOVERNOR BROWNS' SECRET WAR ON CLEAN ENERGY

His family wealth and oil ties were more important than the California economy and environment.

On taking office in 1975, Governor Jerry Brown, son of legendary California Governor Pat Brown, and his allies aggressively wielded power in ways that directly benefited Brown's family, which included killing nuclear power plants.

Former Governor Jerry Brown has actively sought to advance oil and gas interests, including his own. A two-year investigation by Environmental Progress concludes that no American politician has killed more clean energy than Governor Jerry Brown — and in ways that often benefited his own family financially. [74]

San Diego Gas and Electric sought to build five nuclear plants in 1976 a project called "Sundesert," which Brown attacked directly and through the agencies he controlled. Brown's allies at the California Energy Commission (CEC) argued that future demand should instead be met by burning oil and coal. [75]

Continuing throughout the 1970s we saw even more significant activity by Brown's closest allies including: changing pollution regulations to benefit his family's Indonesian oil monopoly; killing the Sundesert nuclear project; and lobbying Mexico's President to approve a natural gas project. Smells like a conflict of interest, that mainstream news media opted not to explore as they may have been tagged as climate change deniers, during the green renewable tidal wave. A major part of Jerry Brown's legacy will be as the fossil fuel millionaire who killed more clean energy, and increased emissions, more than any other in recent history.

Former Governors Jerry Brown and Pat Brown had strong ties to natural gas and oil in Indonesia and Mexico, which established the family's oil monopoly in California.

The Brown family's oil and gas ties extended into Mexico. Brown Sr. did business deals with a Mexican oil and gas family headed by Carlos Bustamante, head of a powerful Mexican oil and gas family, the New York Times reported in a long, front-page investigative article for the Sunday paper in 1979. [76]

Between 1976 and 1979, Brown and his allies killed so many nuclear power plants that, had they been built, California would today be generating almost all its electricity from zero-pollution power plants.[77]

LEAKAGE OR EMISSION DISPLACEMENT TO OTHERS

The use of deceptive techniques to displace emissions, or "leak" them, to other states and countries, is used to achieve the states' low emission goals.

California climate policy, by law, is supposed to account for and avoid greenhouse gas emission displacement, or "leakage" (lack of transparency) to other states and countries. Carbon leakage occurs when there is an increase in greenhouse gas emissions in one country as a result of an emissions reduction by a second country with a strict climate policy.

California's 2017 Climate Change Scoping Plan simply states that AB 32 requires the State's climate policies to minimize emissions leakage, as relocation would shift greenhouse gas emissions outside of the State without the benefit of reducing pollutants that contribute to overall global warming impacts [78].

Each country's share of CO2 Emissions shows that China, India, and "Rest of the World" contribute 56 percent of CO2 emissions, versus the 1 percent that California's 5th largest economy in the world contributes.[79] "Leaking" further emissions to countries with significantly less stringent environmental controls than California is detrimental to the world's environment.

The estimated greenhouse gas emissions from worldwide manufacturing have changed over time.[80] Today, China, Mexico, Japan, and India are major contributors of emissions resulting from the efforts to "leak" emissions for California's demands to locations outside of the state.

Shifting greenhouse gas emissions to other locations also reduces the availability of associated jobs and could impact a local tax base that supports services such as public transportation, emergency response, and social services, as well as funding sources critical to protecting the natural environment and keeping it available for current and future generations [81].

California's green and political interests do not want in-state oil and natural gas drilling, hydraulic fracturing (fracking), pipelines, nuclear, coal, or hydroelectric power plants – or mining for the materials needed to support EV batteries, battery energy storage systems for electrical grids, and the intermittent electricity from wind and solar generation. They prefer to have that work done somewhere else where Californians cannot "see" the emission numbers or any mining degradation to the landscape. They pursue more "leakage" of emissions and environmental degradations at sites outside of California instead. Additionally, the state continues to increase imported crude oil, electricity, and consumer goods from locations with significantly less environmental controls than California and the U.S., i.e., perpetuating the "leakage".

California has long wanted a totally electric vehicle (EV) fleet, which they claim would be clean, ethical, climate-friendly, and sustainable. Of course, those labels hold up only so long as they look solely at activities and emissions within California state boundaries – and not where, i.e., the "leakage" to others associated with the mining, manufacturing and electricity generation taking place at other out-of-site locations. Obviously, that kind of "life cycle" analysis, and the impact on the world's climate would totally disrupt their claims.

Consider copper usage: A typical internal combustion engine uses about 50 pounds of this vital everyday metal the International Copper Association says.[82]. A hybrid car requires almost 90 pounds; a plug-in EV needs 132 pounds; and a big electric bus can use up to 812 pounds of copper. If all 31 million registered vehicles were EVs, or just 1.5 percent of the world's 2 billion vehicles, the California EV fleet would need almost 2 million tons of copper.

But copper ores average just 0.5 percent metal by weight, notes energy analyst Mark Mills.[83] That means 400 million tons of ore would have to be dug up, crushed, processed and refined to get the copper to support the 2 million tons of copper needed by 31 million electric vehicles to replace the current fleet of registered vehicles in California. Almost every step in that process of copper mining in the countries where emission leakage is delegated would also require fossil fuels for the mining activities and transportation of the mined copper that would result in additional emissions to the worlds atmosphere of carbon dioxide and pollutants.

Adequately analyzing the nature and extent of potential greenhouse gas emission leakage from the state is crucial for the success of California's greenhouse gas reduction measures, which must be measured by global emissions, not just California in-state emissions.

There is substantial evidence that the state has stimulated and will continue to induce emissions increases in other locations that are currently unaccounted for in California, allowing California to claim in-state greenhouse gas reductions, while emission increases are "leaked" to other states and countries.

Elected leaders and special interest groups blatantly ignore the California climate policy that is supposed to account for and avoid greenhouse gas emission displacement, or "leakage" to other states and countries. California's greenhouse gas emission reduction program also ignores the increased emissions caused by population leakage to higher per capita greenhouse gas emission states.

A substantial amount of the California economy is supported by high-emission activities at other locations. Many, if not most of these product-based greenhouse gas emissions reflect long-term siting and supply decisions based on lower costs for energy and labor, as well as less stringent environmental and other regulatory restrictions.

Even the wealthiest California companies, including Apple,

Google, and Facebook, choose to locate higher energy consuming data centers and manufacturing to other states and countries with lower energy, housing, and labor costs, and greater emissions from locations with significantly less environmental controls than California.

California's high-tech economy illustrates the greenhouse gas accounting problems created by California's focus only on in-state activities. Apple Inc., Alphabet Inc. (Google) and Facebook are the three largest public companies in California, and, as of the end of 2017, three of the five largest in the United States, based on market capitalization. Each of these firms uses electricity and products made from petroleum derivatives and produces greenhouse gas emissions worldwide to make products, such as smartphones and computers, and to power and cool large online data centers.

With significant manufacturing no longer within the state, California's greenhouse gas reduction accounting also ignores greenhouse gas emissions from products consumed by Californians but manufactured elsewhere and transported back to California from other states and countries. Very little of these greenhouse gas emissions – or the manufacturing jobs that correspond to the production activities that emit greenhouse gas emissions - are in California, the most environmentally controlled location in the world.

Deployment of the exotic minerals and metals to support "clean" intermittent electricity from wind and solar generation, battery energy storage systems for electrical grids, and EV batteries on a massive scale using existing technologies requires obtaining rare, raw materials, such as lithium, nickel, graphite, and cobalt, from politically unstable, socially exploitative, and environmentally damaging sources outside of California.

The Democratic Republic of Congo (DRC) is by far the world's largest producer of cobalt, accounting for roughly 60 percent of global production and includes the use of children in dangerously narrow tunnels.[84] Very little tangible progress has been made to reform these "blood battery" labor practices in the DRC.[85]

Nickel mining has been linked with serious environmental damage and worker disease in the Philippines, Colombia, and Russia.[86] Graphite mining is known to damage crops, homes and personal property from soot deposits, respiratory disease, and polluted drinking water.

There are yet no cost-effective and feasible means for storing the intermittent electricity generated from solar or wind for use when needed. Until a cost-effective and ethically acceptable storage solution is invented, adding intermittent solar and wind electrical generation to any large-scale grid will require an alternative, more reliable electrical generating source to meet peak demands and provide power when solar is unavailable (i.e., cloudy days and at night) or there is insufficient wind to power the state's turbines.

The Scoping Plan calls for the deployment of 4.2 million EVs by 2030 or about 7 or 8 percent of all registered vehicle in the state, up from the 650,000 or 2 percent of the 31 million registered vehicles in 2019.

Californians buy about 2 million new cars per year including more EVs and hybrids than any other state; in 2016, however, the share of new EV sales fell from 5.0 percent to 4.7 percent and the share of new hybrids remained flat.[87] There is no current consumer trend suggesting the elimination of fossil-fueled cars in California especially for the 18 million that represent the Hispanic and African American populations.

Virtually none of these battery-related emissions in the batteries imported into the state from manufacturers in China, Japan or the Tesla plant in Nevada will occur in California. Yet California's climate regulators can claim the greenhouse gas reductions from the diffusion of EVs in the state while ignoring the greenhouse gas emissions required to produce the EV batteries (or other manufactured car components) for these vehicles.

California climate regulators, and elected leaders, completely ignore the greenhouse gas emissions associated with products manufactured outside of but consumed in the state. The absence of transparency and any effective analyses has also precluded any meaningful disclosure of cost-effectiveness, or consumer costs, in California's greenhouse gas reduction efforts.

Exporting manufacturing and production jobs and driving out middle-class jobs to states or countries where housing is more affordable exacerbates income inequality, violates the civil rights of California minorities, and are counterproductive to global reductions of greenhouse gas emissions.

In 2020 California experienced a catastrophic loss from the COVID-19 impact on the economy when the California Marathon Petroleum announced it will "indefinitely idle" it is Martinez Refinery.[88] The decision will remove hundreds of jobs, billions of dollars, and nearly 7 million gallons of gasoline, diesel, and other petroleum liquids per day from the energy-hungry California economy. It will also send fuel prices even higher for minority and other poor families that already pay by far the highest gasoline prices in the continental United States: $1.32 more per gallon of regular than in Louisiana and Texas.

During the 2020 pandemic, airlines and cruise ships were virtually shut down, vehicle transportation was at an all-time low, and the demand for fuels and petroleum derivative products manufactured from petroleum, were at an all-time low, the Northern California refinery, one of the largest in the state has just become a COVID victim.

We have all seen the photos of those foreign tankers with their crude oil cargo parked off the coast of California as the refineries had reduced usage to manufacture products that were in limited demand during the pandemic.

Soon as we recover from the COVID-19 pandemic, the decision to idle Marathon will involve more foreign imports – i.e., "leaking" emissions" from the energy-hungry California economy.

As the economy starts to recover back to near-normal fuel demands for the 5th largest economy in the world, the future looks very bleak for ALL 40 million residents of the state as fuel prices may even go higher for minority and other poor families that already pay by far the highest gasoline prices in the continental United States: $1.32 more per gallon of regular than in Louisiana and Texas.[89]

The near-normal daily energy use for California's 145 airports (inclusive of 33 military, 10 majors, and more than 100 general aviation) is enormous.[90] Currently, this behemoth consumes 13 million gallons of aviation fuels, or one-fifth of the nation's jet fuel consumption.[91] California's 31 million registered vehicles are gas-guzzlers.[92] Consuming 10 million gallons a day of diesel and 48 million gallons a day of gasoline.[93] Collectively, that is about 65 million gallons of various fuels needed daily to run the CA economy, but now with Marathon Martinez out of business the future supply may not be able to meet the demand.

California may take credit for the emission reductions resulting from the Marathon Refinery closure, but the state's energy demands will continue to be met with imported finished products from afar, i.e., more "leakage" of emissions to other manufacturing locations to support the demands of the state. California leaders understand there are more than 700 refineries worldwide of suppliers that can meet global demands, and all are in areas with significantly less stringent environmental controls than California.[94]

CALIFORNIA: A MINOR PLAYER ON THE WORLD'S EMISSIONS STAGE, WITH A BIG STICK

California professes to be the leader of everything and spouts voracious pride of being the State that imports more electricity than any other state.[95] Its dysfunctional energy policies has forced California to be the only state in contiguous America that imports most of its crude oil energy demands from foreign country suppliers to meet the energy demands of the state.[96]

California has unique characteristics from other states in America:

- Largest state at 163,696 square miles.
- Divided into 58 counties and contains 482 municipalities.
- Most populated state with almost 40 million residents, but just 0.5 percent of the world's 8 billion.
- Most unauthorized immigrants at 2.2 million or 20 percent of all unauthorized immigrants in America.
- Most diversely populated state with 39 percent Hispanic, 14 percent Asian, and 6 percent African American.
- Least percentage of Whites at 38 percent versus U.S. average of 64 percent.
- Most number of registered vehicles in America with 31 million.
- Half of all electric vehicles in America registered in California at 650,000 or 2 percent of all registered vehicles.
- Most major airports with 10.
- Most military airports with 33.
- Most general aviation airports with 100.
- An energy "hog" demanding more than 65 million gallons (aviation at 13, diesel at 10 and gasoline at 42) of various transportation fuels daily from suppliers to meet the demands of the states' economy.
- An "energy island" situated between the Pacific Ocean and the Arizona/Nevada Stateline, with no existing pipelines over the Sierra Nevada Mountains.
- Imports the most electricity than any other state at 32 percent.

- The only state in contiguous America that imports most of its crude oil energy from foreign country suppliers.
- Elected representatives perpetuate continued decreases of in-state oil production, to further increase the 58 percent of imported crude oil from foreign countries to meet the demands within the state.
- Elected leaders and special interest groups blatantly ignore the California climate policy that is supposed to account for and avoid greenhouse gas emission displacement, or "leakage" to other states and countries for California to meet the energy demands, and emission reduction targets, of the 5th largest economy in the world.

Politicians and environmental special interest groups that are avoiding any transparency of the state's actions to "leak" emissions to other states or countries for the electrical and crude oil demands of the California economy is blatantly obnoxious and illegal, but it continues to this day.

Even though the state's actual greenhouse gas emissions account for less than 1 percent of the world's anthropogenic greenhouse gas emissions, California policymakers continue proclaiming others will follow their greenhouse gas reduction example.

California represents a miniscule 0.5 percent of the world's population (40 million vs. 8 billion) but plans to bring all activities to zero emissions with the shuttering of power plants leading the way by 2035. Two blatant problems with the plan to decarbonize California are:

1. More "leakage" of emissions and environmental degradation to countries with less environmental controls and non-existent labor guidelines.
2. Just intermittent electricity from wind and solar cannot manufacture the 6,000 products made from petroleum

derivatives that are the basis of daily lifestyles, transportation, and economies around the world.

Even if California, or the entire USA, eliminated all fossil fuel use tomorrow – it would not make an iota of difference for global carbon dioxide levels as:

- The oil and gas industries are not just an American business with a few stateside refineries to service the demands of its residents, but an international industry with more than 700 refineries worldwide that service the fuel and product demands of almost 8 billion living on earth.[97, 98]
- There are 62,500 power plants around the world operating today, all types, generating electricity for the world's inhabitants.[99] Of that total, more than 2,449 are coal-fired power plants. Over half (1,363) of the world's coal power plants (2,449) are in China and India whose populations of mostly poor peoples is roughly 2.7 billion. Together they are in the process of building 284 new ones of the 546 total.[100] [101] They are putting their money and backs into coal – to provide scalable, reliable, and affordable electricity, that will continue to increase emissions to the worlds' atmosphere.

China and India are the two most populous countries in the world. As of 2018, China had almost 1.4 billion people, a figure that is projected to grow to 1.5 billion by 2045.[102] India accounted for approximately 1.3 billion people [103] in 2020 and is expected to grow to almost 1.7 billion by 2045.

Before California weans itself from all fossil fuels, the elected leaders and special interest groups should attain a little more energy literacy by reading the book "Just GREEN Electricity – *Helping Citizens Understand a World without Fossil Fuels*".[104] The book brings simplicity and clarity about understanding a world without fossil fuels. The inventions of the automobile, airplane, and the use of petroleum in the early 1900's led us into the Industrial Revolution and victories in World War I and II. The 'aha moment' for every green advocate is that intermittent electricity from wind or solar cannot produce any of the more than 6,000 products manufactured from petroleum derivatives that did not exist before 1900, that are the basis of daily lifestyles and economies worldwide.

The state's Democratic party leaders of the highly regressive climate schemes have engendered and continue to inflict disparate financial hardships on middle and lower-income workers and minority communities.

The blackouts that hit California in 2020 exposed the fragility of one of the most-expensive and least-reliable electric grids in North America. They also showed that California's grid cannot handle the load it has now, much less accommodate the enormous amount of new demand that would have to be met if the state attempts to "electrify everything."

The push to electrify everything would prohibit the use of natural gas in buildings, electrify transportation, and require the grid to run solely on renewables (and maybe, a dash of nuclear). But attempting to electrify the entire California economy will further increase the cost of energy at the very same time that the state's electricity rates are among the highest in the country and continuing to soar. That will result in yet-higher energy costs for low and middle-income Californians. These costs are before a new multi-trillion-dollar grid will need to be constructed to handle100 percent intermittent electricity from wind and solar only.

The problem of California's poor electric reliability will likely get worse. On Sept. 10, 2018, then- Governor Jerry Brown signed Senate Bill 100, committing California to obtain 100% of its electricity from "clean energy sources" by 2045. Replacement of coal, nuclear, and natural gas generators with intermittent electricity from wind and solar will continue to erode grid reliability.

For decades, California Democrats have argued that major economies can run mostly, if not entirely, on intermittent electricity from wind and solar renewables. "We are in the future business in California and that means we're in the renewables business," said Governor Gavin Newsom in 2016, when he made the case for closing the state's nuclear plant, Diablo Canyon that had been providing emission free and continuous uninterruptible electricity.

While the official numbers are not yet in, 2018 appears to have been a big year for natural gas power plant retirements in California with three in one year.

Natural gas power plant retirements in 2018: [105]

- Encina at 854 MW
- Mandalay at 560 MW
- Etiwanda at 640 MW

Blackouts in the failing state of Democrat-run California have forced Governor Gavin Newsom to admit green everywhere is falling short. The Governor has acknowledged that the transition away from fossil fuels has left California with a gap in the reliability of its energy system.

In September 2020, following blackouts from a lack of generated electricity being available, State officials threw a lifeline to four natural gas power plants along the Southern California coast, deciding the facilities are still needed to provide reliable electricity even as they contribute to the climate crisis.[106] The four plants, that were given a shuttering deferral for the time being, that allows them to continue providing continuous and uninterruptable generated electricity were:

1. The 1, 310 MW Natural Gas Power Plant at Redondo Beach, that was scheduled to be shuttered in 2023.

2. The 823 MW Natural Gas Power Plant at Scattergood in Playa Del Rey, that was scheduled to be shuttered in 2024.
3. The 575 MW Natural Gas Power Plant at Haynes in Long Beach, that was scheduled to be shuttered in 2029.
4. The 472 MW Natural Gas Power Plant at Wilmington, that was scheduled to be shuttered in 2029.

The Redondo Beach natural gas power plant continues to be an important source of clean and reliable electricity, generating 1,310 MW, which is enough power to light more than one million California homes and businesses.[107] [108] Natural gas accounts for 31% of the Los Angeles's power supply in 2017, with much of that generation coming from Scattergood, Haynes, and Harbor.[109]

The state lacks enough reliable electricity generation capacity to run everyone's air conditioner during hot summer evenings. Despite these capacity shortfalls, the state is moving ahead with plans to remove 2,200-MW of reliable electricity from the grid. That is the amount of power produced by Diablo Canyon nuclear power plant, which will be closed in stages in 2024 and 2025.

For more than a decade, California has been closing coal and nuclear power plants. Recently, the state also began closing natural gas-fired plants as part of a continuing effort to fight global warming.

Beginning in 2021, California will be removing 4 hydroelectric power plants, in addition to all those Natural Gas plants being removed:

- Four Dams on the Klamath River, three in California and one in southern Oregon have been identified for removal as soon as 2021 at a cost of $450 million. Iron Gate, Copco #1, Copco #2, and J. C. Boyle Dams make up the Klamath River Hydroelectric Project and removing them will be the largest dam removal project in the

> world. It will also open hundreds of miles of salmon and steelhead spawning habitat. PacifiCorp's 169-megawatt (MW) Klamath Hydroelectric Project (FERC No. 2082) is in a predominantly rural area in southern Oregon. The project generates approximately 716 gigawatt-hours of emissions-free electricity on an annual basis – enough power to supply the energy needs of approximately 70,000 households.[110]

Frightening trends for California, despite the 2020 stay of execution for four natural gas generating plants targeted for shuttering, are more power plant closures in the future. To accomplish its zero emissions target by 2045, the GREEN movement wants the in-state generation from Natural Gas and Nuclear that was about 56 percent in 2018 (46.54 percent + 9.38 percent respectively) to be zero percent![111]

In mid-2019, Berkeley became the first city in the U.S. to ban natural gas connections for all new residential buildings and most non-residential buildings. Since then, some 31 other local governments in California have enacted measures that either ban or restrict the use of natural gas. San Jose requires that all new, low-rise residential buildings and municipal buildings be all-electric. [112] Menlo Park's ordinance requires electric space heating and electric water heating but allows the use of gas for fireplaces and cooking.[113] According to the Building Decarbonization Coalition, about half of the California governments that have imposed restrictions on natural gas are also including electric-vehicle infrastructure to their list of regulations.[114]

Nationally in 2016, natural gas-fired generators accounted for 42% of the operating electricity generating capacity in the United States_with 200 more set to open. [115] [116]

- California's goal is ZERO natural gas power plants to generate continuous uninterruptable electricity.
- California's green goals are to only rely on intermittent electricity from wind and solar and hope that the Northwestern and Southwestern states can generate enough extra power to meet the electricity demands of the 5th largest economy in the world.

If we look at the bigger national picture, it is scary as the Democrats' Green New Deal want 81 percent of our current national generation to go away and be replaced with intermittent electricity from wind and solar (Natural Gas at 38 percent, Coal at 23 percent, and Nuclear at 20 percent). Unlike California that can import electricity from nearby states, America will have no outside sources of electricity and will need to learn how to live with intermittent electricity from wind and solar, or just accept blackouts from a lack of continuous uninterruptible electricity.

Here are the national numbers for electricity generation in 2019 and the Democrats Green New Deal (GND) targeted generation by 2050. [117]

	2019	Democrats GND target by 2050
Natural Gas	38 percent	Zero percent
Coal	23 percent	Zero percent
Nuclear	20 percent	Zero percent

California has built many subsidized solar and wind farms for intermittent electricity generation and retired or canceled many clean-burning natural-gas power plants that have been providing continuous uninterruptible electricity.

All Americans need to be aware of the Democrat's Clean Energy Standard, which would put them in the same position as Californians. Making the entire U.S. reliant upon intermittent wind and solar electricity. The California blackout episodes should be a wake-up call for all politicians thinking of jumping on the green energy bandwagon.

California has had decades to prepare for this. No one failed to predict it. They only failed to prepare for it. The writing has been on the wall since 2001 when the state was hit with a series of massive blackouts and soaring electricity prices. The result was the successful recall of then-Democrat Governor Gray Davis in late 2003. He was replaced by the equally energy policy incompetent Republican Arnold Schwarzenegger.

While there is nothing funny about people losing power during a heat wave, it is still hard not to laugh at a state that is so scientifically backwards. The luddites were thwarted by a lack of wind and clouds.

In preparation for that next heat wave to hit California's growing population, rather than constructing more reliable electricity generation, the California dysfunctional energy policies have seen three natural gas power plants close in 2018, and four more to be shuttered, along with the last nuclear power plant at Diablo that have been providing continuous uninterruptable electricity to the grid.

Now, faced with the electricity supply crisis, caused by the dysfunctionality of the states' energy policies, Governor Newsom has suspended air-pollution regulations, which may increase the use of diesel generators, and worsen air pollution in the inner-city.

Maybe the state of California should listen and learn from the Washington state utility report of July 14, 2020 titled "Wind Power and Clean Energy Policy Perspectives," the utility's commissioners said they "do not support further wind power development in the Northwest." [118]

Many of their conclusions in the Washington state utility report are supportive of the energy literacy messages in two recent books by Ronald Stein and Todd Royal, *Energy Made Easy*, and *Just GREEN Electricity*, both available on Amazon.

- Wind power in the Pacific Northwest will not result in consequential reductions in national or global greenhouse gas emissions attributable to Washington State utilities and will do very little to mitigate the increasing risk of northwest power grid blackouts.
- Wind will unnecessarily contribute to increases in northwest utility retail electricity rates which could erode the economic development advantage low rates has given our region for many years.
- Benton Public Utility District believes the best long-term, sustainable, and environmentally responsible strategy toward meeting the Comprehensive and Economic Trade Agreement goal of 100% clean electricity in Washington State by 2045 could be to transition coal power to natural gas and then natural gas to nuclear.

Even the Japanese government has decided to remove the two remaining wind power turbines it installed off the Fukushima Prefecture due to the unprofitability of the project. Kyodo News reported that the government's decision came despite Japan's goal of raising its offshore wind power generation. To commercialize wind power generation, the operational rate of a turbine must remain at 30 to 35 percent or more, according to the ministry. But the rates of the turbines off Fukushima were only around 4 to 36 percent. At a meeting in Fukushima, industry ministry officials briefed fishermen and other participants about the plan to scrap the wind power turbines, with locals saying the government had wasted taxpayers' money and should conduct a thorough study of why the project had failed. [119]

If the state wants to do a better job at providing continuous uninterruptable power to its residents, they need to build power plants "in" California, which means natural gas, nuclear, or hydro, all of which are the exact power plants that California wants to shutter which is consistent with the California "green" religion that is against coal, natural gas, nuclear, and hydro power plants.

California's renewable portfolio standard mandates that 60 percent of its electricity must come from renewable energy (mainly wind and solar power) by 2030, but few are being constructed to replace the shuttered generating plants!

All you need to do is build more power and nuclear plants and the problem is solved. Yes, it really is that easy.

Into the grandest of fantasies, reality intrudes. And so, it may be that Mother Nature in the form of annual forest fires, will force a key California agency to face reality and modify the overly ambitious and unrealistic intermittent electricity fantasy that has characterized the state's energy planning for a decade. [120]

Can you see the conundrum the state is in by shuttering power plants and being surprised by inevitable blackouts? That is why we constantly refer to the energy policies in California as being dysfunctional!

Mark Mathis from Clean Energy Alliance produced a video in 2020 after the rolling blackouts in California, identifying the numerous other scapegoats for California's failing electrical systems.[121] Scapegoats abound in misspent moneys, bad decisions, priority problems, neglect, and corruption. Among the scapegoats highlighted in the video are:

- Poor forest management that continues to provide unlimited "fuel" for the next fire.
- PG&E not spending billions on upgrading transmission lines.
- $7.5 billion from Cap & Trade goes to rate payer benefits.
- $2.2 billion for rebates for rooftop solar.
- $100 million for solar systems in low-income areas.
- $150 million on battery storage systems.
- $130 million to install 7,500 electric car charging stations.

- The people of California that continue to elect the folks contributing to the dysfunctionality of the electrical system.
- Energy reality has yet to register with the people or their elected leaders.

California's blackouts are a product of its politically determined reliance on intermittent, unreliable renewable electricity that the State is not constructing in sufficient quantity to replace shuttered generation, not a product of a heatwave.

California was incapable of generating enough in-state power for its needs and had to import 32 percent of its electricity needs in 2019. With the state having no plans to replace the capacity of the recent three natural gas power plants that were shuttered in 2018 and the upcoming five power plant closures (four natural gas and one nuclear) with in-state intermittent electricity from wind and solar.

Shuttering in-state power plants, the state will need to import more high-priced electricity from the Southwest and Northwest to fill the void and let residents and businesses pay the premium. This great dysfunctional energy plan to "leak" emissions to other states for California's electricity needs, to help achieve in-state emission reductions, is also based on the hopes that other states can generate enough to replace the shuttered California power plants and will have the extra capacity to add to the grid!

Not only have the shuttering of these gas plants been deferred to a later time, but the zero-emission generating plant at PG&E's Nuclear 2,160 megawatt Generating Plant at Diablo Canyon is to be shuttered in 2024 remains on schedule. The Diablo plant supplies zero emission electricity to 3 million households. [122]

The August 2020 rolling electric power blackouts afflicted as many as 2 million California residents as a heat wave gripped the Golden State. At the center of the problem was that power demand peaks as overheated people turned up their air conditioning in the late afternoon just as solar power supplies cut off as the sun goes down. In addition, output from California's wind farms was erratic since the wind does not blow during hot, dry California high-pressure weather systems.

Currently, California does not have the technology or the budgeted funds to build enough battery energy storage systems for electrical grids to back up and store intermittent electricity from wind and solar.

Until the 2020 summer, California utilities and grid operators were able to purchase extra electricity from the Southwest and Northwest, but when the heat wave stretches from Texas to Oregon there is little available to make up for California's power shortage.

Completely ignored in the reporting is that California has been shutting down a huge source of safe, reliable, always-on, non-carbon dioxide–emitting, climate-friendly electricity—that is, nuclear power. In 2013, state regulators forced the closing of the San Onofre nuclear power plant that supplied electricity to 1.4 million households. By 2025, California regulators plan to close down the Diablo Canyon nuclear power plant that can supply electricity to 3 million households.

California's Brave New World of wind and solar has not only brought shortages of power but the state's consumers incur electricity prices that are 60 per cent above the US average. As in Australia and Germany, an unholy alliance of naïve environmentalists and renewable energy subsidy seekers has driven energy policy.

California's prioritization of climate change policies has led to the most expensive electricity and fuels in the country and laid bare the realities of systemic racial, health, economic, and environmental injustices that impacts those that can least afford expensive energy.

Everywhere renewables are implemented, they have driven up costs and driven down reliability. Germany tried to step up as a leader on climate change, by phasing out fossil fuels and nuclear, and pioneered a system of subsidies for wind and solar that sparked a global boom in manufacturing those technologies. Today, Germany is failing to meet its climate goals of reducing carbon-dioxide emissions even after spending over $580 billion by 2025 to overhaul its electrical generation. Germany's emissions miss should be a "wake-up call" for governments everywhere.[123]

Today, German households pay almost 50% more for electricity than they did in 2006 as power prices in Germany are now among the highest in Europe.[124] Much of that increase in electricity cost is the Renewable Surcharge that has increased over the same period by 770%. Germany has learned that clean energy is not energy in totality, as wind and solar only provide electricity, and more accurately its only intermittent electricity at best. Renewables have also been the primary driver behind the high costs of electricity for residents of Australia and California.

A landmark Australian economics report shows that climate policies and renewable subsidies cost Australian households around $13 billion per year, or $1,300 per household. [125]

California has demonstrated that the state is a slow learner as they continue to follow Germany's and Australia's failed climate goals which should be a wake-up all for governments everywhere.[126] Germany's obsession with intermittent wind and solar has resulted in power prices that are now the highest in Europe, if not the world, for those Germans who are lucky enough to be supplied with it.

Over the last year or so, the Sierra Club, along with the Rocky Mountain Institute and other groups, have been pushing local governments to prohibit natural gas use and force California consumers to rely solely on electricity. In July 2019, Berkeley became the first city in the United States to pass a ban on natural gas hookups in new buildings. Since then, about 31 other local governments in California have passed restrictions or bans on the use of natural gas. Last month, the Sierra Club gleefully announced that the city of Piedmont had committed "to going gas-free."

These restrictions are being labeled as an essential part of California's efforts to slash its greenhouse gas emissions. But in practice, they are regressive energy taxes that will hurt low- and middle-income consumers and in doing so, exacerbate California's poverty problem.

Restrictions on natural gas are popular in the Bay Area and Silicon Valley. Municipalities in Silicon Valley that have passed restrictions include Cupertino, the home of Apple; Mountain View (Alphabet); and Menlo Park (Facebook). Those restrictions may be politically popular, but the regressive effects of the electrify everything push cannot be ignored. Banning the direct use of natural gas for cooking, home heating, water heaters, and clothes dryers, will force Californians to instead use electricity which, on an energy-equivalent basis, costs four times as much as natural gas.

The bans and restrictions on natural gas are being implemented at the same time California's electricity rates are soaring.

As part of global warming efforts, officials want all citizens to switch their natural gas stoves and furnaces to electric models. More than 30 California cities have enacted bans on gas appliances, including the major cities of San Francisco and San Jose. Almost 10% of the state population now lives in an area covered by restrictions against gas appliances in new residential construction.

American cities such as Berkeley, San Jose, San Francisco, Houston, Los Angeles, New Orleans, Albuquerque, and other U.S. cities are moving to ban natural gas as a step toward becoming carbon free in the next few decades placing more electrical demands onto an electricity starved grid. They are about to take one giant step toward Germany's failed climate goals which should be a wake-up call for governments everywhere.[127]

Unfortunately, many people do not realize that climate polices that tax businesses, simply increases the businesses cost to produce goods and services. This tax is therefore passed through to consumers. The so-called "tax on carbon" is no exception. Businesses will pass the cost of these carbon taxes onto the products and services and obviously onto the consumers, making energy more expensive for everyone, not just businesses. Moreover, everything that uses energy — transportation, manufacturing, heating, cooling, and cooking, etc. — also will become more expensive.

California's residents, of which 39 percent are Hispanic and 6 percent are African American, or collectively 18 million of the states' 40 million residents, presents a substantially different story than the aggregated statewide data used by CARB to conclude the state is enjoying a successful boom.

In fact, California has the highest homeless population, and the 5th largest percentage of homeless (behind D.C., New York, and Hawaii, and Oregon) including more than a quarter of a million families, children, and adults.[128] California also has the nation's highest poverty rate, and by far the largest number of Americans living in poverty: about 8 million Californians, and more than 2 million California children, live below the federal poverty level.

More than six million Californians already live in overcrowded housing, representing about 2.5 times higher than the nationwide rate. [129]

Rather than heal the wound by doing everything possible to "reduce" energy costs for 40 million residents, the state is excited to place band-aids on the wound with further housing for the homeless as it continues to do everything possible to further increase the cost of electricity and fuels.

California housing and energy costs for electricity and transportation fuels have risen well above the national average, while income growth for the remainder of Californians has been largely stagnated, particularly for the poorer, less educated workers, and households, and the 39 percent that represent the Hispanic and the 6 percent that are African Americans.

California electricity prices continue to rise higher than other states as documented by the Center for Jobs and the Economy at the California Business Roundtable are:

1. Residential Electricity Prices are 55.8 percent higher than other states.
2. Commercial Electricity Prices are 70.8 percent higher than other states.
3. Industrial Electricity Prices are 117 percent higher than other states.[130]

California also wants residents to transition from gasoline-and-diesel-powered cars and trucks to plug-in electric models. So, when those blackouts occur in the future, not only will your lights and air conditioners fail, but you will not be able to cook your food or drive your car either.

In 1968, Tom Wolfe wrote amusingly about hippy movement in the Electric Kool-Aid Acid Test.[131] Today the fodder for the risible has become, at least in California, the politically inevitable. Ever since Arnold Schwarzenegger found climate religion; the state has been determined to push the green agenda to its fullest extreme. In a succession of regulatory moves, California is accelerating its detachment not only from fossil fuels, including natural gas, but the most reliable no-emissions electricity, nuclear power.

Yet at the same time as the state has subsidized massive increases in intermittent wind and solar electricity, it has also accelerated plans to electrify everything from home kitchens to buses, trucks and, of course, cars. All this at a time when reliable power supplies are increasingly shuttered and constrained. Elon Musk can build all the vehicles he wants --- increasingly outside

California --- but how are they going to be powered and at what cost to the consumer?

Well, we know the cost side all too well. Since 2011 electricity prices have increased five times as fast as the national average. In 2017 alone, they increased at three times the national rate.[132] These prices have been devastating to poorer Californians, particularly in the less temperate interior where "energy poverty" has grown rapidly.[133]

These electricity price increases are no fluke; all countries that have adopted draconian "green" mandates --- Germany, Demark, even resource-rich Australia --- have experienced huge spikes in electricity prices. In Europe, notes one study, reliance on renewables both reduces incomes and boosts rates of household poverty. [134]

A family of four managing on $60,000 a year, or poverty level, needs to make those tough critical decisions to spend their income for utility bills, rent, or food for the family, or for transportation. The Democratic party leader's endless pursuit of emission reductions at any cost remain guiltless about those tough decisions for a struggling family.

California continues to sacrifice reliable electrical power on the altar of the fight against global warming. There is no evidence that state efforts will have the slightest effect on global temperatures, but they will be great for candle and flashlight sales.

California is reaping its own harvest from requiring a wind/solar-rich electricity supply. The State Government requires a 60 per cent renewable supply by 2030, and the state is presently at 33 per cent. To meet demand of 42-44 GWh, California has a capacity of 76 GW, a bit more than Australia. But 27 GW of California's plant is solar (useless for evening peaks) and 7 GW is wind (useless in hot windless days).

California is a failed state, and it is not the fault of the Democrat Party — it is what it is... It is the fault of voters who continue to double down on failure, who keep electing the same leaders with distorted plans for its citizens and have given these policymakers a veto-proof majority in the state legislature. Maybe they should consider Republicans for office as well as Democrats. Just a thought...

AUTOMOBILE MANUFACTURING "LEAKED" TO OTHER COUNTRIES.

The cars manufactured in the two largest countries in the world, China, and India, are also the countries that have the least stringent environmental regulations and the least labor controls for their workforce.

The number of vehicles in the world are at 1.2 billion vehicles on the world's roads_today with projections of 2 billion by 2035 [135]. By some estimates, the total number of vehicles worldwide could double to 2.5 billion by 2050.

The once vibrant automobile manufacturing state of California has "leaked" the manufacturing emissions and labor opportunities to other countries. A summary of the once vibrant major automobile plants in California, with years of operation are as follows [136]:

- General Motors, Oakland, 1916-1964
- Willys-Overland, Maywood, 1928-1954
- Ford, Long Beach, 1930-1959
- Chrysler, City of Commerce, 1932-1971
- Studebaker, Vernon, 1936-1956
- General Motors, South Gate, 1936-1982
- General Motors, Van Nuys, 1947-1992
- Nash, El Segundo, 1948-1955
- Ford, Milpitas, 1955-1983
- Ford, Pico Rivera, 1957-1980
- General Motors, Fremont, 1962-1982
- Nummi, Fremont, 1984-2009

A 6-minute video of the automobile manufacturing "needle" that shows China came from zero production in 1950, to 2019 where it produces more cars than the USA and Japan collectively.[137]

	Automobiles manufactured per year.	
	1950	2019
China	None	28 million
United States	8 million	11 million
Japan	31 thousand	9.8 million
India	15 thousand	5 million

China and India are the two most populous countries in the world. China has almost 1.4 billion people a figure that is projected to grow to 1.5 billion by 2045.[138] India accounted for more than 1.3 billion people and is expected to grow to almost 1.7 billion by 2045.[139] Currently, 2 out of every 7 people on our planet are Chinese or Indian. Both countries are desperate for electricity and the population growth numbers are conservative estimates.

China and India also have minimal environmental regulations which promotes emissions and environmental degradation to the countries landscape, and minimal labor regulations which promotes labor atrocities onto the working class within their countries.

ENERGY IMPORTS

California imports more electricity than any other state and imports most of its crude oil from foreign country suppliers to meet the energy demands of the state.

California was the largest net electricity importer from generating facilities outside the state, of any state in 2019, as documented by the U.S. Energy Information Administration. [140] California's net electricity imports were the largest in the country. California utilities partly own and import power from several power plants in Arizona and Utah. In addition, California's electricity imports include hydroelectric power from the Pacific Northwest, largely across high-voltage transmission lines running from Oregon to the Los Angeles area.

California has the least reliable electrical power system in the nation. Between 2008 and 2017, California was the leading U.S. state for individual power outages with almost 4,297 blackouts in the ten-year period, more than 2.5 times as many as its closest rival, Texas. Power outages are now commonplace in California.[141]

To the detriment of Californians, the costs of expensive imported electricity is being borne by residents and businesses alike, and precisely hard hitting to the disadvantaged residents, particularly Latinos and African Americans that collectively represent 45 percent of the state's population. California households are already paying 50 percent more, and industrial users are paying double, yes, twice the national average for electricity.[142]

An important statistic to remember is the 45 percent of the California population are Hispanic and African American.[143] The median income for Latino households in 2016 was $56,200, $96,400 for white households, and $55,200 for black households.[144] According to several studies, as many as 40 percent of all Californians cannot regularly meet basic monthly expenses. [145]

In 2019, 57 percent of Black families and 50 percent of Latino families with children were poor in terms of net worth, lacking enough financial resources to sustain their families for three months at a poverty level, finds new research from Duke University.[146]

The primary focus of California's climate policies has been toward the oil and gas industry that was virtually non-existent before 1900. Today, California has less than 0.5 percent of the world's population (40 million vs. 8 billion) but targets its energy policies onto an industry that did not exist a century ago.

The oil and gas industry are not just an American business with a few stateside refineries to service the demands of its residents, but an international industry with more than 700 refineries worldwide that service the fuel and product demands of almost 8 billion living on earth.[147]

With in-state crude oil production at an all-time low and going lower with pressure from the Governor, California's dependency on other suppliers has increased imported crude oil from foreign countries from 5 percent in 1992 to 58 percent today of total consumption.[148] California spends more than $60 million dollars a day importing oil, yes, every day, being paid to oil-rich foreign countries, depriving Californians of jobs, careers, and business opportunities.

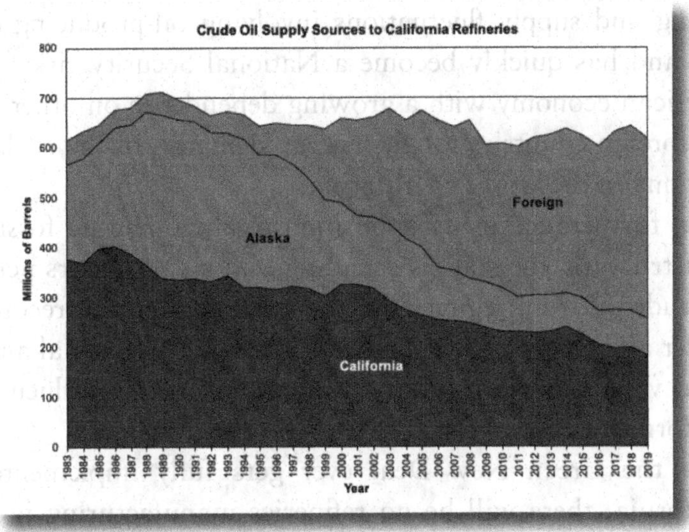

Figure 1-1 Crude Oil Supply Sources to California Refineries

In recent years, while the rest of the country, and other world powers, are rushing to build natural gas power plants, California continues to downsize its natural gas fleet.[149]

As noted by the state's non-partisan Little Hoover Commission, California's reliance on imported electricity and crude oil allows the state to ignore the greenhouse gas emissions from most of these foreign sources, as well as significant portions of the production, refining and manufacturing, and transportation of its energy supplies. [150] State law also precludes in-state providers from entering long-term contracts to buy out of state energy produced from high-emission sources, such as coal.

Opposition to oil pipelines and rail transport has largely succeeded in barring oil imports into California from the lower 48 states and Canada. The major sources of relatively light and clean crude oil that California has been willing to use are produced outside North America in locations the state has no ability to regulate and occur in countries that do not embrace California's environmental, labor, and social equity values.

California is also more vulnerable to political disputes and pricing and supply fluctuations involving oil-producing countries and has quickly become a National Security risk for the American economy with a growing dependence on other states and foreign countries for the energy demands of the 5th largest economy in the world - California.

In furtherance of its opposition to the domestic fossil fuel infrastructure, the state is spending billions of dollars per year on crude oil from regimes that have especially poor records on gender and LGBT rights – as well as many other social and political values that are ordinarily more politically influential in California.

If the Green New Deal ever gets fully implemented in California, there will be no refineries manufacturing in-state. The state would be getting all those thousands of products from the derivatives from oil, and its fuels, from foreign refineries via ships to the states' ports. Governor Newsom may have difficulty suing offshore refineries for their nuisance to society and their emissions to the environment!

At the same time, highly regressive transportation fuel price increases, force generally less affluent, minority-majority inland residents to disproportionately bear the greatest cost burdens for California's climate programs.

FORESTRY MISMANAGEMENT

California policies promote continued accumulation of forest "fuel" for the next fire.

Back in 2012 utility companies lost the authority to manage the greenery within their right-of-way due to Obama-era legislation that changed the management of 193 million acres of national forests and grasslands. Those regulations established a new blueprint to guide everything from logging to recreation and renewable energy development.

The guidelines apply to all 155 national forests, 20 grasslands and one prairie and represent the first major overhaul of forest rules in 30 years. It seemed like a good idea at the time despite the loud opposition voices warning against the new regulations.

Obama-era forestry restructuring regulations waylaid already established preventative maintenance measures that could have averted power outages and raging conflagrations resulting in the need for people to be relocated to safer grounds at the expense of the federal government.

We are all for safety but, for god's sake, teach a man to fish, please. If you don't address the real problem, you're welcoming the "new normal" where fires burn uncontrollably and electrical infrastructure decays that cost more and more money Washington has grown tired of funneling your way will continue unabated.

As California continues to throw good money after bad and baste in the fires of its own Gehenna, the White House has threatened to cut off emergency aid (FEMA) to residents displaced by the recent forest fires.

Environmental advocacy, litigation, and the resulting bureaucratic paralysis have created vast areas of "fuel" in our forests that continue to accumulate from mismanaged forest lands with dense underbrush and stunted tree growth that are uniquely susceptible to catastrophic wildfires that emit far more greenhouse gases from combustion and the subsequent decay.

California Governor Newsom has convinced himself that the devastating forest fires in the state are a direct result of climate change. The governor is oblivious to the facts that USDA Forest Service data documents the greatest number of wildfires occurred from 1928 to 1936 and that today, they are one-tenth of the record. i.e., that is a 90 percent reduction in burned acreage from almost a century ago.[151]

With lightening being the cause of most forest fires, we are fortunate that today, we have fire departments with vast resources of manpower, equipment, and fire-retardant dispensing aircraft that were non-existent before the 1900's. In the past, fires would burn endlessly. Today, the world's best fire management system is in northern Australia, and it is led by Indigenous land managers that manage to minimize the accumulation of unlimited fuel for the next fire.[152]

California has forbidden forest management, and builders continue to build wood tinder box homes adjacent to the stockpile of "fuel" for any fires in the forest, and homeowners do very little to fireproof their homes sitting amongst all that "fuel" for the fire. The perfect storm for fire "fuel" is present in the state:

- California forests get the least rain, with generally 7 to 8 months of summer weather without rain.
- The bark beetle has killed nearly 150 million drought-stressed trees during the last decade.

Even though there are hundreds of millions of dead and dying trees, i.e., fuel for the next fire, in the state's badly mismanaged forests, environmentalists are opposed to culling dead trees and high fire hazard underbrush from the forest lands that cover 33 percent of California, generally citing species protection concerns and climate change.

Rather than facilitate the removal of dead trees and long-term management required to sustain healthy forests that sequester greenhouse gases in trees and plants, California has instead pursued policies that increase the risk of death and property damage from forest fires. Most of the available funding has benefitted the acquisition of rooftop solar and electric vehicles by wealthier residents comprising the top 20 percent of the state's income earners.

California's wealthy, coastal environmental advocates have blocked opportunities to create working and middle-class jobs even when potential industrial expansion would help meet the state's climate goals.[153]

California has vast forest lands which can be managed to protect species and habitats, encourage tree growth to sequester carbon, produce electricity from wood-waste biomass fuels, create substantial jobs in poorer, inland regions of the state by manufacturing wood products, and lower housing production costs with locally produced materials. All this sustainable, prudent economic activity can lead to building three million homes that, as all leading California candidates for Governor agree, must be built to alleviate the state's housing crisis.

A 2018 Little Hoover Commission Report, Rethinking Forest Management in the Sierra Nevada found that wildfires during 2001-2010 directly released huge amounts of CO2 from state forests, and that post-fire emissions could be five times greater than the amounts released during the fire itself. [154] By any measure, wildfires that burn unnaturally hot in mismanaged, over-fueled state forests are producing uncounted greenhouse gas emissions.

The state greenhouse gas emissions subject to the Scoping

Plan include none of these fire-related emissions, despite the fact they could amount to nearly 60 percent of the entire state emissions inventory reported by CARB for 2015.

Among other challenges, restoring California's forests to health and to function as a net carbon sink for greenhouse gas emissions by removing over 100 million dead trees from insect and drought that are now decaying in place and emitting rather than sequestering greenhouse gases.

Identifying, incentivizing with permit streamlining and/or subsidies the use of removed vegetation for economically beneficial purposes, such as structural wood products, lumber products, and biomass power generation, many of which have been in decline due to regulatory constraints and environmental opposition.

Unlike many Scoping Plan policies, the implementation of forest management reforms will produce immediate and long-term greenhouse gas reductions from a resource that cannot be relocated or "shuffled" to other states and countries.

Yet, even if 10 percent of the funding for the High-Speed Rail project, or about $6.4 billion, was redirected to forest treatment, maintenance and the construction of biomass wood product and energy facilities outlined in the Forest Carbon Plan and the Little Hoover Report would produce many times the greenhouse gas emission benefits claimed for the High-Speed Rail project[155].

It is a shared responsibility of all parties as to why we have accumulated so much fuel for a fire and continue to allow its growth?[156] Specifically, the Federal regulations forbid forest management, and builders continue to build wood tinder box homes adjacent to the Federal stockpile of "fuel" for any fires in the forest, and homeowners do very little to fireproof their homes sitting amongst all that "fuel" for the fire.

Rather than taking responsibility for decades of policies that avoided safety precautions like careful and controlled burning, Governor Gavin Newsom is blaming PG&E for not stepping up to solve the mess caused by previous bad public and private management. How will forcing PG&E to shoulder the financial burden of preventing and cleaning up after California fires, and reduce the fuel awaiting the next spark?

About 33 percent of the state is forested, (need to control the fuel) and for decades pre-European settlement growth patterns have been altered to result in a much more dense, overgrown and less fire resilient mix of tress and vegetation than would occur under natural circumstances. Mature trees are stunted in growth, less healthy and more susceptible to insect infestation that has killed well over 150 million trees.

The extent of shade tolerant trees and lower lying vegetation with much less ability to survive fires has greatly increased throughout the forests. Consequently, the amount of carbon dioxide removed from the atmosphere and stored in living forest vegetation has, despite the overgrown nature of the forest lands, been reduced by as much as 25 percent over the last 150 years.[157] Instead, the Governor sues PG&E rather than reduce the cause of the fires, the fuel!

Power companies have done everything they can to oblige with liberal policies from funneling billions of dollars into "renewable" intermittent wind and solar technology to shutting off power for millions of customers, severely cutting their bottom line. But they cannot be expected to fill the gap left behind by irresponsible leadership, especially when the money that they have lost due to California's "progressive" energy policies could have been dedicated to fireproofing efforts and forest management.

The state's mismanagement of past funds has put it in the situation it is in today. Rather than patch the hole in the sinking ship the state would rather bring in a larger sump pump to evacuate the water. The water in this case is its fire maintenance budget and the sump pump represent more FEMA funding.

California continues to encounter environmental and political battles regarding retrofitting and creating defensible spaces around homes and/or thinning forests to reduce the potential "fuel" for the fires. Those choices will set the tone for fixing the problems or funding other problems created by not preventing the original problems in the first place.

Residents escaping the high costs of living in urban areas have created a problem for California's forestry maintenance guidelines. As they move closer to forested areas, animals are seeing their habitats disturbed. Utility services are expected to be brought to them no matter where they live.

This expansion of homes into once forested areas is causing headaches for utility providers who want more customers but do not appreciate the burden minimal forest maintenance has placed on their ability to safeguard their equipment and the public who use it. It has caused them to petition the California Public Utilities Commission and receive the authority to shut down services where it deems those services create a public hazard.

The California Public Utilities Commission (CPUC) sets the rates for consumers that utility companies can charge their customers, NOT PG&E. Recently the CPUC granted the authority to all California utility companies to shut down services where it deems those services create a public hazard. In addition, Governor Newsom's own government forces the company to buy expensive unstable, unreliable intermittent electricity for billions more than less expensive nuclear power and natural gas that provide continuous and uninterruptable electricity.

Today, electric utility companies up and down the state have all implemented Public Safety Power Shutoff (PSPS) events, aka rolling blackouts, as a last line of defense against electric service disruption and possible massive wildfires to vulnerable communities caused by utility equipment.

In October 2019, Pacific Gas & Electric shut off power to some 800,000 customers in 32 counties. Those blackouts were dubbed "public-safety power shutoff events" designed to avoid a repeat of the deadly wildfires. The blackouts eventually affected nearly three million people and PG&E warned that it could continue imposing blackouts for another 10 years.

Powering down swatches of residential areas affects those on life support systems, residents' abilities to recharge EVs to escape the devastation and cell phones that need charging; effectively cutting them off from the rest of the world unless they have ham radios or other highly regulated emergency transmission devices. Some of those devices effect communications between fire fighters and emergency assistance first responders.

It is time for Californians to make the right choice to forego their pseudo-environmentalist ways, at least in this case, and rewrite Obama-era regulations of all the country's forest maintenance and survival rules that have taken away utility companies' control of their right of ways that allow them to provide continuously uninterruptable electricity to all its customers. Or invest in fire hazard suits for all their family members and pets who live in heavily forested areas.

The climate programs "fuel" environmental advocacy, litigation, and the resulting bureaucratic paralysis that have created vast areas of mismanaged forest lands with dense underbrush and stunted tree growth uniquely susceptible to catastrophic wildfires that emit far more greenhouse gases from combustion and the subsequent decay, rather than renewal, of burnt lands.

It is a given the fires are going to come, but we can keep their destruction to a minimum with a modicum of preventive medicine because Washington has grown weary of handing out aid to California over fire mismanagement and energy policies more concerned with green virtue signaling, as Joel Kotkin calls them.[158]

DISPARATE COST EFFECTS ON POORER, LESS EDUCATED RESIDENTS OF ALL RACES AND ETHNICITIES

California policies promoting high cost of electricity and fuels imposes financial hardships onto those that can least afford expensive energy.

California's climate change policies, and their influence on other state legal requirements, such as the California Environmental Quality Act (CEQA) have significantly distorted the California economy and will likely continue to have a disparate effect on Latinos, African Americans, as well as all poorer, less educated residents of all races and ethnicities.

The state's Democratic party leaders relentlessly impose highly regressive climate schemes that result in disparate financial hardships on middle- and lower-income workers and minority communities while providing direct economic subsidies to wealthier Californians in environmentalist strongholds like Marin County. Most of the available funding from highly regressive greenhouse gas -related fees and taxes to assist poorer Californians affected by higher energy and housing costs, has benefitted the acquisition of rooftop solar and luxury electric vehicles by wealthier residents.

The impacts of higher housing and energy costs has had even worse impacts on other, less economically fortunate locations, particularly for income growth among Latino, African American and poorer, less educated workers, and households. Homeownership rates also fell significantly and are lower for the state's minority-majority and poorer, less educated population than environmentally aware Caucasians lined along the coastal regions of California.

California's climate change policies impose highly regressive cost burdens that particularly affect basic living expenses, including housing, transportation, heat, and electricity for the state's historically disadvantaged, and minority majority populations, as well as less affluent and educated residents in all demographic groups.

California's inland regions, where winter and summer conditions are much more extreme than in coastal areas, and where Latino and less affluent households have increasingly clustered to find affordable housing are suffering due to climate change policies. The state's inland population is also required to commute longer distances to work. Commuting delays have regressive effects on the least affluent and can spur higher emissions.

California's climate costs are also disproportionately borne by those living outside the most temperate coastal climate zones in the state's largest employment hubs, such as San Francisco/Silicon Valley, and the western areas of Los Angeles, Orange, and San Diego counties. Only the very wealthy can afford to live in the parts of the state where energy costs for utilities and travel are substantially lower due to the mild climate and proximity to employment.

Adding to California's increased cost of living are expensive energy burdens driven by state government's climate change polices. From California Energy Commission data, average household electricity used in 2018 was as much as 69 percent higher in the hotter and colder interior regions, compared to the milder climate coastal areas. [159] In other words, the lower income and increasingly middle-income workers forced to live in those interior regions must also bear a far greater share of the cost coming from the state's progressive energy policies.

California's big government energy agenda particularly impacts poor, rural, minority, and inland communities. Monthly power bills, for example are 57 percent higher on average during the summer months than in coastal communities. As unemployment rates are higher in the Inland Empire and Central Valley, residents in these communities cannot afford these higher energy costs.

Climate strategies that include intentionally increasing highway congestion in a failed attempt to persuade more people to commute by bus, and highly regressive transportation fuel price increases, force generally less affluent, minority-majority inland residents to disproportionately bear the greatest cost and longest commutes to support California's climate programs.

California's climate programs also reduce the state's ability to generate higher wage jobs for residents without college degrees in manufacturing or other industries that are highly sensitive to energy and housing costs.

California's polices are already driving residents and jobs to other states (Census Net Exodus) with much higher greenhouse gas emissions. [160] State policies also do nothing to address transparency for the importation of products and energy from countries with far less stringent v, environmental, worker protection, and human rights standards. This already well-developed pattern of net domestic out-migration and high- greenhouse gas product and energy importation in lieu of instate manufacturing or generation directly undermines the goal of achieving net worldwide total, not just in-state greenhouse gas reductions.

Since the Global Warming Solutions Act became effective in 2007, fuel prices are the highest in the nation in California due to numerous causes: the state taxes, the state's cap and trade program compliance costs, the low-carbon fuel standard (LCFS) program compliance costs, and the renewable fuels standard (RFS) program compliance costs.

Californians continue to pay almost $1.00 more per gallon of fuel than the rest of the country for transportation fuels.[161] The high cost of fuels, fuels (no pun intended) the growth of our homelessness and poverty populations.

California's substantially higher transportation costs are the result of three components.

1. First are direct regulations. Fuels sold within the state must comply with specific formulations adopted under clean air regulations plus additional provisions adopted under the state's climate change program including additional formulation requirement, and the cost of emission credits for both production and subsequent consumer use of the fuels.
2. Second, California in 2017 increased fuel taxes to what are now the highest levels in the nation.
3. Third, Compliant fuel is more expensive to produce, and noncompliant fuel cannot be sold within the state.

Progressives argue that these are justifiable and absorbable costs because their goal is to force consumer changes to adapt to climate change. As shortages and prices spikes-yet again- appear due to the latest mishap at the refineries, the first response of the political hierarchy has been to say nothing, hoping to weather the latest event with their climate change regulations intact. When price spikes persist, they typically pass the buck by calling for an official investigation to try and change the subject, when they know government policies are the primary culprit for rising fuel prices.

California has a history of having the highest gasoline prices in the country. Why? For one, the West Coast fuels market is isolated from other supply/demand centers. As such, the West Coast is susceptible to unexpected outages of West Coast refineries as it is unable to backfill an unexpected loss in supply by quickly supplying additional product from outside of the region.

The federal excise tax of 18.40 cents per gallon (cpg) on gasoline is the same across all states and has been at this level since 1997, but state and local taxes and fees levied on gasoline vary from state to state. According to the American Petroleum Institute (API), taxes and fees, including the federal excise tax, average 62.25 cpg across the U.S. *but excluding California*, with a high of 77.10 cpg in Pennsylvania to a low of 32.19 cpg in Alaska. California is shown at 81.45 cpg, the highest indicated by API. However, the API's pricing index does not include all the taxes and fees.[162]

In California, the current gasoline taxes and fees applied to gasoline are: [163]

1. **Federal Excise Tax** – This includes an 18.3 cpg component for the Highway Trust fund and an 0.1 cpg component for the federal Leaking Underground Storage Tank Trust Fund.
2. **California State Excise Tax** – This tax revenue is deposited into the State Transportation Fund. In 2017, the state legislature passed Senate Bill 1, the Road Repair and Accountability Act of 2017, which raised the excise taxes on motor fuels effective November 2017 and provided to inflation adjust the tax rate. The current excise tax rate is 50.5 cpg.
3. **Sales Tax** – Sales taxes are applied to the price of gasoline similar to but at a lower rate than general sales taxes. The sales tax rate in any given location includes county, transit, and local sales taxes in addition to the state sales tax.
4. **Underground Storage Tank Fee (UST)** – UST is used to provide revenue for the state's Underground Storage Tank Cleanup Fund. The primary purpose of the fund is to provide financial assistance to the owners and operators of underground storage tanks to remediate conditions caused by leaks, reimbursement for third-party damage and liability, and assistance in meeting federal financial responsibility requirements. The current CA UST tax rate is 2.0 cpg.
5. **Fuels Under the Cap (FUC)** – This is part of California's Cap & Trade (C&T) program that requires fuel suppliers to purchase Allowances (basically a license to emit a

ton of greenhouse gas (GHG)) to offset the GHG from the combustion of the fuel. This fee will vary with the price of C&T Allowances. This provision was applied to gasoline in 2015. At recent C&T allowance prices, this corresponds to 14.3 cpg.

6. **Low Carbon Fuel Standard (LCFS)** – This program that began in 2011, requires suppliers of high carbon intensity (CI) fuels, like the petroleum portion of gasoline, to purchase credits from suppliers of low-CI fuels such as ethanol, biodiesel, renewable diesel, renewable natural gas, and electricity. Through 2015, the cost of the LCFS was low, but as the price of credits has increased and the LCFS standards have tightened, the costs have become more significant. This fee will vary with the price of LCFS Credits and will increase as the annual LCFS standard is scheduled to become more stringent through 2030. At recent LCFS credit prices, this amounts to 22.6 cpg.

The last two items, Fuels Under the Cap (FUC) and Low Carbon Fuel Standard (LCFS), are not considered in the tax and fee category by API or the U.S. Energy Information Administration (EIA) as these items do not strictly use a cent-per-gallon rate, a percentage of price, or may not be collected by the state. The two GHG "fees" are an indirect charge to the fuel on the fuel characteristic and a GHG emission price. However, since the application of both GHG programs mimic a fee or tax and are passed through to the retail consumer, they are included in our definition of taxes and fees.

Contrary to what some may claim, GHG-reduction programs and taxes on petroleum fuels have a cost, and these costs are passed through to consumers.

A few years ago, at an April 23, 2018 hearing before the State Senate Committee on Business, Professions and Economic Development. I testified in support of Senate Bill 1074 (John

Moorlach) called "Disclosure of government-imposed costs," which would have required gas stations to post near each gas and diesel pump a list of cost factors, to include federal, state, and local taxes, as well as costs associated with the state's various environmental rules and regulations.

The Democrat controlled committee was adamant they did not want the public to see all the costs included in the posted pump price. Instead, the Democrat-controlled committee killed the Bill that would make gas pricing transparent from future consideration and would have eliminated the need for an investigation in the first place.[164] Today, we are hearing the same concerns that Senate Bill 1074 (Moorlach) would have remedied. And the dance continues unless the voters begin to consider Republicans, Libertarians, or other alternatives to the current Democrat supermajority at all levels of California state government.

Governor Newsom and Attorney General Becerra have short memories as "their own Democratic" lawmakers already killed transparency of pricing at the pump. The Governor and Attorney general have already forgotten that a 2018 bill, SB 1074, sponsored by State Senator John Moorlach (R) from Costa Mesa and championed by the minority party in the State was effectively bottom drawered by their own ruling party.

Notably absent is any discussion of how the state's existing costs, let alone additional burdens, severely harm lower-income and historically disadvantaged communities and households as the cost of living in California, for such fundamental necessities as housing, transportation, and electricity, is simply dramatically higher than the national average. Under either the Supplemental Poverty Measure (SPM) or National Academy of Science (NAS) based estimates, the number of state residents in poverty is about 8 to 9 million, roughly the population of Austria or Switzerland. [165]

A recent study by the Bay Area Council indicates that many workers must commute over 2 hours from lower-cost houses in the Central Valley to the costly Bay Area and make the same trip home again each day.[166]

The total California system electric generation in 2018 tabulates the following[167]:

- Natural Gas was 46 percent of all in-state power generation, but that will be going down with the shuttering of three natural gas power plants in the Los Angeles area.
- Nuclear power generation was 9 percent of all power generation for California needs, but that will be going to nil with the closure of the last nuclear plant at Diablo Canyon.
- Imported electricity from the Northwest and Southwest accounted for 32 percent of California s needs as the state was incapable of generating its own power needs from in-state sources[168].
- The California Energy Commission (CEC) shows that the 2018 imported electricity from the Southwest was comprised of 17 percent from coal fired power plants. Another case of California successfully "leaking" emissions to others for the electricity demands of the fifth largest economy in the world[169].
- With no wind or solar plants proposed to take up the loss of shuttering a total of four power plants that have been providing continuous uninterruptible electricity, the imports from the Northwest and Southwest MUST rise to fill the void if the Northwest and Southwest can generate that much extra power. If not, California will experience continued power shortages.

The tunnel vision of the Scoping Plan is a fundamentally inward-looking document that requires minute and highly intrusive regulations of Californians to achieve at any cost, in many cases, very small greenhouse gas benefits measured against state goals, let alone the national and international emissions that will ultimately determine future global temperatures. The climate policy leaders have little to no concern about the greater harm to middle- and lower-income, minority, and most vulnerable and historically underprivileged populations.

Since the Global Warming Solutions Act became effective in 2007, California climate policies and related regulatory programs, including the California Environmental Quality Act (CEQA), have contributed to significantly higher costs for energy, housing, transportation, food, and other necessities in the state. These costs are highly regressive because they disproportionately burden residents and households with lower incomes and wealth.

Despite its surprisingly ineffective record, the state still intends to push ahead with policies and seems determined to ignore adverse effects on the large proportion of its population already burdened by higher energy, housing, and other costs, including minority, less affluent, and less educated residents. The real-world effects that the Scoping Plan and related climate policies will have on the millions of Californians who are today in or near poverty, and those who are struggling just to make ends meet, have never been seriously or transparently evaluated.

In 2018, California used about 285 million megawatt hours of electricity, but it imported over 30 percent—90 million megawatt hours—from other locations. California is by far the nation's largest net electrical-importing state.[170] By 2050, state consultants estimate that electrification mandates, including those for trucks, will cause total demand to skyrocket to about 500 million megawatt hours.[171]

With the states' electricity policies focused on ridding itself from coal, nuclear, and natural gas-powered plants, the dubious aspect of the all-electric policy can be summarized in a question: Where will the new "juice" (electricity) come from? California policy leaders have displayed a lack of empathy for struggling residents. Blue collar workers and middle-class families should not be punished by energy policies. Their requests for support should be heeded and responded to appropriately.

In September of 2020, rather than have Californians drive more fuel-efficient vehicles in the decades ahead, Governor Newsom issued an Executive Order to phaseout the sale of gas-powered vehicles by 2035.[172]

The Governor's Executive Order will most likely have California replacing Cuba as the vintage car capital of the world as older vehicles are continuously re-registered each year, that are less fuel efficient, and bigger emission polluters, than state-of-the-art gas-powered newer vehicles that will be banned in the state.[173] A vintage car state will most likely result in greater emissions each year from the aging fleet of vehicles.

The Governor's Executive order will result in more electrical charging demands onto a dysfunctional energy program that has already shuttered one nuclear power plant and three natural gas power plants in recent years and has five more power plants to shutter in the cross hairs – the last nuclear plant at Diablo Canyon and four more natural gas power plants. California HOPES that other states will be able to generate enough power to meet the demands of the state.

The Governor's green actions are supportive of jumping onto the EV train, knowing that EV's have a very dark side of environmental atrocities – and the non-existing transparency of human rights abuses occurring in other countries, both of which are directly connected to the mining for the exotic minerals and metals that are required for EV batteries[174].

Zero and low emission vehicles are generally from the hybrid and electric car owners which are a scholarly bunch; over 70 percent of respondents have a four-year college or post-graduate degree.[175] This likely explains why the average household income of EV purchasers is upwards of $200,000.[176] If you are not in that higher educated echelon and the high-income range of society, there may not be an appetite for an EV.

The Public Policy Institute of California (PPIC) shows that no race or ethnic group constitutes a majority of California's population. according to the 2018 American Community Survey. Latinos surpassed whites as the state's single largest ethnic group in 2014.[177]

In a recent Los Angeles Times article, citing Edmunds data, The number of battery-electric models available more than doubled from 2018 to 2019, but EV sales budged in the wrong direction. In response to the major efforts by manufacturers, the horrific EV sales data shows that only 325,000 electric and plug-in hybrid vehicles were sold in the U.S. in 2019, down from 349,000 in 2018. Half of all EV's in America are in one state – California. The rest of the country seems to be less enthralled with EV's.[178] Are EV carmakers driving off a cliff?" [179]

Those EV numbers represent a dismal 2% of the 17 million vehicles of all types sold in the United States in 2019. California sales were more than 10 percent of the nation, as California vehicle sales have exceeded 2 million for three straight years[180].

California has 31 million registered vehicles.[181] These millions of vehicles were consuming 10 million gallons a day of diesel and 42 million gallons a day of gasoline[182]. In 2019-20 and 2020-21, it was estimated that fuel consumption accounted for $7.2 billion and $7.5 billion, respectively for taxes. The fuel tax is one of several sources of revenue for the transportation program[183].

The Governor's goal to reduce fossil fuel driven vehicles will also reduce the billions of dollars of fuel tax funds for road maintenance from the resultant reduction of fuels taxes.[184] There would also be a reduction in fees for the environmental compliance programs such as the cap-and-trade program, the low-carbon fuel standard program, and the renewable fuels standard program, that are currently dumped onto the posted prices at the pumps.

Governor Newsom may have forgotten that whatever type of vehicles use the roads, there are huge funding requirements for both California's transportation infrastructure, and for the environmental compliance programs that have come from the gas pumps. California has almost 400,000 miles of roadways that are heavily dependent on road taxes, the same tax base that will be diminishing in the decades ahead.[185] So, in addition to buying new vehicles, owners may think they are saving money by not buying fuel, but lookout – here comes that VMT (Vehicle Mileage Tax) program for vehicle miles traveled on highways for the state to get back the revenue they are losing at the gas pumps.

The VMT sound like a logical idea – requiring the users of the highways to pay the fees to maintain those highways. The challenge will be how to implement that great idea which may require odometer annual readings! The VMT concept may be another book all by itself!

The Governor's dance continues!

COST OF LIVING BURDENS

In the state that has the highest poverty rate of any state in America, the rising cost of electricity and fuel in California disproportionately impacts lower- and middle-income families who lack the disposable income to absorb the extra costs.

Environmental groups like the Sierra Club issued bold proclamations against racism, but they still push policies that, in the name of fighting climate change, only lead to higher energy and housing costs. .[186] This destroys the aspirational poor.[187] Many businesses, including small firms, must convert from cheap natural gas to expensive, green-generated electricity.[188] This policy is adamantly opposed by the state's African American, Latino, and Asian-Pacific chambers of commerce[189].

The Scoping Plan nowhere acknowledges or considers how California's cost of living burdens disproportionately harm economically vulnerable and historically disadvantaged populations in the state, including Latinos (now the largest California ethnic group at 36 percent), African- American and black households at 9 percent, and residents who do not have college or graduate degrees. All these groups are more adversely affected by the high housing, energy and other costs resulting from state climate and related policies that constrain new home construction, raise the cost of housing that can be built, and boost energy prices for both electricity consumption and transportation fuels well above the national average.

Energy in California and the world has transformed the horse-and-buggy days of less than two hundred years ago into today's global economy, supplying the insatiable fuel and product demands of militaries, airlines, cruise ships, supertankers, container shipping, and trucking infrastructures.

In the span of two centuries, man has virtually eliminated weather-related deaths, greatly extended life expectancies, and drastically reduced childbirth-related fatalities[190].

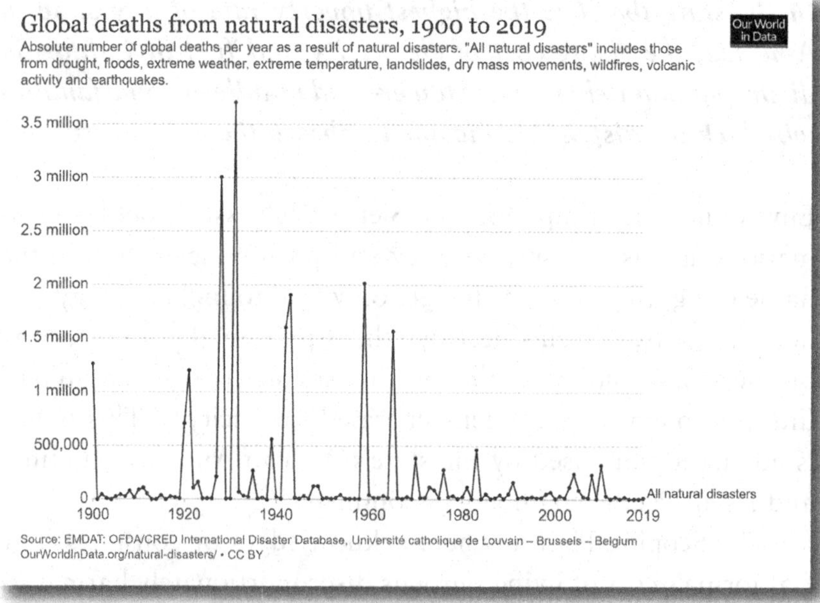

Figure 1-2 Deaths from climate and non -climate catastrophes

Due to regulatory constraints and costs, including state climate policies that increase energy prices and require significant and expensive energy efficiency installations, as well as California Environmental Quality Act (CEQA) permitting reviews that are heavily weighted against lower cost housing development in suburban areas, California failed to build sufficient housing to meet demand in almost all areas of the state.[191] The average cost of a home in California is about 2.5 times higher than in the rest of the U.S., and several times higher in coastal urban communities, including the Bay Area, the western portions of Los Angeles County, and in Orange and San Diego counties. As the Legislative Analyst's Office (LAO) documents, the average rent in the state is also about 50 percent higher than in the rest of the country and much higher in coastal urban areas.[192]

Effective climate regulations require transparency and accountability to California consumers and voters. The real consequences of high cost, dense housing, the decline of working- and middle-class employment, slower job growth in those portions of the state that do not have the Bay Area's resources or demographics, and lower homeownership must be quantified and disclosed for all income and ethnic groups in the state.

No state advertises its green credentials more than California. These policies largely accomplish the goals by "leaking" emissions to other locations with significantly less environmental control than California. This hurts the economy, driving up housing costs and narrowing opportunities for working-class people while not even doing much for the environment, has not discouraged the state's environmental overlords.[193]

These policies have been devastating to poorer Californians, particularly in the less-temperate interior, where "energy poverty" has grown rapidly.[194] Alternative energy sources like recycled natural gas and nuclear could lessen the pain, but the climate purists who dominate California policymaking find only intermittent electricity from wind and solar acceptable. Even some climate-change advocates caution that overreliance on intermittent, weather-dependent energy will push costs so high and lead to such massive land and environmental impacts that the public will turn against such policies.[195]

The biggest losers in California's green policies are minority entrepreneurs and blue-collar workers. Most of the roughly 18,000 drivers working at the Ports of Long Beach and Los Angeles are Latino, while the 4,500 in Oakland include large contingents of Indian Americans. Some 1,800 small and sole proprietors, who dominate the trade, have little wherewithal to absorb the huge costs associated with electric trucks, forcing them out of business, in favor of larger rivals.

A Massachusetts Institute of Technology (MIT) report suggests, overreliance on intermittent electricity from wind and solar will continue to impose costs, including for massive new (and environmentally unfriendly) battery-storage capacity; it will also threaten reliability, particularly without energy from other sources, such as nuclear plants.[196] Virtually every place that has tried to base its energy on a short-term shift to renewables—Germany, Demark, even resource-rich Australia—has experienced huge spikes in energy prices. In Europe, notes one recent study, reliance on renewables both reduces incomes and boosts rates of household poverty. [197]

California has the highest poverty rate of any state in America. A growing element of this problem is the cost of electricity; rising electricity prices disproportionately impact lower- and middle-income families who lack the disposable income to absorb the extra costs.[198]

California, with its suffocating cost of living and huge population, is home to an inordinate number of households receiving public assistance. Over $103 billion a year is going toward welfare for more than 3.6 million residents.[199] The Golden State's spending on the financially needy is more than the next two on the list combined.[200] New York, at number two, paid out $61.4

billion in 2015, while Texas, in the third spot, spent $35.4 billion, according to U.S. Census Bureau data.

In addition to upgrading the grid to meet the renewable mandates, California utilities must be ready to deliver lots more electricity to meet new demand due to the bans on natural gas. They must also generate more juice to meet new demand that will be coming from the transportation sector. In June, the California Air Resources Board adopted a rule that requires commercial trucks sold in the state be electric starting in 2024 and for the commercial fleet to be all-EV by 2045.

Restrictions on natural gas use will result in even-larger electricity bills for California residents and further exacerbate the state's poverty problem. Since 2019, 31 local governments in California have enacted regulations that limit or prohibit natural gas consumption in buildings. The policies were adopted in the name of climate change and decarbonization. Proponents of the restrictions claim that prohibiting the use of natural gas will help in the effort to achieve net-zero carbon emissions and that slashing building-level greenhouse emissions is essential in meeting that goal. California has a statewide goal of net zero emissions by 2045.[201]

While the number of bans is still relatively small, they are a harbinger of further efforts to restrict, or prohibit, the use of natural gas in California and other states. Proponents of the bans want California regulators to impose a statewide building code that would require all new buildings in the state to be all-electric.[202] In November 2019, Brookline, Massachusetts voted to prohibit natural gas use in new buildings.[203] Other cities in Massachusetts, Washington, Maryland, and Michigan are considering similar_restrictions[204].

The California electric grid (like the U.S. electric grid) is heavily dependent on generators that burn natural gas. Roughly a third of the electricity used in California is produced by burning natural gas, but more and more are targeted for shuttering.

Furthermore, achieving that 80 percent reduction will eventually require retrofitting nearly all the homes and commercial buildings that now consume natural gas so that they only consume electricity. The increase in electricity use will require upgrading the electric grid with new generators, wires, poles, and transformers, all of which will have to be paid for by consumers. As California plans to shutter four natural gas power plants and its last nuclear power plant, expensive imported electricity data from the Energy Information Administration (EIA) shows that California's electricity rates are soaring.[205]

The August 2019 "common interest agreement" agreement between the California Public Advocates Office and Sierra Club to help the California Public Utility Commission (CPUC) ban the use of natural gas will inflict more costs onto those that can least afford more expensive utility costs as electricity is four times more expensive as natural gas on an energy-equivalent basis.[206]

Banning the use of natural gas imposes a regressive energy tax on low- and middle-income consumers. Prohibiting the direct consumption of natural gas in furnaces, stoves, clothes dryers, and water heaters forces consumers to buy electricity, which in California is four times as expensive as natural gas on an energy-equivalent basis. To perpetuate the problem, the state imports more electricity than any other state[207] as it has been incapable of generating in-state electricity.

While the total number of local governments that have banned or restricted the use of natural gas remains relatively small, the fact that this is happening at all requires attention, because California has long been a trendsetter. The local bans on natural gas use are likely a harbinger of future policy moves by California and other states that have pledged drastic action on climate change. To achieve its climate goals, California will have to slash its use of hydrocarbons by 80 percent or more. That will require shuttering and dismantling nearly all the state's natural gas infrastructure, including its natural gas power plants, a move that will cost billions of dollars — and most of those costs will have to be borne by ratepayers.

Many of the cities that have restricted the use of gas are in the heart of Silicon Valley and are homes to some of the world's biggest technology companies[208]. Cities that have restricted the use of gas include:

- ***Cupertino***, home of Apple.
- ***Mountain View***, home of Alphabet.
- ***Menlo Park***, home of Facebook.
- ***San Francisco***, home of Visa.
- ***Los Gatos***, home of Netflix.

Also, on the list: Los Altos Hills, known as the wealthiest town in America, where the median household income is $244,000 per year[209].

Wealthy localities have erected barriers to entry. The California governments that are restricting or prohibiting the use of natural gas are far wealthier than the state and national averages. Many of them are also in the heart of Silicon Valley and are homes to some of the world's biggest technology companies, including Apple, Facebook, and Alphabet. (Sources: U.S. Census Bureau, Sierra Club, Silicon Valley Business Journal, San Mateo Daily Journal; Graphic: FREOPP)

Renewable-energy policies are a key driver of the higher costs. In 2008, then-Governor Arnold Schwarzenegger signed an executive order that required the state's utilities to obtain a third of the electricity they sell from renewables by 2020.[210] In 2015, Governor Jerry Brown signed a law that boosted the mandate to 50 percent by 2030.[211] In 2018, California lawmakers imposed yet another mandate SB-100 that requires the state's electric utilities to procure at least 60 percent of their electricity from renewables by 2030, and to be producing 100 percent "zero-carbon" electricity by 2045.[212]

Today, due to the lack of in-state electricity generation from wind and solar, California's dysfunctional energy policies have resulted in the State importing more electricity than any other state_to meet the electricity demands within the state.[213]

If California wants to avoid increasing the number of its people living in poverty, it must strive to keep electricity and fuel energy affordable. Furthermore, if the state's policymakers want to slash greenhouse gas emissions, they must find ways to reduce greenhouse-gas emissions by increasing the supply of reliable, low-carbon electricity, such as that generated by nuclear power, instead of by increasing energy costs for California's most economically vulnerable residents.

California's sustainable climate change measures need a major update to achieve not just "local" but net "global" reductions of greenhouse gas emissions and allow for affordable electricity and fuels to achieve greater equity and upward mobility for all Californians.

Progressives conclude that the poor "can't afford to live" on what they are being given and use these conditions to push for higher assistance levels. When that fails, they then push to increase regulations even more, through rent control, restrictions on driving, rules to force people out of their cars, and others that make the supply and cost conditions even worse, And when more regulations fail to fix the problem, they propose a government take-over of that part of the economy under the assumption that this time, the government will figure out how to run it better and provide the services at a level that people will actually want.

Looking back to the 1970's when former Governor Jerry Brown and his allies killed so many nuclear power plants that, had they been built, California would today be generating almost all of its electricity from zero-pollution power plants while likely having the lowest cost of electricity today. Now we have the highest cost of electricity in the country. California energy policies are a major failure, and unfortunately the upcoming Biden administration is doubling down on the same polices. A major part of Jerry Brown's legacy will be as the fossil fuel billionaire who killed more clean energy, and increased emissions, more than any other in recent history.

Brown's selfish action to protect his family wealth brings to mind that government involvement in our daily lives recalls the most terrifying nine words in the English language:" I'M FROM THE GOVERNMENT AND I'M HERE TO HELP."[214]

CHAPTER TWO

GERMANY

- ***The first country to go Green now has the highest cost of electricity in the world.***
- ***With only 1 percent of the world's 8 billion population, its residents endure the cost burdens of Germanys environmental policies.***

INTRODUCTION

Back in 2009, then U.S. President Barack Obama echoed this fatalistic statement: "(we) four years to save earth to avert eco-disaster," which was originally exclaimed by leading NASA scientist and climate expert James Hansen.[215] One can only imagine what he would say about Donald Trump's attempt to save humanity after multiple Nobel Peace Prize nominations.[216] Current Trump and U.S. haters have warned America is becoming a "failed (climate) state."[217] What would they say about Germany? They are following the advice of white males like James Hansen, so why is Germany's energy policies, electrical prices, and national security in peril?

Germany is the first country to make the futile attempt to go green, decarbonize, and have zero emissions by an identified date, off false proclamations. Germany is not alone, as an upcoming chapter will highlight Australia's fanciful road to economic ruin and destruction by putting their faith in solar panels and wind turbines for generation of intermittent electricity since they are keeping up with the German's quest for a carbon-free society.[218]

No other industrialized country has followed Germany's led to completely phase out coal-fired and nuclear-powered generation of electricity. A plan to avert disaster according to the German government was put in place after the Fukushima nuclear disaster in 2011.[219]

Germany had 17 nuclear reactors in operation generating electricity in 2011, but now only has seven still online.[220] Despite an electrical generation shortfall – Germany plans to decommission all its remaining nuclear reactors by 2022.

The Fraunhofer Institute is a German state-funded organization who has researched German energy policies and found:

> "55.7 percent of net electricity generated (in 2020) came from renewable sources including wind and solar (both intermittent electricity), biomass and hydro. Coal accounted for almost 20 percent, nuclear and natural gas with about 12 percent each. (Coal, nuclear, and natural gas are) continuous and uninterruptable electricity."[221]

Ominous is the word for these policies since Germany is preparing to "close all 84 of its coal-fired power plants by 2038."[222] Cleaner sources from solar panels and wind turbines is the justification for the closures without accounting for the intermittent and unreliable nature of renewables.

Questioning German energy policies should be the number one concern for their own national security in America, the European Union and NATO. After nuclear and coal are phased out of existence where does reliable, continuous, uninterruptable electricity come from to meet demand? This question has never been answered. More importantly, renewables at this time cannot power a modern, industrialized society like Germany.[223] The sun, wind, and water have never powered any thriving state, country, or continent, because they were never meant to have that responsibility.

The deeply troubling part about Germany's decision to vacate the responsible use of coal-fired power plants and zero-emission nuclear power is that it is all predicated on the questionable notion of anthropogenic (man-made) global warming. The British medical journal *The Lancet* for years now has published questionable data and notions proclaiming global warming is the number global killer of humanity when nothing could be farther from the truth.[224]

The continuation of 11 million children in the world dying every year from preventable causes of diarrhea, malaria, neonatal infection, pneumonia, preterm delivery, or lack of oxygen at birth as many developing countries have no or minimal access to the thousands of products from oil derivatives enjoyed by the wealthy and healthy countries is horrendous.[225] Those awful infant mortalities are directly related to the lack of access to the thousands of products from oil derivatives enjoyed by the wealthy and healthy countries, not from global warming.

New data has shown: "anthropogenic global warming will be only one-third of current midrange projections, well within natural variability and net-beneficial to life and health."[226] Then what are the real-life consequences of Germany attempting a clean energy transition?

With diminishing supplies of continuous uninterruptible electricity generation, Germany now has the highest electricity prices in the world.[227] The international consulting firm McKinsey & Co. has deemed the German government and people's green energy policies a significant threat to their national security through NATO, and energy readiness.[228] For McKinsey to be this harsh and unrelenting in their criticism of German energy policies is unprecedented. This is a crushing example of German citizens voting in and supporting green policymakers while ignoring baseload electrical grid reality.

The leading German newspaper *Der Spiegel* agreed with the consulting firm's findings while adding the term "disaster" to the so-called green energy transition (termed *Engerwiende*).[229] Energy policies based on intermittent electricity (mainly solar panels and wind turbines) comes from the destructive, misguided, and false notion the earth is disastrously warming over mankind's activities referred to as anthropogenic global warming.[230] None of these announcements have ever come true from the global warming/climate change totalitarian movement.[231]

With barely 1 percent of the world's 7.6 billion people, German residents endure the cost burdens of Germany's environmental policies that do nothing to affect the climate becoming hotter or colder.[232] Actually, these policies are making failed politicians like former U.S. Vice President Al Gore wealthier; to the tune of over $200 million dollars since leaving office in 2000.[233] So why do the Germans keep making the same decisions by advocating for intermittent electricity to power a modern, technologically-advanced country like Germany? The sun and wind have inherent difficulties they have never overcome in recorded history since the power turns off when the sun does not shine, or wind does not blow.[234]

Germans believe wind, solar, and battery energy storage systems for electrical grids are solutions that will allow them to decarbonize their entire economy. The German government led by Angela Merkel has gone so far as to sign on with Mayors of "12 major cities with over 36 million residents" to divest from fossil fuel companies, and only invest in a "Green and Just Recovery from COVID-19 Crisis."[235] Fashionable, faddish recovery plans will imperil Germany and the European Union (EU).[236]

All this green pandering without realizing the recovery from the global pandemic is only taking place because of large, multinational energy companies. *The Wall Street Journal* exclaimed in April 2020: "Big Oil to the Coronavirus Rescue," over the thousands of life-saving devices and pharmaceuticals used by hospitals, doctors, and first responders, which come from crude oil and petroleum to fight the COVID-19 pandemic.[237] Refineries manufacture the derivatives from oil that are needed to make more than 6,000 products that are essential to the medical industry, electronics, communications, transportation infrastructure, electricity generation, cooling, heating, manufacturing, and agriculture.[238] Virtually every aspect of daily lives and lifestyles to fight the pandemic. Those thousands of products the world did not have before 1900 are made from petroleum derivatives, which are also needed for aviation, trucking, automobiles, and cruise ship segments that were brought a standstill over the coronavirus in 2020-2021.

Berlin with Merkel's backing is swaying these 12 Mayors and cities away from the truth; none of them are divesting from fossil fuels this century. They are more carbon-intensive than ever. Unknowingly, or naively they are using deep earth minerals and fuels to make transportation fuels and thousands of products from oil derivatives and supporting the likes of Exxon Mobil, British Petroleum, and Saudi Aramco for their cities, countries, and continent's existence. Understand this – over 6,000 products originate from a barrel of crude oil, and without them most people would likely die.[239] Our entire modern existence for fuels and products comes from a barrel of crude oil.

Every part of what makes a German wind turbine that produces intermittent electricity comes from crude oil.[240] German solar panels cannot absorb the sun's rays for generation of intermittent electricity without a 42-gallon barrel of crude oil refined during the petrochemical process.[241] In a head scratcher for Germany's dreams of a green country (no one can define exactly what that means by the way) solar panels are now becoming some of the dirtiest forms of electricity over the environmental degradations occurring during the mining processes for exotic mineral and metals around the world, to the toxicity of disposing of used panels.[242]

Massive amounts of conventional energy from fossil fuels are used to build the energy intensive hardware in solar panels, wind turbines, battery energy storage systems from oil derivatives for electrical grids and home use along with all electrical vehicles having an enormous carbon footprint. The environmental challenge from renewable waste is currently unimaginable.

Journalist Emily Folk of *Renewable Energy Magazine* brings the entire waste and ecological impact of renewables a step further by admitting:

> "When talking about renewable energy, the topic of waste does not often appear (over the supposed) pressures of climate change and alleged urgency to find alternative (electricity) sources making people hesitant to discuss possible negative impacts of renewable energy."[243]

Solar panels and turbine blades are ending up in landfills transported to and serviced by fossil fuel vehicles and machinery. Germany's reliance on renewables and a just, green society only grows the financial statements of multinational energy companies, waste management conglomerates, and transportation firms.

The U.S. state of Washington understands this reality when a leading electrical provider the Benton Public Utility District, wrote in a lengthy report of the ecological impacts and electrical generation ineffectiveness of wind turbine energy to electricity farms:

> "Wind power has emerged as a popular choice for helping to meet greenhouse gas emission reduction goals, reasonable questions continue to be raised regarding its ability to cost-effectively contribute to the powering of modern civilization and how

> the lifecycle environmental and ecological impacts compare to other types of technologies."[244]

If a medium-sized U.S. state's regulators and utility understands the destructive nature of renewables displacing hydrocarbon's, and do not support further wind power development in the Northwest state of Washington, then who is behind this rush to the energy catastrophe for German citizens?[245]

Partly it is the passion of activists, politicians, producers, and investors for green electricity to reduce emissions and the fossil fuel divestment crowd supported by international governments, multinational financial institutions and investors seizing the day for global warming/climate change.

The above-mentioned mayor's and cities control over $295 billion in assets and call on multi-trillion-dollar pension funds to do the same. Damn the costs or return on investment for everyday working people or younger workers looking to make their mark, buy homes, get married, start families, and live better than their parents. Those days are bygone to environmental-totalitarians, and their lust for power and divestment from the supposed evils of fossil fuels. Activist investors and their proxies hijack good corporate governance. Add the western media feeding the onslaught against hydrocarbon production puts all oil, natural gas, petroleum, coal, and ultimately nuclear power companies and firms at risk of going out of business, to the detriment of world economies.[246]

Environmental totalitarians fail to explain how the world will function without the 6,000 products that are made from oil derivatives. Germany has incorporated this form of environmental nihilism into their national existence, which is multifaceted, petty, vicious, and has global implications beyond German borders. Germans are following old, tired white billionaires such as Bill Gates and the evil George Soros who want to shut down the world over a climate crisis using the slogan a "revolutionary moment," as justification.[247]

Instead, the Germans should be listening to 49 NASA climate scientists who wrote an open letter to NASA debunking the "dangerous climate change from CO2 hypothesis" (which Merkel, Soros, Gates, Gore, and western environmentalists believe) – who wisely pointed out NASA "is hyping unsubstantiated and unverified claims about climate."[248]

Will the Germans ever begin paying attention and learning the truth about the dangers of believing climate change alarmism hype and renewables being the answer for stable electricity that cannot provide any of the oil derivatives to make the products that are the basis of economies and societies worldwide? The verdict is unclear, but one thing is certain: German emissions are increasing over their rejection of nuclear power. Now they are re-committing and opening back up shuttered coal-fired power plants since industrial solar panel and wind turbine farms for electricity are unable to meet German electrical generation needs.[249]

For now, it does not look promising for German and European aspirations, which are like China and India's overwhelming use of coal for electricity over German plans for a green future leading to a carbon-free European Union.

GERMANY'S ILL-FATED RENEWABLE DREAMS

Globally recognized U.S. historian, Dr. Victor Davis Hanson unknowingly and succinctly captured Germany's energy madness by stating:

> "The problem with conventional wisdom is not that it is always wrong. The rub is that most "experts" unthinkingly and habitually mouth its validity until they ensure that it becomes static,

> unchanging, and immune from reexamination and dissent – an intolerant religious orthodoxy that finally becomes dangerous."[250]

Dr. Hanson has used common sense wisdom to describe the global warming/climate change fanatics; more akin to an ideological cult gripping German consciousness. The conventional wisdom of the German global warming/climate change (GWCC) ideology believes climate change will be reversed using solar panels and wind turbines for intermittent electricity and that electricity alone will save the earth. Germans falsely believing in their own elitist nobility is nothing new. Under current and future technological constraints for solar panels and wind turbines – German pretensions have gone to new heights – making the "great green reset" without merit or the possibility of it ever-taking place.[251]

Even worse, German, or global warming/climate change-troglodytes cannot answer how to replace the over 6,000 products that come from a barrel of crude oil.[252] A post-COVID-19 world based on intermittent electricity from renewables will lead to economic malaise and electrical grid blackouts.[253]

The facts under today's technology dictate the severely overstated impact of global warming/climate change. The U.S. Congressional Budget Office studied the economic impact of climate change and found over "30 years it will lower GDP output in 2050 by 1 percent."[254] High degrees of uncertainty came with the 1 percent lower GDP claim and the range of outcomes varied. Under normal projections using econometrics and regression analysis this means 1 percent is not statistically relevant. This study was not worth U.S. taxpayers time or money, and Germany is following the same path.

Winston Churchill implied the German people were worshippers of absolute power and allowed themselves to be ruled, as if they were children being led by adult, even if it meant their selfish interests were fulfilled."[255]

Exactly what the Germans are doing following global warming/climate change is similar to the way National Socialism seduced them decades ago. Apropos to the German people's bloodlust for renewables again when Churchill implied how Valhalla, Pericles the Greek statesman, George Washington in a positive sense and the negative energy choices in line with Attila the Hun and Tamerlane."[256]

For the German's, they must ask – peace or war for your energy policy choices? But it is fatalistic, a choice of Freud or Marx, with Freudian-Marxism winning in their psychotic energy policies. Common sense and the requirements for civilization is then no longer a useful tool for everyday choices keeping humanity alive, or the complex, interconnected world of energy.[257]

All choices, which invite Russian domination via the still uncompleted (chapter written in September-October 2020) Nord Stream 2 natural gas pipeline.[258] This energy imbroglio inextricably links the Germans to Russian Nord Stream 1 & 2 natural gas pipelines – since renewable-zealotry cannot power a modern, German society, or any country at this time.[259]

Try displacing the Russians out of domestic Germany during freezing months when the wind is not blowing, and the sun is not shining. Then, engineering and physics undoubtedly reveal there is not enough battery energy storage systems for electrical grids to store the trillions of cubic feet of natural gas needed to heat Germany. Under current technological constraints battery energy storage systems for electrical grids cost billions to achieve working status and cannot keep electrical grids working for more than minutes at a time.[260]

The type of wisdom that says the wind and sun will replace fossil fuels and hydrocarbons to power a modern, German society with electricity only, without the thousands of products from oil derivatives should ask Merkel and her acolytes in Berlin why they need Russian gas if plentiful German wind and sun can decarbonize their country's electricity? Extrapolated outwards to European security, NATO stability, and the U.S. desperate to leave foreign obligations. Germany cannot defend itself from a Russian invasion without the U.S. military and U.S. nuclear arsenal.

Germany's dangerous vision of just electricity from wind turbines and solar panels keeping Germans safe and Europe out of Russia's grip borders on dangerously naïve, or insidious moral relativism producing a national quagmire not seen since World War II.[261] Choosing between U.S. natural gas from Pennsylvania, or the frozen Artic tundra where Russian natural gas is located should not be a difficult choice.[262]

Philosopher, theologian, and author C.S. Lewis described the grip moral relativism, the belief of nothing and everything all at once, can descend upon the human condition and its cogency in the arena of energy. He implied that wise old men who were once wise, but now in power face the cardinal problem of reality; the soul, the solution of knowledge, and by wit self-discipline and virtue are thrown out the window if profitable off the backs of taxpayers and the global poor."[263]

Reality, conventional wisdom, common knowledge, are traits missing from Germany electricity discussions. German policymakers and voters hoping for green electricity solutions are subduing reality "to the wishes of mankind."[264] Simply put, just green electricity is a failure from California to Germany.[265] Even the British for all their brash, Brexit votes have failed to meet the requirements of a carbon-free, renewables-based electricity country.[266]

The EU with Germany being the largest economy within the bloc takes the brunt of the blame from the *European Environmental Agency* (EEA) for failing to achieve any of its environmental goals outlined in *European outlook – state and outlook: 2020.*[267] Every environmental goal based on a just a green world powered by renewables and put into existence by the global warming/climate change exploiters does not produce the intended results according to the EEA. The EU-led environmental movement with Germany in the driver's seat continues to selectively promote their version of renewables leading to electrical baseload generation grid failure.

Have the Germans asked what will be done with solar panels, wind turbines and toxic battery energy storage systems for electrical grids in 10-20 years once they need to be replaced and tossed into landfills? The negative environmental impact is worse than any 1-1.8-degree Celsius rise in temperature equated to global warming/climate change.[268]

Lithium and cobalt imported from nations and continents with atrocious environmental abuse records will tarnish Germany for decades.[269] Conditions now so malicious causing a rare U.S. bipartisan approach approving President Trump's executive order declaring U.S. mining of rare earth metals a national emergency via Presidential Executive Order.[270] We should all be asking how millions of tons of toxic waste will be reprocessed in Germany or elsewhere if they are not broken down safely for disposal?[271]

Casting aside rare earth environmental and national security impacts – Germany's green movement – emboldens labor and working conditions that "face allegations of abuse, including land rights infringements, corruption, violence or death over the past 10 years."[272]

If Germany should be importing U.S. natural gas to counter Russian aggression, then add abundant lithium deposits in California's Mojave Desert to the list.[273] Casting aside renewables for German electrical needs and the scale of mining needed to begin worldwide for rare earth minerals for 100 percent renewable energy to electricity doable, how do Germans overcome "only 3.8 percent of global energy" originates from solar panels, wind turbines, biomass, ethanol, algae, and other acceptable renewable sources?[274]

A lesser known, but equally deadly renewable is biomass. Any entity – individual, non-profit, non-government organization, corporation, etc., - should understand the complete destruction to forests for biomass power plants existence.[275] So who profits then? The United Nations, western environmentalists, and most western governments by claiming green virtue erroneously.

With current science undoubtedly showing burning global forests for electricity "is disastrous for biodiversity, generates more emissions than coal, and isn't carbon neutral," the UN persists forward promoting biomass.[276] Why? Because otherwise they would need to disavow the 1997 Kyoto Protocol that credits electricity from biomass as carbon neutral.

Forests are being cleared to pay power companies incredibly large subsidies to burn wood pellets for electricity. This is currently happening in Britain, the EU, and in nations craving for western and UN environmental credibility by claiming zero emissions off the discredited Kyoto Protocol.[277] This faux environmental treaty is now creating more emissions from biomass forest cutdowns in the U.S., Canada, Russia, Eastern Europe, and Vietnam. The UN has never condemned biomass for being a "climate destabilizing carbon accounting loophole."[278] The UN receives their billions while hurting the environment and Germany buys into their environmental schemes.

Wind turbines environmental footprint are just as bad, maybe worse than biomass. This pictorial displays the number of materials it takes to make just one wind turbine, which Germany relies on to meet their green electricity goals.

Did you know that every part of a wind turbine depends on steel? [279] Global steel production is dependent on coal as 70 percent of steel is made from coal.

- Generator is 65 percent steel and 35 percent copper.
- Tower is 90 percent steel.

- Blades are held in place with steel.
- Foundation is made from steel and reinforced concrete.

Far left documentary filmmaker Michael Moore documented the complete destruction of biodiversity and the environment in his early 2020 documentary *Planet of the Humans.*[280] Unsurprisingly, the climate cultists and financial beneficiaries of renewables – such as the German government – hated the film.[281] The Merkel-led government is wiping out entire German forests and habitats for a so-called inevitable transition to electricity only from a wind-powered existence that will likely never take place in the next 20-40 years.

Merkel and most Germans are deeply worried about global emissions however, they are doing more harm than good to the environment by using wind turbines just for electricity. This generation's wind turbine needs "45 tons of rebar and 481m3 of concrete, (making) its carbon footprint a massive 241.85 tons of CO2."[282]

All across Germany "millions of acres of forest have been clear-felled and great swathes cut through others, to allow some 30,000 (wind turbines) be speared across Deutschland for tons of CO2 global warming emissions."[283] Begin with the assumption that all 30,000 wind turbines need to be replaced.

Then the behemoth costs for disposal are unknown. German wind companies, utilities and ratepayers are plagued with high costs throughout the entire value chain.[284] Feed-in requirements, which forces utilities and power providers to purchase expensive wind electricity is set to expire.

The inevitable course of action for many operators will be to shut down old wind turbine farms and let the courts decide their fate. Ancestry turbines are no longer valid for what the wind industry is producing. Blades "so large as to exceed the height limits placed on turbines by Germany's planning rules.[285] In the Lower Saxony region of Germany approximately 1,000 of 1,691 wind turbine site-farms are currently unavailable for repowering, according to Olaf Lies (SPD) the Lower Saxony Minister of Energy and Environment.[286]

Wind farm sites will be deemed unprofitable and abandoned. Germans will have paid billions to men like Al Gore, Bill McKibben, Tom Steyer, Michael Bloomberg, Bill Gates, and EU environmental regulators for toxic, rusting, useless, wind farms.

A study published in *Nature Communication* – who agree that the only way to stop climate change is with renewables – conceded the mining and use of renewables will annihilate biodiversity.[287] The study soberly stated:

> "Mining potentially influences 50 million km2 of Earth's land surface, with 8 percent coinciding with Protected Areas, 7 percent with Key Biodiversity Areas, and 16 percent with Remaining Wilderness. Most mining areas (82 percent) target materials needed for renewable energy production, and areas that overlap with Protected Areas and Remaining Wilderness contain a greater density of mines (our indicator of threat severity) compared to the overlapping mining areas that target other materials."[288]

Counteracting the global warming/climate change-advocates the study also said:

> "Mining threats to biodiversity will increase as more mines target materials for renewable energy production and, without strategic planning, these new threats to biodiversity may surpass those averted by climate change mitigation."[289]

Simply put, more problems will be caused by mining for rare

earth minerals and metals to support solar, wind, EV batteries, and battery energy storage systems for electrical grids than what they seek to mitigate or eliminate. Namely, undefined global warming and/or climate change that is vastly overdue for questioning instead of being taken at face value.[290] The two are virtually interchangeable in the mind of the western environmentalists or governments that uses the term(s) as instruments of fear for control and votes.

Abandoning industrial wind and solar intermittent electricity renewables should be the goal since they are pointless, intermittent, toxic, cause higher emissions, are overly expensive, and underdeliver what they promised.[291] Considering their undeniable reproach to humanity isn't it time to cast them asunder? Delusional bordering on fanatical or evil is how renewables for a German future should be viewed.

It takes a special kind of stupid to continue believing the wind and the sun will power, continuously and uninterruptedly, a small Bavarian farm, much less Berlin or Hamburg. But foreign policy, national security, and geopolitical concerns are bigger problems than unreliable and expensive electricity for German ratepayers and businesses. Placating the global warming/climate change-crowd who believe renewables are the answer needs to be understood from a German perspective before they completely destroy their country and the EU by replacing fossil fuels (mainly coal, natural gas, oil/petroleum) and nuclear, with solar panels and wind turbines which would eliminate from societies and economies the 6,000 products that are now made from oil derivatives.

WORLD WAR III IS IN THE OFFERING IF GLOBAL WARMING/CLIMATE CHANGE-IDEOLOGUES FLOURISH.

The well-meaning, but leftist Catholic Pope – Pope Francis from the secure, comfort of the Vatican believes the teachings of Saint Francis of Assisi from the 13th century can be extrapolated forward to possibly outlaw private property and capitalism based on "trickle-down" economic policies.[292] Pope Francis in his fervent heart takes climate change as one of the highest issues for mankind to tackle.[293]

Whether or not the Pope wants to confront the global pandemic within the Catholic Church of Priests sexually abusing vulnerable children and adults did not cross his mind musing on the theme of human fraternity in his latest encyclical. This is the type of person in the global warming/climate change-movement, you the reader, needs to grasp must be defeated if you want reliable, uninterruptible electricity and continue to benefit from those thousands of products made from oil derivatives that did not exist before 1900.

The section heading offers the conviction that World War III is a possibility if wind and solar backed up by battery energy storage systems for electrical grid becomes the dominant reality. The first part of this section will be the final facts connected to physics, chemistry, engineering, and electricity reality when it comes to solar panels, wind turbines, and large-scale batteries for electrical grid storage.

Solar panels can convert the sun to electricity at about 33 percent, and solar panels at best can convert energy to electricity at a 26 percent effectiveness rate; wind turbines can receive and convert at a 60 percent rate and turbines effective rate is at best 45 percent. Germans are only capturing 45 percent of wind to electricity from wind turbines. This is the reason why they have the first or second most expensive electrical rates in the world depending how they are measured.

Tesla has the world's largest battery factory in the world located in Nevada to store electricity for the sun and the wind. It would take over 500 years to make enough batteries to store only one day of daily U.S. electricity needs. Over decades of subsidized technology, possibly trillions of taxpayer monies squandered, and still only 3 percent of the world's electricity needs are being met from solar panels and wind turbines. Germans, Americans, the UN, and others are making Elon Musk a billionaire many times over based on technology that does not work as advertised from his solar panel endeavor, battery energy storage system for electrical grids, and electric vehicle companies. All are built from nonrenewable exotic mineral and metal materials when considering a "100 MW Wind Farm to deliver electricity to 75,000 homes (not an insignificant number) requires 30,000 tons of iron ore, 50,000 tons of concrete, and 900 tons of plastic and for solar panels the numbers increase 150 percent." The world and particularly Germany will need a "2000 percent increase in rare earth metals (lithium, cobalt, copper, iridium, dysprosium) to achieve a green future," and endure the worldwide environmental degradation and humanity atrocities associated with the mining for those materials in developing countries.[294]

Since China dominates the supply chains for the minerals and metals needed by renewables, Germany will be giving money through mining operations for rare earth minerals and metals to the Chinese Communist Party (CCP), which is hostile to Germany, and Berlin is now pushing back against the CCP's egregious ways.[295] That is the definition of a national security risk to Germany, NATO, the EU, and global stability. When this does not have to be the case since there is now an almost inexhaustible worldwide supply of hydrocarbons (oil, petroleum, natural gas, coal).[296]

It is roughly the same cost to drill one oil well or use hydraulic fracturing (fracking) in the exploration and production process (E&P) of hydrocarbons, as it is to build one wind turbine.[297] A wind turbine produces the equivalent of one barrel of oil per hour and when backed up with battery energy storage systems for electrical grids costs over $200 to deploy whereas an oil well produces ten barrels of oil per hour and costs less than $0.50 cents to store in a propane tank.[298] Math, physics, and chemistry highlights the national security and foreign policy nightmare for the Germans, EU, and U.S.-led NATO over using the limited worldwide supply of exotic minerals and metals to support wind and solar renewables instead of hydrocarbons.

President Biden's vision for America does not include EU-energy-bailouts, coming to Germany's rescue, or enlarging NATO. He will allow New York City to continue floundering over poor land use practices and anti-economic development that the city's Democratic politicians keep invoking.[299] Does Berlin's leadership believe Joe Biden, or the U.S. Congress will rescue Germany from Russian domination, or their own egregious energy choices?

None of this risky, energy behavior is happening outside the first-world confines of western environmental organizations and government leaders who back them. Likely why former American enemy now coveted ally Vietnam – which is continually threatened by China's aggressive behavior in the South China Sea and all of Asia – has announced a $5.09 billion liquefied natural gas power and terminal project to be developed by Exxon Mobil.[300] Power generation will begin in 2026-27, and initial capacity will be 2.25 gigawatts (GW), and can be expanded to 4.5 GW by 2029-30. The terminal will have a capacity of six million tons of liquid natural gas (LNG) per year – including large amounts of U.S. LNG.

Outcries of racial inequity, environmental injustice, and social accountability are missing from the Reuters piece and Vietnamese government. They are desperate for reliable electricity to grow their country and enhance national security while disengaging and drawing closer to western allies for protection from China. Same moves Germany should be making to reverse their actions when it comes to using Russian natural gas. But where are the chants in Hanoi from protesters crying out: "fossil fuels kill the planet! "clean air over capitalism!" Those sentiments are not reality in Asia. Only from wealthy, secure westerners.

What is now called the "Asian Century" by the *Financial Times* causes realpolitik in force, energy, and economics to rule the day over sloganeering, and false pandering.[301] Nothing will slow development of Vietnam for electricity and access to all those products made from petroleum derivatives to enhance their lives in their quest for first world status to compete with and defend themselves against China. The Institute of Energy of Vietnam is currently working:

> "On a master power development plan and has compiled a list of 22 LNG power plants with a combined potential capacity of up to 108.5 GW, the first of which will be operational by 2023."[302]

I dare Greenpeace, or individuals like Greta Thunberg, Al Gore, Tom Steyer, et al., to march against fossil fuels in major Vietnamese cities. At best anyone engaging in said marches will be arrested, and rot in a 3rd-world hellhole masquerading as a jail. At worse they will be shot and/or executed. The hypocrisy and billions these westerners receive to tout their environmental credentials is mind-numbing when they are such cowards.

The hypocrisy with Bill Gates, Al Gore, Bernie Sanders, Tom Steyer, Leonardo DiCaprio and other climate activists is that they speak about the changes we need to change every aspect of our lives to fight global warming but they're living one way for themselves and imposing another set of austerity on the rest

of us. The World's Wealthiest Consume 20 Times More Energy Than the World's Poorest.[303]

But there has been an even larger geopolitical event taking place in Europe and Germany's sphere of influence. In late September 2020, Egypt, Israel, Greece, Cyprus, Italy, and Jordan "established the East Mediterranean Gas Forum (EMGF) as an intergovernmental organization."[304] Formal status is now accredited to the forum, the U.S. and France want observer status, and the "group seeks to promote natural gas exports from the eastern Mediterranean and that Israel hopes will strengthen ties with Arab neighbors."[305]

The East Mediterranean Gas Forum (EMGF) was established to counter Russia's Nord Stream 1 & 2 along with the Turk Stream (country of Turkey) natural gas pipelines.[306] All countries are bitter rivals at one time or another with Turkey and Russia. Turkey a NATO member is locked in drilling rights disputes with EU members Greece and Cyprus. Germany continues to sit on the sidelines and not take sides since their wind turbine and solar farms do not work. They need Russian natural gas more than European and NATO security interests. While environmental billionaires continue profiting, the average European and Mediterranean citizen is in danger of Russian, Chinese, and Turkish hegemonic ambitions dominating their countries and individual lives. Where is Germany you might ask in these geopolitical disputes? Germany is doing absolutely nothing while sermonizing on ambiguous climate change.[307]

Try having another war in Europe or the Mediterranean between NATO members Greece & Cyprus on one side and Turkey quietly backed by Russia on the other. It does not take a climatologist to understand emissions will rise if a shooting conflict takes place. Germany could end this nonsense, bring Turkey back into NATO's fold, and cripple Russia by unilaterally importing and using U.S. natural gas for electrical generation over wind turbines, solar panels, and biomass. In addition, electricity can help the militaries around the world function, but electricity alone cannot make any of the military equipment: aircraft carriers, battleships, destroyers, submarines, planes, tanks and armor, trucks, troop carriers, and weaponry.

U.S. natural gas production and gross exports set new records in 2019 before COVID-19 hit.[308] Texas saw its production rise 15 percent and Pennsylvania output rose 10 percent. The U.S. is one of Germany's closest allies. For decades ahead the U.S. has more natural gas than they can possibly use. But the pull of the billionaires of the environmental movement does not care about German citizens or its safety; only taxpayer-backed profits for their solar panel and wind turbine farms, which forces German utilities to purchase grossly overpriced wind and sun electricity for their ratepayers.

Regional politics where Germany sticks its head in the sand or wrings its hands over infrastructure and transportation costs does nothing to alleviate the tension between Turkey, Eastern Mediterranean countries, and neighbors caught in the crossfire. If the EMGF project somehow stalls, expect Russian and Turkish gas to rule German and EU energy security with an iron fist.

While Germany toils away with blubbering climate change rhetoric and funding the Kremlin's military by purchasing their natural gas – the U.S. shale revolution continues – in ways Germany never thought possible. Literally, American fracking has changed the world.[309] The Germans will have to decide one way or another if they want to be part of this revolution or the renewable debacle that puts them in Vladimir Putin's crosshairs?[310] Putin hates shale natural gas and oil, because it clips his geopolitical and economic power.[311] The Russian economy has zero diversification since Putin favors crony capitalism that support his ambitions.[312] It's only real source of revenue and job growth is from fossil fuels. That is, it, nothing else.

Revolutions are about more than ideas or Marxist political movements they are now about energy. They are nothing without access to abundant, affordable, reliable, scalable, and flexible energy and electricity. The U.S. provides that for any nation abundantly – the question for Germany is which side will they choose? With Germany overreliance on renewables this is now a dire emergency for the German people, the entire EU, and NATO.

Thomas Donilon, national security advisor to former U.S. President Barack Obama said: "a stronger hand in pursuing and implementing its international security goals," is what the shale revolution affords.[313] U.S. Secretary of State Mike Pompeo said: "the shale revolution has provided the United States with a flexibility in international affairs that it had not had for decades."[314] The Brookings Institution agreed with this bipartisan approach to energy by stating:

> "In the modern era, no other commodity (oil, natural gas, coal i.e., fossil fuels/hydrocarbons) has played such a pivotal role in driving political and economic turmoil, and there is every reason to expect this to continue."[315]

Previously, the Middle East has been central to world oil supplies. Saudi Arabia backed by OPEC has been the leader to balance global oil supplies. If any type of price spike or disruption occurs, the Saudi's have historically brought spare capacity online quickly. Think of the Saudi's historically in the oil era, as the "central bank of world oil."[316] That is no longer the case. If the U.S. is eager to leave Europe, and retrench at home – why does anyone believe they will continue staying in Germany? The U.S. is bringing home thousands of military personnel from German bases for the first time in decades.[317]

Berlin is rapidly learning they are on their own. The U.S. security umbrella is fading away. A Biden administration will likely exacerbate this trend. This is critically important, because of energy security and reliable electricity. Germany always knew the U.S. would come to its rescue – whether it entailed defeating Russian troops – or backing up German renewable folly. Not anymore. Intermittent electricity from wind and solar renewables are now a life and death crisis for Germany and Angela Merkel. What about the nature of the U.S. security agreement to the Persian Gulf and Arab nations?

Completely different now that U.S. engagement is no longer needed to fuel daily energy needs. Once heavily dependent on the Middle East, the U.S. in 2019 only imported about "11 percent from the Persian Gulf," whereas Canada is now the number one supplier of American fossil fuel imports.[318] The oil sands of Alberta, Canada is a larger influencer than Saudi Arabia. Germany is stuck in the middle. Turn to Putin for natural gas while relying on white male billionaires like Bill Gates for intermittent electricity from wind and solar renewables to enhance your green credibility, or embrace Roughnecks drilling for oil and natural gas in Texas, Pennsylvania, and North Dakota?

The U.S. on the other hand has decided that Europe and Germany may no longer be worth the headache. If earlier in this chapter was termed the "Asian Century" then the geopolitics of oil and particularly natural gas are playing out on the world stage.

In March 2015, a supertanker with oil left from the U.S. Gulf Coast, crossed the Panama Canal, and was welcomed by Sinopec, one of China's two major oil companies. More oil flowed that year to Shenzhen, China where this time it was American LNG. While the life-and-death U.S.-China rivalry is real, energy supplies cut barriers and the overarching need for petroleum and heating oil supersede political rhetoric. The Asian pivot is not led by diplomatic soft power coming from the U.S. State Department or German diplomats from Berlin and Brussels, but energy from Texas, Pennsylvania, New Mexico, North Dakota, Oklahoma, and Louisiana shale formations.

German interests want the U.S. tied up in the Strait of

Hormuz. It keeps the U.S. military closer to the EU. The U.S. being present in Asia leaves German, EU, and NATO behind. Trade wars, commonality of energy interests, and coronavirus blame are cast asunder for abundant energy resources. Germany can pursue wind turbines for intermittent electricity all they want, but the U.S will supply South Korea (number one purchaser of U.S. LNG), Japan, and more importantly, India to counter China.[319] India's purchasing of U.S. LNG, oil, petroleum, and petroleum products brings both nations closer to block China's ascendant rise in Asia. A contentious past is washed away, because of the shale revolution.

Germany better quickly learn that nothing is changing the world the way the U.S. shale natural gas and oil/petroleum revolution has to world politics, soft power, realist balancing, and geopolitics. Otherwise, Germany will become a Russian client-state more than the leader of the EU.

IN CONCLUSION: GERMANY HAS DECISIONS TO MAKE

The geopolitical, human rights, and environmental concerns need to be comprehended before launching a gargantuan increase in mining, destroying biodiversity, wastefully clear-cutting forests, demolishing child protection laws, and creating waste issues no one understands, all for the benefit of environmental billionaires. So far, these issues are not being understood by Germany and westerners with interests in Europe.

U.S. oil production overtook Russia and Saudi Arabia in the fall of 2018 for the first time in over four decades. What does this mean for Germany? Better choose if you believe renewables solve climate change, to rebuilding coal-fired power causing Russian natural gas to not have as big an influence over the German economy. Does China step in to fill the void if the U.S. continues European retraction the way they have in Iran and elsewhere globally via their multi-trillion-dollar Belt and Road Initiative? Germany should ask at the highest levels of government and engaged citizenry, how long will U.S. production gains last?

The COVID-19 pandemic, or German lectures on climate change will not stop exploration and production (E&P). Only changes in U.S. environmental and regulatory policy. The U.S. and Germany are virtually alike when it comes to the destructive nihilism that plagues both countries from their environmental lobbies and interest groups. A Biden administration will likely waste trillions on a Green New Deal the way Germany has in their disastrous quest for renewables to counter overblown climate change without understanding the consequences. Environmental opposition is the most potent force hampering western countries from growth and balancing the China threat. Germany will be forced to eventually decide – the U.S. and NATO, or China?[320]

To that end would Germany choose to ban fracking. If they do, expect a flood of Russian natural gas thru Nord Stream 1 & 2 and Turk Stream. Imported Russian hydrocarbons brings Russian influence, and more Russian invasions of countries in their historic ring of influence such as the Baltic States and all of Eastern Europe. Is Germany comfortable with Russian troops on their border again?

Negative impacts would be felt globally. Which is why giving into renewable fantasies and the George Soros-backed environmentalists is so dangerous. Let Mr. Soros spend and give away his money as he chooses, he lives in a free country. But understand the dangers and totalitarian nightmare you are inviting in by supporting solar and wind for electricity. Mr. Soros, Mr. Gates, all environmental billionaires, and government leaders, et al., are never inconvenienced, only the average German citizen and European.

Noted energy historian, author, and vice chairman of IHS Markit, Daniel Yergin writes in his new book *The New Map: Energy, Climate, and the Clash of Nations*:

> "The shale revolution has transformed the world oil market and is changing concepts of energy security. 'OPEC versus non-OPEC,' the arrangement that has defined the world oil market for decades, has been overtaken by a new paradigm, the 'Big Three', the United States, Russia, and Saudi Arabia."

How Germany chooses will define their country for the remainder of this century. Renewables and climate change versus German energy security. Baked moral equivalence within energy better not overtake Germany the way it has for over 150 years from the Kaiser to Merkel.

The world needs to comprehend that energy is more than intermittent electricity from wind and solar. Ever since the discovery of the versatility of products available from petroleum derivatives, and the beginning of manufacturing and assembly of cars, truck, airplanes, and military equipment, the world has had almost 200 years to develop clones or generics to replace the crude oil derivatives that are the basis of medications, electronics, communications, tires, asphalt, and fertilizers.

The social needs of our materialistic societies demand continuous and reliable electricity from coal, natural gas, or nuclear electricity generation backup. Otherwise, the manufactured products derived out of the petrochemical process that begins from a barrel of crude oil and is the basis for modern life and thriving economies ceases to exist.

The two co-authors of this book and our second book released in June 2020 titled, *Just Green Electricity – Helping Citizens Understand a World without Fossil Fuel* explains that no one has the answer at this time as to how we can continue to make products and move things without fossil fuels and the derivatives from petroleum.[321] Most of all the discovery for Germany and humanity has been the arrival of an "aha moment" – electricity alone from wind and solar cannot replace the fuels and products that are manufactured from petroleum, which is the basis for all modern life and vaccines fighting the coronavirus.

*Special note...this chapter is dedicated to my best friend growing up Wade Gilham who passed away a few years ago, and his older brother Preston Gilham, whose new book – *Swagger* – gave me the inspiration to write this difficult and well researched chapter.

CHAPTER THREE

AUSTRALIA

- ***The second country to go Green now has the second highest cost for electricity.***
- ***With only 0.33 percent of the world's 8 billion population, its residents endure high costs for an unstable electrical grid.***

INTRODUCTION

Australia is currently the second country after Germany to attempt a transition away from fossil fuels to intermittent electricity with solar panels and wind turbines. A cautionary tale comes from two international energy companies struggling with their net zero emission plans/transition – Royal Dutch Shell and British Petroleum (BP) – both are laying off over 20,000 people.[322] Other European Energy giants: "Norway's Equinor, France's Total, and Spain's Repsol have all made similar commitments to hit net zero by 2050."[323] The U.S.' largest oil and gas company Exxon Mobil has gone the other way by doubling down on additional exploration and production for oil, petroleum, and natural gas.

Climate change alarmists bordering on an ideological cult are scaring energy companies with bad press and social media green virtue shaming into pledging no emissions from all-of-the-above mentioned energy conglomerates.[324] Expect tens of thousands lost jobs and hundreds of billions in lost business, revenues, and benefits to assuage the global warming/climate change advocates.

Australia is learning the reality of trying the technologically impossible task of achieving net-zero emission electricity.[325]

Australia has some of the highest electricity prices in the world via overreliance on renewables; causing economic floundering, and China is feeling emboldened over Australian weakness to take advantage of them on the international stage and Asian hemisphere. And environmentalists could care less.

Australia's burgeoning renewables leading to decarbonization plans sets the stage for China to continue manipulating Australians and world governments with their dubious announcements about relying on wind and solar for electricity. China uses renewables to manipulate naïve western environmentalists (particularly U.S advocates of wind power) and gullible Australians.[326] Be under no starry-eyed illusions – China is a growing, hegemonic, duplicitous power that will use any means necessary – such as using renewables for geopolitical gain to intimidate and conquer all of Asia.[327]

Under current technological constraints Australia will not power their economy with the wind and the sun intermittent electricity for this simple reason: "the wind doesn't always blow, and the sun doesn't always shine."[328] Peak power demand coinciding with continuous and uninterruptable on-demand electricity does not align with fluctuating weather and climates. Australia has gone further than most countries attempting country-wide decarbonization by heavily using industrial wind farms. Grid reliability in Australia is critically deficient, and the notion of renewables being as cost-effective or lower than fossil fuels or nuclear for electricity is simply false.[329]

Could it be Australia is betting on a Joe Biden U.S. Presidency to kick start renewables by implementing the American Green New Deal? The former U.S. Vice President and current U.S. President has said, the "Green New Deal (is) a crucial framework for meeting climate challenges we face."[330] Mr. Biden cannot answer exactly what are the climate challenges, and admits renewables, which are the cornerstone of Australian and western climate ambitions cannot move forward without hydraulic fracturing (fracking) for fossil fuels.[331]

The Australian and global energy landscape will continue experiencing renewables only being able to provide no more than 9 percent for electrical generation.[332] The International Energy Agency's *World Energy Outlook 2020* counters the claim that renewables are the new, dominant source for electricity and can replace hydrocarbons under decarbonization plans.[333] The outlook within the report confirms growing fossil fuel use moving forward.[334] Additionally, Australia is in the same hemisphere as India. For decades ahead India is building and expanding use of coal-fired power plants.[335] Australia does not have any answers for how to counter India and China's growing emissions from hundreds of new coal-fired power plants.[336]

Australian policymakers, western environmentalists and anyone interested in health, wealth and prosperity should understand renewables are only for electricity and cannot produce the over 6,000 products that come from a barrel of crude oil.[337] Without those products – ranging from plastics to pharmaceuticals – our modern, thriving, technologically engaging world would likely end. The co-author of this book, Ronald Stein, is an expert explaining why these products do not exist without hydrocarbons, and how there is no current replacement for them.

Each of Ron's chapters within this book, and our other two books: *Energy Made Easy* and *Just Green Electricity* should be read, pondered, and understood for how crude oil is the greatest source of health, wealth, and material prosperity the world has ever known. Outside of Alex Epstein's *The Moral Case for Fossil Fuels* no one does a better job than Ron explaining the role of crude oil in our everyday, affluent lives the past 200 years.[338, 339]

If hydrocarbons are not the answer, and renewables are not breaking single digits for global electricity consumption then battery energy storage systems for electrical grids must be the answer. The Australian government believes this is true, but battery storage technology is developing too slowly to make a difference.[340] Global media outlets, western political parties, and environmentalist billionaires such as Michael Bloomberg, Bill Gates, and Tom Steyer would scoff at this assertion.

But factually, by 2023 battery storage is expected to double in the U.S. and will do the same in Australia.[341] The EU is following suit. All want more battery energy storage systems for electrical grids. So that must be a good thing, right? On the contrary, take a large, modern city, say New York City or Sydney, Australia, and understand this: every battery energy storage system on planet earth combined is a pittance of what is needed to power a modern city.[342] New York City could use every electrical grid size battery in existence, and would not have enough storage to keep the lights on in the Big Apple for one hour.[343] Elon Musk is trying to make technologically advanced batteries, but so far, the technology has not caught up to the hype or ambition over the soothsaying truth of physics and economics.[344] Now imagine the amount of rare earth minerals and metals needed for just Sydney or Melbourne, Australia to run off the wind and the sun backed up by batteries for electrical grids.

Jobs, jobs, jobs, and the lifestyles and economies that thousands of products from oil derivatives are what most people want. Green jobs in Australia are a myth, the same way they are in Germany and California. Australian workers and unions in the country are sick and tired of political gamesmanship for unreliable wind and solar over fossil fuels and nuclear electricity jobs.[345] South Australia's promotion and usage of subsidized wind turbine farms for intermittent and unreliable wind power has been tough on their economy, jobs creation, uptick in electricity prices, and hampered job growth. South Australia is now like the U.S. state of Wisconsin.

Based on extensive research and data without relying on modeled assumptions:

> "Wisconsin's 10 percent renewable target has knocked a (negative) billion- dollar hole in economic activity; killing off 10,000 real jobs every year; and failed to deliver any sign of 'green' jobs."[346]

Australian trade unions wisely studying the promises made to them of additional green energy employment installing turbines and solar panels has fallen short. Reliable and affordable power supplies when not in place kill off jobs in manufacturing, mining, mineral processing, transportation, construction, and any sector in the Australian economy that depends on plentiful, inexpensive, continuously uninterruptible electrical power generation.

There is a multi-trillion-dollar price tag attached to Green New Deals without explaining how to overcome the wind not blowing, the sun not shining, and the environmental and human atrocities associated with mining billions of tons of rare earth minerals in 3rd world countries.[347] The law of physics cannot be overcome no matter how much political rhetoric.[348] A Green New Deal with renewables for just intermittent electricity is unrealistic for Australian energy policies.

U.S. estimates range from $18-$29 trillion, and those are conservative estimates since this does not factor in the loss of over 6,000 products that come from a barrel of crude oil – including every part in solar panels and wind turbines.[349] By comparison the U.S. gross domestic product (GDP) is approximately $20 trillion a year. Can Australia weather ridding fossil fuels from their economic and lifestyle equation? Doubtful for decades ahead, and in the short-term could lower Australian life expectancies since pharmaceuticals come from crude oil, and national security risks associated with China's domination of the supply chains for the minerals and metals needed by renewables.[350]

Assuaging billionaire's environmental credentials for profitability seems like the worst Faustian bargain (deal with the

Devil) a person, country or continent could make.[351] Each day Australia depends on the Faustian bargain of renewables for intermittent electricity is another death-knell being sound for their country. These bait-and-switch politicians are employing in the west and Australia promising green jobs, and taxpayer monies to wealthy environmentalists who advocate for intermittent electricity from renewables deprives economies of gainful employment, grid instability, and national security implications. Australian locals are shunned, components outsourced, and suppliers bankrupted – however, the billionaire environmentalist could care less – but why?

THE BILLIONAIRES BANKRUPTING AUSTRALIA FOR THEIR GAIN!

Roger Pielke is a faculty member at the University of Colorado since 2001. He teaches on public policy and governance in the areas of science, innovations, and sports. His degrees are in mathematics, public policy, and political science. Mr. Pielke is a Democrat just like Michael Shellenberger – both environmentalists I trust for one good reason. They are honesty brokers when it comes to the environment and energy policy. Both give clear and sober analysis building towards consensus and workable environmental and energy policy solutions.

Mr. Pielke is an energy and science wonk of the first degree for good reason. He truthfully states to his own detriment in early 2020: "How Billionaires Tom Steyer and Michael Bloomberg Corrupted Climate Science."[352] These are the very billionaires who are corrupting Australian energy policy and killing their country's economic competitiveness for power and money. Billionaires have unbelievable wealth, prestige, and access for absolute power to corrupt absolutely – and climate change coinciding with climate science and literature backing assertions benefiting billionaires and renewables go together.

Mr. Pielke argues for scientific integrity working together for solutions towards lowering emissions and mitigating the effects of exploding energy and electrical growth from China, India, and Africa for the remainder of this century.[353]

Here is the power of billionaire environmentalists and the foundations and governments that support them. Popular Mechanics now believes based on the International Energy Agency's findings:

> "It's Official – Solar is the Cheapest Form of Electricity in History!"[354] Australia employs industrial solar farms for electricity. South Australia (SA) ratepayers are expected to absorb a $2.4 billion USD increase to connect chaotically intermittent wind and solar farms in South Australia to the main Australian grid since the South Australian electrical grid is collapsing over intermittency issues from the wind and the sun.[355]

Australia continues building solar and wind farms despite the fact power rationing is in place when air conditioners run continually during hot summer months, and the daily spot pricing for electricity is among the highest in the world.[356] This causes Australian coal-fired power plants to be unprofitable and then supposedly they need to be closed for intermittent electricity from the wind and the sun, which benefits billionaire-owned renewable wind and solar farms. It is insidious and evil, because it kills the working family trying for a better life and opens Australia to being taken over by China. But the billionaire-environmentalist benefits handsomely when energy losses mount.

These out-of-control costs are never factored into how much electricity prices are for the wind and the sun. Billionaires own Popular Mechanics and most global newspapers, magazines, internet sites, television stations, satellite conglomerates, and streaming services delivering this type of news that solar is the least expensive form of electricity over fossil fuels.

They also own supposedly do-gooder foundations – such as the Rockefeller Foundation (John D. Rockefeller made all his money off oil, petroleum, and natural gas) – who ironically, are trying to destroy fossil fuels since they are heavily invested in climate change and renewables. What is referred to as Big Green, Inc., and includes the founders of Hewlett Packard, Ford Foundation, Gates Foundation, George Soros, and Michael Bloomberg – all are billionaires trying to destroy Australia and the foundation of freedom found in western nations in the post-World World II, U.S.-led, liberal order.[357]

These are the same types of fascists seen 80 years ago in Germany, which Australia is modeling in their quest for renewable electricity. Does anyone realize it was Hitler and the Nazi Party who made the first serious overtures for wind power?

Ruper Darwall, author of *Green Tyranny: Exposing the Totalitarian Roots of the Climate Industrial Complex* wrote:

> "German obsession with renewable energy (intermittent electricity) originates deep within its culture. Few know today that the Nazis were the first political party to champion wind power, Hitler calling wind the energy of the future."[358]

The roots of the environmental movement are human-hating, desolate environmental nihilism tracing its roots to Hitler, Henrich Himmler and the Nazi high command. A blood and soil ideology, which ruined Europe, and started World War II.

But these people and their acolytes want devastation – and that starts with destroying Australia's power grid by replacing fossil fuels and nuclear with intermittent electricity from solar panels and wind turbines.[359] Climate science has been entirely corrupted by billionaires who seek control and suffering for average people to line their pockets by using an obscure, technical term called Representation Concentration Pathway (RCP8.5).

Representation Concentration Pathway (RCP8.5) has gone from obscurity to fame using apocalyptic climate-rhetoric centered around one part of climate science.[360] It has had devastating consequences to climate policy discussions and detrimentally accelerating ridding the world and Australia of fossil fuels, and all those products from petroleum derivatives that were not available before 1900.[361]

A story began in 2012 about Tom Steyer, the California billionaire who made his fortune from an investment management firm focused on coal-fired power plants, and Michael Bloomberg. *The New York Times* reported in November 2012 Tom Steyer convening a meeting of environmentalists and leading Democrats after he made his billions from coal-fired power plants.[362] Mr. Steyer has never disavowed how he made his billions or given money back to anyone now that he touts renewables and the horrors of climate change. Steyer concluded the focus of the question of how to scare the masses with climate change and double his fortune was "how do you make climate change feel real and immediate for people?"[363]

Pocketbook and behavioral economics working together. Steyer then convinced two other billionaire rent-seekers of U.S. taxpayer dollars – Hank Paulson, Republican CEO of Goldman Sachs and Secretary of the Treasury under President George W. Bush, and Michael Bloomberg to each contribute $500,00 towards a project that made climate change a potential devastating threat to global economies and multinational corporations.[364]

These contributions by billionaires funded imaginary expert studies with a heavy political bias by the Rhodium Group and Risk Management Solutions. The flawed report using the obscure Representation Concentration Pathway (RCP 8.5) term was published in 2014, titled, *Risky Business: The Economic Risks of Climate Change in the United* States.[365] Since 2014 this is now widely reported as factual for why the earth is warming beyond repair. The donors and advisors are a Who's-Who's of U.S. foreign policy and investment management elites with heavy ties to other European centers of power and environmental activism overly influenced by the U.S. Democratic Party.

The report made significant errors, because it made Representation Concentration Pathway (RCP 8.5) bereft of any type of climate policy from governments, which is not true. It also took the United Nations Intergovernmental Panel on Climate Change (IPCC) and dismissed all outcomes from the report and changed "RCP 8.5 to RCP 2.6 (as well as RCP 4.5 and RCP 6.0) to come at no cost."[366] The modeling experts who created these scenarios said this was inappropriate and did not endorse this new RCP 8.5 usage.[367]

Climate science was bastardized by billionaires; no wonder people do not trust climate change rhetoric whether it is true or not. The horror is what if the climate is inexorably changing for the worse, and the climate apocalypse crowd is no more believed than the "boy who cried wolf." The faux *Risky Business* report was nothing more than a money and power grab meant to scare people.

But Paulson, Steyer, and Bloomberg took a road show approach and made 11 talks hyping their funded report at such places as the annual meeting of the American Geophysical Union (AGU) in San Francisco.[368] The largest annual gathering of climate researchers. The billionaire triumvirate went somewhere they would be celebrated instead of debating experts such as Paul Driessen, Dr. Judith Curry, or Gail Tverberg.[369] This set off a cascade of flawed climate research, and "the delusion that experts are wise enough to redesign society."[370]

Flawed papers and research based on the obscure Representation Concentration Pathway (RCP 8.5) are now the norm, which has accelerated the efforts to eliminate natural gas and nuclear for intermittent electricity from wind turbines and solar panels. The prestigious journal *Science* used the flawed *Risky Business* paper to erroneously push RCP 8.5.[371] It has been cited 413 times according to Google Scholar on October 26, 2020 when this section was written. Hardly any criticism or peer review – outside of Roger Pielke – stops this horrendous mistake moving forward that is now setting electricity policy globally.

Science then published a more extensive study in 2017 "where the abstract brazenly announced its methodological error."[372] This analysis concluded the "United States would see a 10 percent hit to its economy under the most extreme version of Representation Concentration Pathway (RCP 8.5)" without stopping to ask if anyone even understands what exactly is RCP 8.5, and how is it going to drop the U.S. economy over 10 percent.[373] This 10 percent loss is the most extreme loss per the U.S. National Climate Assessment. Not one extreme climate prediction the last fifty years has ever been proven true.[374]

What's maddening are the amount of climate experts who know the methods produced in this billionaire-influenced paper peddled as science is patently false. According to Gary Yohe, the Huffington Foundation Professor of Economics and Environmental Studies at Wesleyan University; Professor Yohe is an active participant in climate assessment for many years and states over flawed methodologies influencing global electricity policy:

> "States and urban areas have adopted this approach, as well as the National Academy of Sciences (NA) and the National Climate Assessment (NCA) of the United States."[375]

Even more damning are how few raise concerns since they want precious research dollars that billionaires provide to non-profits, foundations, and western universities to produce results they want to use for political gains and monetary advantages in the global marketplace. For a moment, imagine Donald Trump or any leading U.S. Republican or western conservative working with the Heritage Foundation or the Koch Brothers to vehemently oppose and skew data against any type of balanced climate research.

The billionaire triumvirate of Bloomberg-Steyer-Paulson has not stopped with *Risky Business*; recently the climate-billionaires have corrupted a suspect group called Climate Impact Lab (CIL). These folks are the same ones who put together the *Risky Business* report but have included universities and the Rhodium Group. In place of sourced, factual research seeking to bring billions out of poverty we have a well-funded endeavor to distort climate science. The media, politics, and energy policy are made the worse for wear over academic literature failing the public it is funded to serve by using flawed Representation Concentration Pathway (RCP 8.5) climate modeling.

The Climate Impact Lab (CIL) are peddling climate nonsense at the highest levels since global media outlets continually trumpeted their studies. A few examples are: "1.5 million more people die in India by 2100 due to extreme heat by climate change."[376] This favorite canard for climate enthusiasts is next: "Rising sea levels could swamp major cities and displace almost 200 million people, scientists say."[377] Or this one – "Rise in Climate-Related Deaths Will Surpass All Infectious Diseases."[378] The real-world science has shown that weather related deaths have virtually been eliminated in wealthier developed countries with access to heating and ventilating made available from oil derivatives.

Bjorn Lomberg's book, *False Alarm: How Climate Change Panic Costs Us Trillions, Hurts the Poor, and Fails to Fix the Planet* and Michael Shellenberger's book, *Apocalypse Never: Why Environmental Alarmism Hurts Us All* – both books published in 2020 would factually prove all three of the Climate Impact Lab (CIL) headlines untrue. Along with the thousands published daily promoting billionaire-approved climate

Armageddon's, which simply are not true, overstated, certainly overblown, and keeping continents in poverty, high infant mortality, and shorter longevity so they can make more money and gain additional power. Australian has bought into all of this for their energy and electricity national policies, further enriching mostly white, mail, western-based billionaires.

Australia is following all these policies and climate change prognostications into the economic, national security, and energy abyss. A former U.S. Navy Seal now U.S. Congressman Dan Crenshaw rebukes former U.S. Vice President Joe Biden over his demonization of fossil fuels, which provide over 6,000 daily products for Mr. Biden and his family.[379] U.S. President Joe Biden who believes the U.S. can "transition from the oil industry."[380] President Biden has yet to comprehend the limitations of just electricity from renewables to support worldwide economies and lifestyles[381]

What that does for Australia's foreign policy and ability to counter China is anyone's guess. Biden believes the entire U.S. economy, U.S. militaries, and freedom of the world can be powered by chaotically intermittent electricity from wind and solar. The greatest source of protection and prosperity for the entire western hemisphere and Asia is in jeopardy over billionaires prospering further.

Professor Pielke is correct, and his analysis should be trusted, because he understands and you, the reader should as well, that our educational institutions have let all of us down. Scientific peer review or for that matter, any peer review in literature, science, the arts, and especially public policy that relates to energy should catch glaring methodological errors. The best example for the purposes of energy is how renewable electricity is not energy in totality – it is only electricity – and intermittent at best. Question my rationale, then read Vaclav Smil's *Energy and Civilization*. Professor Smil is respected, read, and frequently quoted by Bill Gates. For Australia to trust the sun and wind to provide electricity and none of the products that support lifestyles for their modern economy and have their media lap it up the way a thirsty dog needs water is despicable.

Simply put, solar electricity is double the costs of natural gas and nuclear, wind uses hundreds of tons of raw materials, are land-grabbing behemoths, noisy, and do not work as advertised.[382] Put both together produce outrageously high electricity prices, which slows down economies, hurts the average citizen immensely, and is a national security risk for Australia, Germany, and any rich country relying on renewables for electricity and wealth.

Climate wealth coupled with media advocacy for global warming where the solution is toxic, emission-causing solar panels and wind turbines is a failure of their government, global universities, multinational corporations, and the billionaires behind the destruction of climate science that is intended to serve humanity.

IS AUSTRALIA COMMITTING ENERGY AND ELECTRICITY SUICIDE?

The answer is yes, they are committing energy and economic suicide leading to Asian and likely global instability.[383] But is there a right to sovereign misrule? In the west since freedom is a cherished choice – again, the answer is yes. At what cost for Australia and Asia since Australia is currently undertaking an "ideology of de-industrialization."[384] If the above section condemns billionaires, and climate science leading to global catastrophe by the likes of China, who in the west believes fossil fuels and nuclear are evil? Two examples are enlightening for how clean electricity exploiters are driving Australia and the west towards herculean levels of ruin.

A Professor of Anthropology in the Boston Review said: "Zimbabwe and Puerto Rico provide models for what we might call pause-full electricity." The good professor continued his palaver by saying the west had created a vast fossil fuel infrastructure – say the over 6,000 products that comes from a barrel of crude the professor relies upon daily – and it was time to "pause-full electricity." Consuming electricity continuously and uninterruptedly with "planet-destroying fossil fuels and nuclear power" we need to save the planet by giving up our demand for on-demand electricity like Zimbabwe and Puerto Rico.[385]

Of course, the good professor never mentioned if we give up 6,000 products, on-demand electricity, or the life-saving products in the fight against COVID-19, which comes from crude oil then what?[386] This type of professor should be fired, prosecuted, and sentenced to decades of prison time for endangering our entire planet for extinction based on what his dangerous rhetoric insinuates.

Patently untruthful lies and outright dangerous sentiments can never be taken lightly – particularly when the *Boston Review* is domiciled within MIT and Harvard University. Add Professor of American History Robin D.G. Kelly of U.C.L.A, then well-known climate activist and, *New York Times* best-selling author Naomi Klein says, "(the) Boston Review is so good right now," and we know western wisdom has gone off a cliff.[387]

The other side of the climate debate shows restraint and the bold-faced truth. Legendary German environmentalist and professor, Fritz Vahrenholt stated in early October 2020 on German television this about current environmental policies Germany and Australia are rushing towards:

> "Climate science is politicized, exaggerated, and filled with fantasy and fairy tales. The Paris Accord is already dead. Putin says its nonsense. The Americans were out under Trump. The Chinese do not have to do anything. It is all concentrated on a handful of European countries. The

> European Commission is massively in on it. And I predict that they will reach the targets only if they destroy European industries. (He then added) Germany is a country in denial when it comes to the broader debate on climate science."[388]

Said Professor continued on television unabashedly "characterizing Europe's recent push for even stricter emissions reduction targets to madness akin to Soviet central planning that was doomed to failure."[389] It should be noted Professor Vahrenholt holds a Ph.D. in chemistry, started his career at the Federal Environmental Agency in Berlin, then went to the Hessian Ministry of the Environment, from 1984-1990 served as State Secretary for the Environment and from 1991-97 was Minister for Energy and Environment in Hamburg before becoming a professor.[390]

Australia's future is similar to the U.S.' largest state, California. California's electrical grid can no longer provide reliable electricity when the wind blows swiftly and its reliance on solar panels and wind turbines has caused the state to have the highest electricity prices in the U.S. Renewable shortfalls in electrical generation are destroying California's economy and their electrical grid, which will take trillions to repair or replace when the entire system collapses. During a brutal heat wave like what Australia experiences during their summer months, Governor Gavin Newsome (U.S. Democrat) said:

> "The state's (California) transition away from natural gas and nuclear generation are contributing factors to the state's rolling (electrical grid) blackouts. The elimination of fossil fuel products and the shift to solar power, windmills and other forms of green energy has led to 'gaps' in the energy grid's reliability."[391]

The U.S. Environmental Protection Agency (EPA) delivered a common-sense, stinging rebuke to California policymakers, voters continuing green energy voting patterns, and Governor Newsome when former EPA Administrator Andrew Wheeler wrote a letter stating:

> "California's record of rolling blackouts – unprecedented in size and scope – coupled with recent requests to neighboring states for power begs the question of how you expect to run an electric car fleet that will come with significant increases in electricity demand, when you can't even keep the lights on today."[392]

California is the 5th largest global economy, but just 4 percent of the world's population, with arguably the greatest research universities in the world: Cal Tech, Berkeley, Stanford, and U.C.L.A, to name a few. Still, it cannot figure out electrical grid scale batteries, and/or how to build and operate solar panels and wind turbines that are as effective as coal, natural gas, and nuclear power generation.

Australia needs to stop kidding itself along with the entire west who believe just electricity from renewables and batteries are the answer to ridding the world of fossil fuels. Not hardly, and if this continues, electricity suicide in Australia and the west will continue unabated. The Australian government could rightly counter global warming is running amuck, and something must be done, right?[393] Again, not hardly.

Nobel Laureate William Nordhaus won the award for groundbreaking work on the economics of climate change accurately shows:

> "The best current research shows that the cost of climate change by the end of the century, if we

> do nothing, will be less than 4 percent of global GDP."[394]

Different outcome than the billionaire-funded *Risky Business* report concluded. Who will you trust? Who will the Australians trust? The fate of global peace over Chinese dominance could rest on those two questions. Best-selling environmental author Bjorn Lomberg, mentioned in the previous section, agrees with Mr. Nordhaus' conclusions, but takes them a step further adding the pithy comment: "instead of seeing incomes rise to 450 percent (of current levels, this section written on October 27, 2020) by 2100, they might only increase to 434 percent."[395]

Australia can still refute the understanding or the climate skeptic that climate change is accelerating at an alarming rate. So why not fill the country with solar panels and wind turbines, and regress away from the products made from the industry they wish to shutter – the oil industry, which successful people from all walks of life say is the only way to stop the planet warming uncontrollably. The problem with that logic is when you study the predictive long-run global temperature chart it has several glaring omissions. It does not record the well-documented and historical Roman and Medieval Warming Periods.[396]

These two periods, the Roman and Medieval Warming Periods, had temperatures as warm, or warmer than what is currently touted as anthropogenic global warming/climate change. More importantly, this was before billions were on the earth, no fossil fuels in use to produce CO2 emissions and the products that led us into the industrial revolution. Are the Australian people willing to bet their country and lives on charts, which omit evidence pointing towards climate change being a lesser threat than advertised? Without balancing and working with western allies towards a robust approach against the Chinese Communist Party – the Chinese government's tyranny – and unlimited ambitions will be unchecked against Australia and the Asian hemisphere.

Hope is a funny thing, and Australian politics proved this true when the 2019 elections saw the center-right Liberal-led coalition of Prime Minister Scott Morrison win re-election despite all opinion poll predicting a clear path to victory for the climate alarmism/renewable promoting Labor Party. People still vote with their pocketbooks, and have a limited, somewhat suspicious feeling about climate change being at the center of governing platforms.

Environmental activism is an issue that can easily backfire and did for the Australian voter in 2019 when a leader political commentor remarked: "How to lose the unlosable election: be anti-coal."[397] Ordinary people understand that economic growth and sustainable livelihoods do not happen without the thousands of products from fossil fuels.

From one barrel of oil, 46 percent goes to making gasoline, and the other 54 percent goes to making products. Life without oil for Australia, as well as the world, would be devoid of such products as: [398]

- Medicines: Most over the counter medications, homeopathic products and vitamins are derived from benzene, a product of petroleum.
- Cosmetics: Makeup and shampoo that has oils, perfumes, waxes, and color are all produced with the help of petrochemicals.
- Plastics: Almost all plastics are made from petrochemicals, from your iPhone to that bottle of water. It is 4 to 5 percent of the total petroleum consumption.
- Synthetic rubber: Thousands of products rely on rubber such as shoes, tires, wet suits, breast implants, and gloves.

- Cleaning products: All those ingredients you cannot pronounce in the ingredient list, all of which are very poisonous.
- Asphalt: There are over 11 million miles of paved road in the world. Asphalt aka bitumen is the glue that binds the minerals together.

THE BATTLE AHEAD FOR AUSTRALIA

Ideological renewable electricity advocates, and global warming/ climate change enthusiasts have tyranny as their goal; unless Australia changes, they will be ruled by these types for the remainder of the century. If Australia does not resist this Nazi-like movement then a transformation into an environmental plutocracy will occur. It has taken place in California and quickly Germany is being led towards another downtrodden path. This political clique will be run by billionaires, environmental extremists and "unionized state bureaucracies" since elected leaders can no longer manage what they vote into existence.[399] The administrative state in Australia and the west is more powerful than ever and they could care less about the people who put them in power.

The environmental zealots are well-funded and motivated to destroy the west into a progressive system with feudal, climate change overlords as the Australian citizenry's masters. Oh, they will pass out crumbs: basic universal income to cover skyrocketing electric prices backed with social justice blathering to scare the electorate. Ultimately, environmental-justice Australia will be remade into like the U.S. state of California, which is attempting to radically transform America, the west, and Australia into its progressive image.

Several factors have turned California into a failed-energy state based on the unrealistic notion of renewables.[400] As the world keeps adding natural gas and nuclear power plants to provide continuous uninterruptable electricity, California is eliminating natural gas and nuclear and relying on wind and solar for intermittent electricity. This causes the state to continue importing electricity from neighboring states – a dysfunctional failing grade for an energy policy on the world stage.[401] Why would Australia follow this path unless it benefitted environmentalists and their billionaire-backers?

The Labor Movement, which the Australian Labor Party backs assaults voters with climate change scare tactics, then uses fear to rake in billions to influence elections and gain political dominance. These mobilized climate armies reach into every locale and institution within respectable democracies. Then Australian environmental zealots will attempt to completely unionize police, firefighters, and all first responders bankrupting Australian cities and states. California Democratic Congressmen and Congresswomen have introduced legislation in the U.S. Congress to that effect.[402] Australia should take notice and stop anything resembling these actions at the federal level.

Australians should be cautious about the use environmental extremism by enacting laws like the California Assembly Bill 32, the "Global Warming Solutions Act," empowering administrative state fiefdoms to thrive and dominate all aspects of housing, transportation, and emissions. Australians should consider this before allowing unelected environmentalists to run their country:

> "California's environmentalists destroyed California's forests. Any attempt to deflect this catastrophe onto climate change is sophistry. Densely packed, tinder dry forests will burn like hell, and that is exactly what happened. It does not matter one bit if (Australian or California) summers are slightly dryer and slightly hotter. They'll still burn."[403]

What Green New Deals accomplish for California and eventually Australia destroy forests, but what Green New Deals truly achieve are that they destroy suburbs and push high-density housing "based on the largely unchallenged assumption that suburban sprawl results in higher per capita greenhouse gases."[404] Joel Kotkin of Chapman University has unequivocally proven this theory is wrong. Go to JoelKotkin.com to read years of articles on this subject, plus numerous books.

Consequences of excessive environmental policies will make everyday products from petroleum derivatives– the basic necessities of life unaffordable for the majority of the population. Life becomes an "era of limits" that negates staples such as housing, transportation, construction, health care, water, and electricity beyond reach for most Australians and people on earth except the billionaires and the leftist government workers who want this to take place.[405] The limits of austerity will be pushed on Australians unbeknownst to the average voter before it is too late.

Energy capitalism is destroyed once these policies are enacted. Imagine public education beholden to teacher unions, California-style, where over $900 million a year is collected and spent on U.S. elections fomenting climate change scare tactics for U.S. Democratic politicians. Identity politics become energy racism and is absolved of wrongdoing with billionaire-owned foundations pushing climate change and renewables for intermittent electricity for their profit and power.[406] That is how this works. It is truly that simple and disgusting. Environmental billionaires own the most powerful corporations on earth with tens of billions at their disposal. How much does it take to buy an Australian election? A billion here, a billion there and you have a new government for Australia serving their net worth, and not the voters and citizens of Australia.

The battle for Australia is the battle for Asia and who controls the Pacific Ocean and ultimately the west coast of the U.S. These environmental organizations, foundations, and left-leaning governments are plutocrats hell-bent on Australian destruction. Progressive feudalism using Green New Deals, solar panels, wind turbines, electric vehicles and each component justified for usage over climate change wants to destroy western political economies and capitalism. They want to reside over the destruction of wealth that energy prosperity brings, "exponentially increasing their own wealth and power."[407] Then the Australian population is controlled, instead of merit, hard work, and character being the determining factors in Australia's success or downfall.

Australia should learn capitalism back-stopped by abundant, affordable, scalable, reliable, and flexible electricity is not dangerous and is the only engine of progress available for this island-nation to block Chinese ascendance. The counter to this aggressive form of capitalism in the eyes of the environmentalist is an eco-topia for Australia and the west. Is it possible?

THE FUTURE OF ENVIRONMENTALISM IN AUSTRALIA

If Australia does not eliminate the modern-day, rent-seeking environmentalists, their country will end up like the U.S. state of California. A country whose energy policies are run thru the filter of social justice. Just once, could someone define what is social justice, and how it helps the downtrodden? Australia will electrify everything while eliminating generating capacity by shutting down coal-fired power plant and nuclear energy, like California, in the name of climate change. And of course, the gasoline-powered transportation sector must be excoriated and pushed out of business with no replacement or plan of how to replace the transportation sector?[408]

Forest fires in the interior will be blamed on climate change – again it all comes back to the ill-defined notion of climate change – when so many questions exist about what exactly is today's definition of climate change?[409] Daring to be a climate skeptic will bring personal and legal opprobrium like the Nuremberg trials.

Insane energy policies, political dysfunction, impoverishing Australians by shuttering their entire economy is what awaits them if they follow California or billionaire-funded western environmentalists. The country has few habitable areas and needs abundant, affordable, and reliable electricity that fossil fuels and nuclear provide. How many navigable rivers, streams, or temperate climates does Australia actually have? Science, physics, engineering, and common-sense electricity policies are what the Australian people need.

The immense changes Australia has undergone since the Japanese were on the precipice of invading the country in World War II is remarkable. Coal, oil, petroleum, and nuclear and the thousands of products from petroleum derivatives that were not available before 1900 are why Australia is a modern, 1st-world economy. Nowhere does environmentalism play a role in that transformation.

Reasonable environmentalism no longer exists. Even the quaint odes of "mother earth" are thrown aside when the heart of western foreign policy wisdom from the *Council on Foreign Relations* stupidly states, "California is a Preview of Climate Change's Devastation for the Entire World."[410] If Australia believes the Chinese Communist Party cares at all about climate change then climate change hell exists in their citizen's minds and governmental hierarchies and they "have made the transition from awesome to awful."[411]

Joel Kotkin of Chapman University in California has a chilling comparison of modern environmentalism to one of the greatest evils the world has ever known since recorded history by stating:

> "Modern environmentalism is largely a California product. To some, particularly the ecological Left (U.S. Democratic Party), environmental rapine is to California what the legacy of slavery has been

> to the South (former Confederate states in the U.S. during the Civil War).[412]

The Sierra Club and radical Friends of the Earth have sprung from this environmental-slavery-inspired movement. This led to discredited though still celebrated book by Paul Ehrlich *The Population Bomb*, which wrongly predicted global starvation. Not to be outdone with wrongheaded claims was Rachel Carson's nonsensical *Silent Spring* advocating against pesticides.

Ask the over 600 million people in Africa without electricity or 2 billion globally without reliable electricity for more than 1-2 hours a day how stable their food supply is without pesticides. Better yet, take Bill Gates, Tom Steyer, Michael Bloomberg, and hippies like former California Governor Jerry Brown and drop them into those regions of the world and let us see how long they last?

An obscure book that gained wide acceptance in 1975 and today influences Australian energy policies was by an American named Ernest Callenbach who published a book titled, *Ecotopia*. This book believes the best way to happiness is by regulating and deeply restricting – even limiting – transportation, all fossil fuels, airlines, shipping, and how many children family, woman or man can have. Nuclear families are considered hostile to eco-purity since they consume. Is this what Australia wants? Keep embracing industrial wind farms for intermittent electricity and this is exactly what you will vote into power (no pun intended).

All this garbage comes from these types of books, which are embraced by today's eco-zealots. An imposing regulatory response will limit every aspect of Australian lives if western environmentalism is continually embraced. Continually growing longevity and the Australian economy will grind to a halt the way COVID-19 has already done to the known world. These people want environmental lockdowns to be a normal way of life.[413] My co-author of this book has suggested California and eventually Australia will look like Cuba with a rickety electrical grid, vintage cars, and a society stuck in the 1950s.[414]

Seemingly, no major media outlet, foundation, non-profit, or respectable university mentions how these environmental policies predicated on intermittent electricity from renewables being the answer for global warming, and electricity policies that limit rather than expand humanity have been disastrous for workers and the middle class.[415]

Climate change is now the excuse for incompetent policy-maker and life-killing energy edicts. This is not Divine Will, if only it was, we are witnessing man-made disasters for profit and political repression excused at the expense of the uninformed and gullible voter. It is the worst practice of corporate cronyism, and Australia is falling for it hook-line-and-sinker unless they change.[416]

This legacy of dread that Australia will experience is seen in Germany and California. The working class more than the rich are decimated and men like British socialist James Heartfield want to ban/restrict children due to a "carbon legacy" where this viewpoint is endorsed by Sweden's Lund University and Oregon State University – both western based. When Australia does not comply with environmental edicts then "experts" for the masses own good are brought in from America, the European Union, or the United Nations.

I beg and hope Australia rejects this autocracy and develops natural gas power plants, and nuclear generated electricity. and continues to enjoy the products and fuels from petroleum, and sensible national security. Asia and the world are counting on them. Otherwise, autocracy becomes more ubiquitous than G'day mate![417]

CONCLUSION

The billionaires, plutocrats, and corporations lining up for the taxpayer subsidized-handout industry of renewables for intermittent electricity based on global warming/climate change is an insatiable, ideological affair. Closely linked to politics and every election whether the Prime Minister or the local dog catcher, inventive ways to preserve their income to the tune of billions and growing is the shibboleth of this growing movement. Renewable intermittent electricity we are told for Australia along with Germany, are to serve the public safety over emission-belching fossil fuels.

Puzzling how emissions-free electrical generation from nuclear is left off the list. Buzzwords abound production tax credits, feed-in tariffs, renewable portfolio standards, carbon-capture, cap and trade for carbon, the list never stops. All to dupe the Australian taxpayer and voter. When formerly reputable sources such as Bloomberg – billionaire Michael Bloomberg's information/media conglomerate – ecstatically announces "wind and solar are cheapest sources of electricity in most places," there is a major problem for unbiased energy policies.[418]

Lazard global investment firm added in their recent *Levelized Cost of Energy Analysis* (LCOE 14.0) finds the levelized cost of electricity for solar and wind is lower than existing gas and coal when subsidies are included.[419] In other words, this is a useless analysis, because without the subsidies they skyrocket electricity more than currently taking place in Germany, Australia, Denmark, Minnesota, Vermont, and California.

No wonder why global finance firms are turning against fossil fuels because they cannot rake in the same taxpayer handouts they can with renewables.[420] If each one of these multinational banks and investment management firms truly cared about the environment instead of virtue-signaling, then stop using the over 6,000 products that come from a barrel of crude oil. If using fossil fuels troubles you so deeply, then find another way to make these products from another source other than reliable hydrocarbons.

Good luck, as we have had more than 200 years to find generics or clones to the oil derivatives that have made today's lifestyles and economies. This constant canard is barked to the heavens by the carnies who peddle renewables as well – that fossil fuels receive more subsidies than renewables. That is patently false, a lie, and Australians need to understand the truth.[421] Renewables receive and need vastly more subsidies than fossil fuels or nuclear, and those huge subsidies trickle down to the public through accelerating electricity rate increases.

Stop telling the Australian people how cheap wind and solar intermittent electricity are when making comparisons to coal, oil, petroleum, natural gas, and nuclear generated electricity. We will know this is the case when renewables overtake fossil fuels and nuclear for electrical generation, but renewables remain incapable of providing continuous uninterruptable electricity, nor the oil derivatives that are the basis of today's living standards.

What you receive is what you pay for and does not negate the difference in price versus value. Wind and sun power may be free, but at what cost and what is the value of the unreliability both produce. Comparing a source of electricity that is weather-dependent (sun and wind) to what is available anytime, anywhere, means there is no comparison.

When some report from an investment bank, billionaire, environmental agency, foundation, non-profit, anyone who says the wind and sun are the same as fossil fuels is being intellectually dishonest. Even more so for Australia, because they are in the headwinds of the biggest geopolitical standoff of the century currently taking place between China and India.[422] Australia is stuck in the middle. They have chosen India over China, and the zenith of retorts against renewables comes the moment Australians realize they better have plenty of coal-fired generation to keep the lights on, the grid humming and their military at high-alert against the Chinese menace.

Realizing the climate change backed by renewables are moral grounds for political gain is like the Maoists employing "Red Guards" during the Chinese Communists' Cultural Revolution in the 1960s.[423] Trade ideological pure communism for Eco brown shirts who want to ruin climate skeptics of all stripes in Australia and elsewhere.[424] Then what – jail them?[425] Rid companies, families, and individuals of all their assets?[426] How about ruining reputations on social medial or other media outlets?

Australia's flirtation and seemingly embrace of the global warming jihadists agenda will restrict every facet of modern life while nothing happens to the likes of progressive-socialist, U.S. Congressman Alexandria Ocasio-Cortez (Democrat, N.Y.) who wore over $14,000 of clothes for a *Vanity Fair* cover photo shoot.[427] Congresswoman Ocasio-Cortez will be the first to rage against nuclear families, single family homes, and prosperity from nuclear and fossil fuels and the thousands of products we get from petroleum derivatives that we not available before 1900. Her type preaches austerity and practice the fruits of capitalism.

Greens and the wealthy who use them for gain are a clerisy that will ruin Australia and throw the entire Asian hemisphere in disarray and eventually a soul-crushing regional war. This is not a Tolkien fantasy – unless Australians wake up, and recapture government electricity policy towards abundant, affordable, scalable, reliable, and flexible coal, natural gas, and nuclear they will commit economic holocaust. Hopefully, this eco-nightmare does not become a reality soon. The world is watching.

CHAPTER FOUR

ENVIRONMENTAL DEGRADATION

- ***Environmental degradations resulting from the mining for the minerals and metals to support solar, wind, and EV batteries are occurring worldwide.***
- ***We get very excited when we "see" EV's powered by batteries and clean intermittent electricity being generated from wind and solar. What we do not see is the environmental degradations from the origins of the minerals and metals around the world required to support and manufacture all that clean electricity.***
  - IntroductionElectric Vehicle Batteries
  - Wind Turbines
  - Solar Panels

INTRODUCTION

For the clean electricity we "see' in electric vehicles, industrial wind and solar farms, and battery energy storage systems for electrical grids this chapter will discuss the worldwide environmental degradations occurring to access the materials mined for the construction of EV batteries, wind turbines, and solar panels.

Green energy uses materials from some of the worst ecological and human rights offenders that exist on earth.

This chapter will comment on the following aspects of each supposedly clean source of intermittent electricity from EV

Batteries, Battery Energy Storage Systems for Electrical Grids, Wind Turbines, and Solar Panels:

- Environmental degradation "leaked" to other Countries.
- Longevity
- Land Use
- Wildlife and Habitat
- Public Health and Community
- Water Use
- Life-cycle global warming emissions
- Recycling

Many in the world believe we are facing climate change caused primarily by fossil fuels where irreversible damage to the planet and humankind is imminent unless we act now and, curtail or quit all fossil fuel use. This book and chapter, along with our second book, ***Just Green Electricity*** will enlighten you on understanding a world without fossil fuels.

With billions living in abject poverty in underdeveloped lands, the authors believe this book will make you look at intermittent electricity from wind and solar, fossil fuels, nuclear energy, and the 6,000 products made from petroleum derivatives that were not available before the 1900's in a new and fresh way.

The rage these days from the Green New Deal, the Paris Accord, and the recent Democrats Clean Energy Climate Policy are all focused on renewable energy to replace our demands from fossil fuels. But wait - renewable energy from wind and solar is only renewable ELECTRICITY! At best, that renewable electricity is intermittent as it depends on mother nature's wind and sunshine to produce any electricity[428].

As will be discussed later in the book, China, and India — two of the world's most populous countries — are rejecting the use of intermittent electricity from wind and solar for scalable, reliable, affordable, abundant, and flexible electricity from coal.

Yes, of all the methods of generating electricity, coal is the most prolific source of emissions, but it is affordable and abundant in economically depressed developing countries like China and India. Even China, with thousands of coal-fired power plants, already has 46 nuclear reactors in operation and 11 more under construction to provide continuous uninterruptible zero-emission electricity.[429]

With the movement to reduce the world's dependency on the oil and gas industry, the Democratic Clean Energy Climate policy for America remains unconcerned with the United Nations warning that the raw materials used to produce electric car batteries and batteries for battery energy storage systems for electrical grids are highly concentrated in a small number of countries. This coupled with their extraction and refinement pose a serious threat to the environment. The unintended negative consequences of the shift to the exotic minerals and metals used to produce the parts for industrial wind and solar electricity, and for all green batteries, are that they are highly concentrated in a small number of countries. Their extraction and refinement pose a serious threat to worldwide ecological degradation and heinous human rights abuses.[430]

The climate cult is fearful of sharing that all the mineral and metals needed to make wind turbines, solar panels, EV batteries, and battery energy storage systems for electrical grids are mined and processed in places like Baotou, Inner Mongolia, Bolivia, and the Democratic Republic of Congo, mostly under Chinese control. All these areas have minimal to nonexistent labor, wage, environmental, reclamation, worker health, and safety regulations in place. The mere extraction of those exotic minerals and metals presents social challenges, human rights abuses, and environmental degradations worldwide.[431]

The United Nations (U.N.) trade body, United Nations Conference on Trade and Development (UNCTAD) has issued a new report breaking down some of the unintended negative consequences of the shift, which include ecological degradation as well as human rights abuses.[432] The U.N. even Warns of Devastating Environmental Side Effects of the Electric Car Boom.[433]

Before 1900 the world had relatively none of the 6,000 products now made from petroleum derivatives, which resulted in no medications, electronics, cosmetics, plastics, fertilizers, and transportation infrastructures. Looking back just a few short centuries, we have come a long way since the pioneer days. The two prime movers that have done more for the cause of globalization than any other: the diesel engine and the jet turbine, both get their fuels from oil. Road and air travel now dominate most people's lives. Also, before 1900, the world had very little commerce and without transportation there is no commerce.

The current passion of activists, politicians, producers, and investors to implement a world with only intermittent electricity are oblivious to the unintended consequences of a world without fossil fuels. The signatories to the green movement have failed to imagine how life was without that industry that did not exist before 1900 when we had, NO medications and medical equipment, NO vaccines, NO water filtration systems, NO sanitation systems, NO fertilizers to help feed billions, NO pesticides to control locusts and other pests, NO communications systems, including cell phones, computers, and iPads, NO vehicles, NO airlines that now move 4 billion people around the world, NO cruise ships that now move 25 million passengers around the world, NO merchant ships that are now moving billions of dollars of products monthly throughout the world, NO tires for vehicles, and NO asphalt for roads, and NO space program.

The world also had virtually no military aircraft carriers, battleships, destroyers, submarines, planes, tanks and armor, trucks, troop carriers, weaponry around the world before 1900. Both WW I and II were won by the Allies as they had more oil than the Axis Powers of Germany, Italy, and Japan to operate their military equipment and move troops and supplies around the world.

Post 1900, after the Industrial Revolution and WW I and II, we now have medications, electronics, cosmetics, plastics, fertilizers, transportation infrastructures and more than 6,000 products that are made from the derivatives of crude oil.[434] Including every part in solar panels.[435]Same goes for wind turbines.[436] Crude oil also allows various fuels to the world to operate planes, trucks, construction equipment, merchant ships, cruise ships, and automobiles.

ELECTRIC VEHICLE BATTERIES

Environmental degradation "leaked" to others for EV's.

Batteries powering electric vehicles are forecast to make up 90 percent of the lithium-ion battery market by 2025.[437] However, battery production uses a lot of energy, from the mining extraction of raw materials to the electricity consumed in manufacturing, and assembly. The bigger the electric car and its range, the more battery cells are needed to power it, and consequently the more carbon produced.

Battery production causes more environmental degradation damage than carbon emissions alone. Consider dust, fumes, wastewater, and other environmental impacts from cobalt mining in the Democratic Republic of the Congo.[438] Then add water shortages and toxic spills from lithium mining in Latin America.[439] Altered ecosystems hurt local communities in Central Asia due to nickel mining in Russia.[440] Mining for rare earth minerals significantly impacts air pollution in northeastern China.[441]

A lithium EV battery weighs about 1,000 pounds. While there are dozens of variations, such a battery typically contains about 25 pounds of lithium, 30 pounds of cobalt, 60 pounds of nickel, 110 pounds of graphite, 90 pounds of copper, about 400 pounds of steel, aluminum, and various plastic components.

Looking upstream at the ore grades, one can estimate the typical quantity of rock that must be extracted from the earth and processed to yield the pure minerals needed to fabricate that single battery:

- Lithium brines typically contain less than 0.1% lithium, so that entails some 25,000 pounds of brines to get the 25 pounds of pure lithium.
- Cobalt ore grades average about 0.1%, thus nearly 30,000 pounds of ore to get 30 pounds of cobalt.
- Nickel ore grades average about 1%, thus about 6,000 pounds of ore to get 60 pounds of nickel.
- Graphite ore is typically 10%, thus about 1,000 pounds per battery to get 100 pounds of graphite.
- Copper at about 0.6% in the ore, thus about 25,000 pounds of ore per battery to get 90 pounds of copper.

In total then, acquiring just these five elements to produce the 1,000-pound EV battery requires mining about 90,000 pounds of ore. To properly account for all the earth moved though—which is relevant to the overall environmental footprint, and mining machinery energy use—one needs to estimate the overburden, or the materials first dug up to get to the ore. Depending on ore type and location, overburden ranges from about 3 to 20 tons of earth removed to access each ton of ore.

This means that accessing about 90,000 pounds of ore requires digging and moving between 200,000 and over 1,500,000 pounds of earth—a rough average of more than 500,000 pounds per battery.[442]

According to the U.S. Geological Survey, Chile is home to 58 percent of the world's lithium reserves, with Australia accounting for almost a fifth of the usable deposits. However, when it comes to production, Australia takes the lead.

Most cobalt reserves are concentrated in the Democratic Republic of Congo (DRC) – with 50 percent of the world's reserves. The African country contributes almost two-thirds of the global cobalt production.

Manganese has almost a third of its global reserves, as well as production, concentrated in South Africa. China, meanwhile, has emerged as a key player in the trade of natural graphite by putting out nearly 70 percent of the global supply, despite having just a quarter of the world's usable deposits.

Batteries powering electric vehicles are forecast to make up 90 percent of the lithium-ion battery market by 2025. They are the main reason why electric vehicles can generate more carbon emissions over their lifecycle – from procurement of raw materials, to manufacturing, use and recycling – than petrol or diesel cars.[443]

Longevity of EV's

Electric car batteries[444] are only warrantied for eight years and 100,000 miles which is half of the time it would take to start saving money on gas.[445] Furthermore, replacement batteries will cost anywhere between $6,000 to $9,000.[446] Potentially large price tags await electric vehicle buyers, which pushes the payback date out even further.

On the downside, EV's rely on large battery packs that tend to degrade slightly with each charge and discharge cycle, and eventually lose their ability to fully charge which directly impacts the cars mileage range. Replacing electric vehicle batteries costs a great deal of money, which is one reason EV's tend to be more expensive than the average car to insure.

If the weather is too cold, the battery's range and performance declines; however, the longevity of the battery is typically not affected. Hot temperatures, on the other hand, can cause the battery to degrade more quickly. For these reasons, most electric vehicles are equipped with a thermal management system that keeps the batteries at a healthy temperature.

When it comes down to it, every electric car battery will eventually face degradation and an inability to provide the full charge and range when new, but many sources state that very few electric car batteries have depleted to the point of needing to be replaced. As it stands, manufacturers aim to engineer batteries that will provide electric cars with long life and top performance.

Land Use for EV's

Lithium is found in many places on our planet, including in our oceans. However, the extraction and processing take scientific expertise, clever engineering, and investments to produce industry-grade lithium.

Some people would find visiting mines all over the world a fascinating experience and many more wonder about lithium mines and lithium mining in general.[447] As production of electric cars increases, and mobile technologies take over the technological world, the cost of lithium used in the electric batteries of both industries, is rising along with the demand, and already is unable to meet it in time.

Lithium has historically been produced from two sources: brines and hard rock mining. Producing lithium from brines remains the most efficient and cost-effective process. The cost-effectiveness of brine operations forced even large producers in China[448] and Russia[449] to develop their own brine sources or buy raw materials from brine producers.

The effectiveness of producing lithium carbonate from salt brines is so favorable that most hard rock mining operations have been priced out of the market. Lithium brines are currently the only lithium source that can support mining without significant other credits from tantalum, niobium, tin, etc.

These brines contain lithium derived mainly from the leaching of volcanic rocks and vary greatly in lithium content, largely because of the extent to which they have been subject to solar evaporation.[450] They range from highly concentrated lithium deposits in the high altitude salars of Chile[451], Argentina[452], Bolivia[453], Tibet and China[454] where lithium concentrations can be very high; to mid-level brines like Silver Peak, Nevada and Searles Lake, California (a former location of lithium production); to lower concentration brines like the Great Salt Lake, Utah. The lower concentration brines have modest evaporation rates and dilution is constant due to a large volume of freshwater inflow and small lithium concentrations varying between 30 to 60 ppm.

Once the lithium is recovered, by-products include saleable compounds such as potash or boron and the chemicals used can be recycled. Lithium recovery from brines may lead to a significant carbon footprint reduction because of a nearly zero-waste mining method.

Wildlife and Habitat for EV's

The extraction of lithium has been expanding in recent years due to a growing demand in electric vehicles.[455] While lithium extraction may be promoted as good for the environment in supplying a "green" technology, there are potentially severe environmental and social impacts associated with the development of this sector.

The Lithium Triangle region, where Chile, Argentina, and Bolivia meet, is the only region in the world where two species of High Andean flamingo breed and feed. Mining for other commodities such as copper in Chile have already impacted the

Andean Flamingo wetland habitats, where some globally significant breeding sites are now largely abandoned.

As environmental impacts can underpin social impacts, it is not just flamingos and biodiversity that are affected by the imminent lithium boom in the Andes. Water scarcity has led to conflict between national governments and mining companies, as well as with local communities, where in some instances indigenous peoples have been forced to abandon their ancestral settlements.

These water scarcity impacts are predominantly related to the intense use of water in extreme desert environments. Lithium mining, through abstraction and evaporation across the *salares* (salt-flat basins where the lithium is concentrated over millions of years), can negatively impact groundwater hydrology.

Public Health and Community for EV's

The hype these days is to stop using those dirty fossil fuel driven cars and trucks and convert everyone to those clean electric vehicles. But wait!

Tesla Motors' "dirty little secret" is turning into a major problem for the EV industry—and perhaps mankind.[456] If you think Tesla's Model S is the green car of the future, think again. The promises of energy independence, a reduction in greenhouse gas emissions, and lower fuel costs, are all factors behind the rise in the popularity of electric vehicles. Unfortunately, under scrutiny, all these promises prove to be more fiction than fact.

Recently, the Environmental Protection Agency (EPA) and the U.S. Department of Energy (DOE) undertook a study to look at the environmental impact of lithium-ion batteries for EV's.[457] The study showed that batteries that use cathodes with nickel and cobalt, as well as solvent-based electrode processing, have the highest potential for environmental impacts, including resource depletion, global warming, ecological toxicity, and adverse effects on human health. The largest contributing processes include those associated with the production, processing, and use of cobalt and nickel metal compounds, which may cause adverse respiratory, pulmonary, and neurological effects in those exposed.

Before you jump onto the EV train, those EV's have a very dark side of environmental atrocities and a non-existing transparency of human rights abuses associated with mining for the exotic minerals that power the EV's.[458]

Today, 20 percent of cobalt is mined by hand. The mere extraction of the exotic minerals cobalt and lithium used in the batteries of EV's presents social challenges[459], human rights abuse challenges, and_environmental challenges.[460] Not only are working conditions hazardous, but living conditions are abysmal, with workers making such meager wages that they are forced to live in abject poverty; and, whether on-duty or off, regularly exposed to out-of-control pollution and many other environmental issues that cannot be ignored.

If one looks under the hood of a "clean energy" battery driven EV, the dirt found would surprise most. The most important component in the EV is the lithium-ion rechargeable battery which relies on critical mineral commodities such as cobalt, graphite, lithium, and manganese. Tracing the source of these minerals, in what is called "full-cycle economics," it becomes apparent that EV's create a trail of dirt from the mining and processing of minerals upstream.

The key minerals used in today's batteries are cobalt, of which 60 percent is sourced from one country, the Democratic Republic of the Congo (DRC)[461], and lithium, of which more than 50 percent is sourced from the Lithium Triangle in South America, which covers parts of Argentina, Bolivia, and Chile[462].

The cobalt mined by children and adults in these horrendous conditions in the DRC in Africa enters the supply chains of some of the world's biggest brands. There are no known "clean" supply chains for lithium and cobalt, yet the richest and most powerful companies in the world continue to offer up the most complex and implausible excuses for not investigating their own supply chains.

Proponents of EV's might counter by saying that despite these evident environmental and social problems associated with mining in many third world countries, the case remains that EV's help reduce carbon dioxide emissions associated with the internal combustion engines run on gasoline and diesel fuels. According to the reigning climate change narrative, it is after all carbon dioxide emissions that are supposedly threatening environmental catastrophes on a global scale.

For the sake of saving the world, the climate crusaders of the richer nations are willing to ignore the local pollution and human rights violations involved in mining for minerals and rare earths in Africa, China, Latin America and elsewhere.

While one might question the inherent inequity in imposing such a trade-off, the supposed advantages of EV's in emitting lower carbon emissions are overstated according to a peer-reviewed life-cycle study comparing conventional and electric vehicles[463]. To begin with, about half the lifetime carbon-dioxide emissions from an electric car come from the energy used to produce the car, especially in the mining and processing of raw materials needed for the battery. This compares unfavorably with the manufacture of a gasoline-powered car which accounts for 17 percent of the car's lifetime carbon-dioxide emissions. When a new EV appears in the showroom, it has already caused 30,000 pounds of carbon-dioxide emission. The equivalent amount for manufacturing a conventional car is 14,000 pounds.

Once on the road, the carbon dioxide emissions of EV's depends on the power-generation fuel used to recharge its battery. If it comes mostly from coal-fired power plants, it will lead to about 15 ounces of carbon-dioxide for every mile it is driven—three ounces more than a similar gasoline-powered car. Even without reference to the source of electricity used for battery charging, if an EV is driven 50,000 miles over its lifetime, the huge initial emissions from its manufacture means the EV will have put more carbon-dioxide in the atmosphere than a similar-size gasoline-powered car driven the same number of miles. Even if the EV is driven for 90,000 miles and the battery is charged by cleaner natural-gas fueled power stations, it will cause just 24 percent less carbon-dioxide emission than a gasoline-powered car. As the skeptical environmentalist Bjorn Lomborg puts it, "This is a far cry from 'zero emissions'[464].

As most ordinary people mindful of keeping within modest budgets choose affordable gasoline or diesel-powered cars, experts, and policy advisors the world over have felt compelled to tilt the playing field in favor of EV's. EV subsidies are regressive: given their high upfront cost, EV's are only affordable for high-income households. It is egregious that EV subsides are funded by the average taxpayer so that the rich can buy their EV's at subsidized prices.

The developed countries have enjoyed the benefits to human activities, lifestyles, and prosperity afforded by fossil fuels and the thousands of products made from petroleum derivatives for the last couple of centuries, but almost half the world — over three billion people — live on less than $2.50 a day. At least 80 percent of humanity, or almost 6 billion, lives on less than $10 a day.[465]

Other nations and continents living in abject poverty without electricity realize prosperous areas such as California, and the European Union are buying into new green deals, renewable futures, and zero-carbon societies are left with the dystopic reality of mass homelessness, filth, and rampant inequality that increasingly characterize the Green New Deal core values.

Water problems affect half of humanity.

In 2005, the wealthiest 20 percent of the world accounted for 76.6 percent of total private consumption. Incredibly, the poorest 40 percent of the world's population accounts for 5 percent of global income. The richest 20 percent accounts for three-quarters of world income.

The determination not to know or to look away when the facts assail our beliefs is an enduring frailty of human nature. The tendency towards group thinks and confirmation bias, and the will to affirm the "scientific consensus" and marginalize sceptics, are rife in considerations by the so-called experts committed to advocating their favorite cause. In the case of EV's, the dirty secrets of "clean energy" should seem apparent to all but, alas, there are none so blind as those who will not see.[466]

The exotic minerals of lithium and cobalt are both extremely limited in their supply and available locations, compared to crude oil that can be found in almost every country and ocean and at various depths. The limitations of supply, and the minable locations, for these in-demand commodities present a very serious challenge as to how to continue the EV revolution when those supplies begin to diminish.

As demand for rechargeable batteries grows, companies have a responsibility to prove that they have ethical supply chains, a priority when implementing green policies, and are not profiting from the misery of miners working in terrible conditions like those in the DRC. The energy solutions of the future must not be built on human rights abuses.[467]

When a company has contributed to, or benefited from, child labor or adults working in hazardous conditions, it has a responsibility to remediate the harm suffered. This means working with other companies and governments to remove children from the worst forms of child labor and support their reintegration into schools, as well as addressing health and psychological needs.

Non-existent proactive environmental regulations and human rights abuses are both on the dark side of green technology.

Water Use for EV's

One of the biggest problems associated with lithium mining is water.[468] The South America continent's Lithium Triangle, which covers parts of Argentina, Bolivia, and Chile, holds more than half the world's supply of the metal beneath its otherworldly salt flats. It is also one of the driest places on earth. That is a real issue, because to extract lithium, miners start by drilling a hole in the salt flats and pumping salty, mineral-rich brine to the surface.

Then they leave it to evaporate for months at a time, first creating a mixture of manganese, potassium, borax, and lithium salts which is then filtered and placed into another evaporation pool, and so on. After between 12 and 18 months, the mixture has been filtered enough that lithium carbonate – white gold – can be extracted.

It is a relatively cheap and effective process, but it uses a lot of water. In Chile's Salar de Atacama, mining activities consumed 65 per cent of the region's water. That is having a big impact on local farmers – who grow quinoa and herd llamas – in an area where some communities already must get water driven in from elsewhere.

Life-cycle global warming emissions for EV's

Every major carmaker has plans for electric vehicles to cut greenhouse gas emissions, yet their manufacturers are, by and large, making lithium-ion batteries in places with some of the most polluting grids in the world. Beneath the hoods of millions of the clean electric cars rolling onto the world's roads in the next few years will be a dirty battery.

A Reuters analysis found that global automakers plan to spend a combined US$300 billion on EVs over the next decade as car companies are betting big on EV's. In the last year, carmakers made some major cash commitments[469]:

- **Audi** is accelerating EV spending to €12 billion by 2024 and plans to offer 30 electrified (20 fully electric) vehicles by 2025.
- **BMW** is funneling €10 billion into new battery-cell contracts for its upcoming electric cars. It had not launched a new all-electric car in seven years, but three new ones are coming online by 2021.
- **Hyundai** just committed US$17 billion for electric and driverless cars by 2025 (less than half of that will go to EVs, so roughly US$8 billion).
- **Fiat Chrysler** has committed to investing €9 billion to launch more than 30 electrified cars by 2022.
- **Volkswagen** plans to spend €60 billion on rolling out 75 fully electric models and 60 hybrid vehicles over the next five years.
- **GM** announced a US$2.3 billion joint venture with South Korea's LG Chem to build an EV battery factory, in addition to spending US$3 billion to build an electric pickup factory in Detroit as part of its plan to add 20 new battery-electric and fuel-cell vehicles by 2023.
- **Ford** in 2018, said it plans to spend US$11 billion by 2022 to produce 40 new electrified cars.
- **Nissan** is pumping US$9 billion into China alone to bring more EVs to that country and plans to introduce more than 20 electric models by 2022.
- **Toyota** earlier in 2019, said that by 2025 all models will have electrified versions. It is spending US$2 billion on developing EVs in Indonesia alone through 2023.
- **Daimler** in 2018, announced plans to buy €20 billion worth of battery cells for its EVs by 2030. Its entire Mercedes product range will be electrified by 2022.
- **Volvo** will launch a new electric car every year through 2025, when it will phase out gas-only car sales entirely.

> Volvo told Corporate Knights it does not disclose its spending on EVs.

In addition to the huge EV battery, there is an average of 1,000 parts made of plastic, or nearly 50 percent of an EV's volume on the average electric vehicle. Hint — plastic is made from hydrocarbons, but it is not the only petroleum product used on electric vehicles. Tires, belts, hoses, all electrical wires are coated in plastic including battery wires, power steering fluid, brake fluid, antifreeze, coolant for air-conditioning, transmission fluid, all plastics in the engine compartment which are too many to begin to list as most newer cars motors are shrouded in plastic. Carbon fiber, fiberglass, most fenders, many grills, windshield wipers, sealants around windows and undercarriage, some side panels, and all paint. Steering wheel, kick panels, air bag, dashboard, carpet, door handles, switches, most parts of the seat that are not leather, center console. EV's cannot exist without the products made from oil derivatives manufactured from fossil fuels.[470, 471]

Replacing hydrocarbons with green machines under current plans—never mind aspirations for far greater expansion—will vastly increase the mining of various critical minerals around the world. For example, as previously stated, but worth reiterating, a single electric car battery weighing 1,000 pounds requires extracting and processing some 200,000 to over 1,500,000 pounds of earth—a rough average of more than 500,000 pounds per battery. Averaged over a battery's life, each mile of driving an electric car "consumes" five pounds of earth. Using an internal combustion engine consumes about 0.2 pounds of liquids per mile.

By 2021, capacity will exist to build batteries for more than 10 million cars running on 60 kilowatt-hour packs, according to data of Bloomberg NEF. Most of the EV battery supply will come from places like China, Thailand, Germany, and Poland that rely on non-renewable sources like coal for electricity.

"We're facing a bow wave of additional CO2 emissions," said Andreas Radics, a managing partner at Munich-based automotive consultancy Berylls Strategy Advisors, which argues that for now, drivers in Germany or Poland may still be better off with an efficient diesel engine.

Just to build each car battery—weighing upwards of 1,000 pounds in size for sport-utility vehicles—would emit up to 74 percent more C02 than producing an efficient conventional car if it is made in a factory powered by fossil fuels in a place like Germany, according to Berylls' findings.[472]

A single electric car contains more cobalt than 1,000 smartphone batteries; the blades on a single wind turbine have more plastic than 5 million smartphones; and a solar array that can power one data center uses more glass than 50 million phones.[473]

Yet regulators have not set out clear guidelines on acceptable carbon emissions over the entire life cycle of electric cars, even as the likes of China, France, and the U.K. move toward outright bans of combustion engines.

Recycling for EV Batteries

As the popularity of electric vehicles starts to grow explosively, so does the pile of spent lithium-ion batteries that once powered those cars. Industry analysts predict that by 2020, China alone will generate some 500,000 metric tons of used Li-ion batteries and that by 2030, the worldwide number will hit 2 million metric tons per year.

With more electric vehicles on the road and fewer gas-guzzlers, drivers burn fewer fossil fuels and put out fewer planet-heating gases into the atmosphere. But as electric vehicles become more popular, they are posing another environmental challenge: what to do with their batteries once they are off the road as batteries are starting to pile up into a gargantuan problem.[474]

We will inevitably need to recycle many of the batteries but harvesting useful materials from used lithium-ion batteries from electric vehicles remains tedious and risky.

If current recycling trends for handling these spent batteries hold, most of those batteries may end up in landfills even though Li-ion batteries can be recycled. These popular power packs contain valuable metals and other materials that can be recovered, processed, and reused. But very little recycling goes on today. In Australia, for example, only 2–3 percent of Li-ion batteries are collected and sent offshore for recycling, according to Naomi J. Boxall, an environmental scientist at Australia's Commonwealth Scientific and Industrial Research Organization (CSIRO). [475] The recycling rates in the European Union and the US—less than 5 percent—are not much higher. [476]

More than 1 million electric vehicles were sold worldwide in 2017. It is estimated that those cars alone will ultimately result in 250,000 tons of discarded battery packs. If those were to end up in landfills, they had run the risk of going through a process called "thermal runaway," which is basically a chemical reaction in the battery that can cause it to heat up, potentially to the point of burning or exploding which is the reason why TSA prohibits spare lithium-ion batteries in checked baggage when you board a plane.

But exploding landfills are not the only reason to avoid dumping old batteries. They can stay useful long after being taken out of a vehicle. Much like your cellphone, over time, the battery in an electric car will not be able to hold a charge for as long, reducing the vehicle's range. So, drivers get a new battery or a new car. But the used battery *can* typically hold and discharge up to 80 percent of the power it did when it was brand-new. What sucker wants to buy a used battery?

Government subsidies for EV's

Both federal and state governments have generous handouts for electric vehicles. The federal tax credit extends up to $7,500. Throw in state subsidies, and that figure can easily top $10,000.[477]

Furthermore, utilities that stand to benefit from drivers plugging in for an electrical charge are spending tens of millions of dollars on EV charging stations and billing the costs back to all ratepayers. And let us not forget, EV drivers do not pay any gas tax, which is literally highway robbery since the federal gas tax is supposed to pay for the Interstate Highway System that the EV's are driving on.

Who is benefitting from this government-forced benevolence? The people who least need help from other taxpayers and ratepayers. According to research from the University of California at Berkeley, 90 percent of the tax credits accrue to America's top income quintile. [478]

Nearly half of all EV sales reside in one state - California. [479] Count EV subsidies as another case of concentrating benefits to the elite and dispersing the costs among the rest of the residents, many of which cannot afford the basic needs of daily living.

EV subsidies are inadvertently supporting the "leakage" to foreign countries for the mining and extraction of the minerals and metals required to support the manufacturing of EV batteries, and the environmental degradation of land in those foreign locations. Carbon leakage occurs when there is an increase in greenhouse gas emissions in one country as a result of an emissions reduction by a second country with a strict climate policy.

As currently structured, the federal tax credit applies to the first 200,000 electric vehicles per manufacturer, and then a phase-out of the credit begins. Two automakers, Tesla, and GM reached that threshold and are in the phase-out stage. Sen. Debbie Stabenow (D-Mich.) is leading the charge to extend a $7,000 tax subsidy for an additional 400,000 vehicles per manufacturer.

Other policymakers want to lift the cap altogether and extend the tax credit permanently. Some iteration could wind up in a larger tax-extenders package. None of the current elected policymakers have any concern about those that can least afford a higher cost of living that are the ones bearing the cost of this giveaway to the elite of society.

Polling consistently shows that EV subsidies are wildly unpopular, and most Americans do not even want to give a nickel to pay for someone else's car purchase. Extending the tax credit and increasing the cap would be an economic loser to most residents and a political loser to current elected policymakers.

WIND TURBINES

Environmental degradation "leaked" to others for Wind

In addition to being weather-dependent, intermittent and unreliable, wind turbines cover vast areas of land; affect scenic views and local wind flow, temperature and moisture; kill bats and birds of prey with no penalties under migratory bird or endangered species laws; have relatively short life spans and require massive amounts of raw materials, especially for ocean turbines, compared to coal, gas, hydroelectric or nuclear plants; involve enormous air and water pollution in faraway countries where most of the mining, processing and manufacturing are done, before turbine parts are shipped to America; and more.

The volume of wind turbine waste is projected to soar in years to come, with mining and manufacturing waste, service waste, and end-of-life waste being the major sources. It is estimated that there will be *43 million metric tons* just of blade waste worldwide by 2050. China is projected to be responsible for generating 40 percent of the waste, followed by Europe (25 percent) and the USA (19 percent).[480]

Despite the hype of the promoters and policymakers,

intermittent electricity from wind turbines is not clean and not green. The climate cult is fearful of sharing that all the mineral products and metals needed to make wind turbines, solar panels, and EV batteries are mined and processed in places like Baotou, Inner Mongolia, Bolivia, and the Democratic Republic of Congo, mostly under Chinese control.[481] The Chinese have minimal to nonexistent labor, wage, environmental, reclamation, and worker health and safety regulations. Amnesty International has documented children and adults mining cobalt in narrow man-made tunnels along with the exposure to the dangerous gases emitted during the procurement of these rare minerals.[482]

It's not too far-fetched to postulate that wind turbines are leading to early childhood death and a return to child slave labor, which existed in the former Chicago meatpacking plants truthfully exposed in Upton Sinclair's *The Jungle.*

Renewables are finally being brought out into the global debate, and asked if they are socially and environmentally worth more than natural gas or nuclear energy? Far-left filmmaker Michael Moore's new documentary *Planet of the Humans* has "unmasked the power and money behind the renewables scam."[483] This documentary film has infuriated the global environmental movement.[484] The film devastatingly reveals the grotesque ecological impact emanating from wind turbines and solar panels.[485] This clearly reveals the hypocrisy coming from men such as California coal-investing-billionaire – turned environmentalist Tom Steyer – who only cares about renewables to enrich his interests and investors.[486]

The worldwide ecological destruction from the mining of precious minerals leave lands uninhabitable and worthless for plants and trees. Renewable taxpayer handouts have stripped landscapes for the list of the sixteen components needed to be mined in numerous developing countries to build wind turbines such as: aggregates and crushed Stone (for concrete), bauxite (aluminum, clay, and shale (for cement), coal, cobalt (magnets), copper (wiring), gypsum (for cement), iron ore (steel), limestone, molybdenum (alloy in steel), rare earths (magnets; batteries), sand and gravel (for cement and concrete), and zinc (galvanizing).

Never discussed by the sponsors of the Green New Deal or Paris Accords are the questionable and non-transparent labor conditions and loose, or non-existing, environmental regulations at the mining sites around the world for the products and metals required for renewables.[487] To meet the goals to go "green" will most likely cause a rare earth emergency as those "green" goals require a massive worldwide increase in mining for lithium, cobalt, copper, iron, aluminum, and numerous other raw materials.[488]

Longevity for Wind

The lifespan of the average wind turbine is 20 to 25 years, according to the U.S. Energy Information Administration (EIA).[489] As wind turbines grow older, their utilization rates become even worse, dropping at a rate of 1.6 percent each year – which eventually requires the turbine to be repowered.[490]

While new wind farms are going up, America's first generation of wind farms are reaching retirement age[491], like the Xcel Energy's Ponnequin Wind Farm on the border of Colorado and Wyoming[492]. The farm of 44 turbines recently retired at the average age of only 18 years old. In Iowa, wind turbines are reaching the end of their lives even faster as MidAmerican Energy plans to repower turbines constructed in 2004, merely 14 years after they were installed.

To make matters worse, these cost-estimates attribute lengthy 30-year lifespans to every power plant – not just wind turbines – even though coal, nuclear, natural gas, and hydroelectric plants can generate electricity for more than 50 years, compared to some wind farms that are only lasting 14 to 18 years.

Because these reports only look at a 30-year window, they

fail to account for the additional spending necessary to repower a wind turbine and extend its life. According to a report conducted last year, investments in repowering these turbines will exceed $25 billion by 2030 in the United States. Please remember this are additional emissions, pollution, fossil fuels, and rare earth minerals needed for repowering the turbine.

By not factoring in this additional repowering spending, these reports not only underestimate the true cost of electricity from wind but overestimate the cost of power plants capable of generating electricity for more than 30 years. In other words, these cost-estimates tell us that electricity from wind would be the cheapest source of energy if all power sources produced electricity for a similar period... but they do not as we are not comparing apples to apples.

Additionally, because wind turbines can only produce energy when the wind is blowing, they generate electricity less frequently than other generation sources. In Minnesota, wind farms produced electricity only 34.67 percent of the time in 2016. During June 2020 in Australia there were lengthy periods when the combined output of every wind turbine connected to the Eastern Grid struggled to top 400 MW (5.1 percent of total capacity).[493]

The short lifespan of wind turbines is rarely discussed, yet it has a massive impact on the cost of electricity for Minnesota families and businesses. Worse yet, wind energy investments were and still are completely unnecessary to meet our energy needs because energy consumption in Minnesota has been essentially flat since 2006, and existing energy sources could meet the entirety of those demands.

Wind energy is indeed very costly, even without fuel expenses. This is because wind facilities require more frequent upgrades than other power plants and produce energy at far lower rates.

Across the country, wind farm clean-up is loosely regulated by lease agreements and a patchwork of local rules. There are no binding federal regulations to mandate cleaning up after a wind farm is shuttered. Unlike coal mines, wind farms are not required to set aside funds for clean-up.[494]

If cost-estimates are going to be calculated correctly, accurate lifespans, reclamation costs, and capacity factors of different power plants need to be accounted for along with the massive land use requirements for industrial onshore and offshore wind farms. A wind farm replacement for The Diablo nuclear energy facility[495] would require about 75 percent more land, or 1,500 acres to have a capacity of 2,000 megawatts but wait – the nuclear plant generates electricity continuously and uninterruptedly while the wind turbines only generate intermittent electricity whenever the wind is blowing.

In addition, the sites most suitable for wind, may not be near civilization which would require transmission lines to deliver the electricity to where people are populated so one needs to add in the cost associated with the 10 or more years it takes to permit new transmission lines.

Wildlife and Habitat for Wind

Wind farms are groups of turbines that are as tall as 30-story buildings, with spinning rotors as wide as a passenger jet's wingspan. Blades can reach speeds of up to 170 miles per hour at the tips, creating tornado-like vortexes. According to the Fish and Wildlife Service, as many as 500 golden eagles are killed by collisions with wind towers, power lines, buildings, cars and trucks each year.[496] The American Bird Conservancy estimates that the wind industry killed over 2,000 eagles in three decades and continue to be giant avian killing fields.[497]

As with all energy supply options, electricity from wind can have adverse environmental impacts, including the potential to reduce, fragment, or degrade local habitats for wildlife, fish, and plants.[498] Furthermore, spinning turbine blades can pose a threat to flying wildlife like birds of prey and bats. Due to the potential impact that electricity from wind can have on wildlife, and the potential for these issues to delay or hinder wind development in high-quality wind resource areas, addressing impact minimization, proper siting, and permitting issues are among the wind industry's highest priorities.

Industrial wind turbines—those giant generators of electricity—are the greatest new threat to golden and bald eagles. But the eagles are hardly the only ones threatened as condors, owls, hawks, falcons, and bats all fall prey to the turbines' mighty blades.[499] By occupying large areas of migratory habitat, wind turbines have also emerged as one of the greatest threats[500] to large, threatened, slow-to-reproduce, and high-conservation value birds like hawks, eagles, owls, and condors.[501]

In 2017, the former President Obamas' administration finalized a rule that lets wind-energy companies operate high-speed turbines for up to 30 years — even if it means killing or injuring thousands of species protected under the Bald and Golden Eagle Protection Act and the Migratory Bird Treaty Act. Under the new rule, industrial wind may acquire an eagle "take" permit from the U.S. Fish and Wildlife Service (USFWS) that allows the site to participate in the nationwide killing of up to 4,200 bald eagles annually, under incidental "take" permits without compensatory mitigation.

It is shocking that industrial wind can legally obtain permits from the USFWS to kill those majestic bald eagles. I cry foul! I wonder if the renewable industry is proud of those new jobs being created also include those that need to clean up the mess from those creatures chopped up by the industrial wind generator blades and from those fried from the heat from the industrial solar panels?

In addition to the spinning turbine blades that pose a threat to flying wildlife like birds of prey and bats, those wind turbines are also accountable for human fatalities. Just through 2020, wind turbines have been responsible for 218 human fatalities as reported by the Caithness Windfarm Information Forum.[502]

To go along with all the hysteria to support renewables, astoundingly, with all the world's efforts to protect life, industrial wind is "legally" killing hundreds of thousands of top predator birds like eagles and hawks, and decimating bat populations every year. Without any regard for how this affects the ecosystems in their areas it is appalling that society has given the Industrial windfarm industry a FREE get-out-of-jail card.

Residents may be oblivious to their elected policymakers' acceptance of animal cruelty toward birds of prey, but more and more of the public are loudly rejecting this atrocity toward "taking", which is a nice word for killing, bald eagles.

Public Health and Community for Wind

"Researchers estimate the U.S. will have more than 720,000 tons of blade material to dispose of over the next 20 years, a figure that doesn't include the newer, taller higher-capacity versions," National Public Radio reported last year.[503] "It's a waste problem that runs counter to what the industry is held up to be: a perfect solution for environmentalists looking to combat climate change" with clean intermittent electricity.

The environmental risk of renewables is not limited to solar panels, either. Those monstrous turbine sails that spin, on view-spoiling acres, and bird-killing windmills must be retired, too.[504]

Promises, promises. We are constantly being blown away with the growing capacity of industrial wind to provide renewable intermittent electricity, but they've yet to produce anywhere near their projected capacity. Compounding their lack of production, is the intermittency of what they produce.[505]

Maybe world leaders should listen and learn from the Washington state utility report of July 14, 2020 titled "Wind

Power and Clean Energy Policy Perspectives," the utility's commissioners say they "do not support further wind power development in the Northwest."[506]

Many of their conclusions in the Washington state utility report are supportive of the energy literacy messages in two recent books by Ronald Stein and Todd Royal, Energy Made Easy, and Just GREEN Electricity, both available on Amazon.

Windmills are prone to violent storms, tornados, and hurricanes. Blades fly off in all directions leaving 'green' power consumers in the lurch for weeks, if not months.

There is the landfill legacy being created by wind turbine blades, with the wealthier developed First World cynically using the Third World as its dumping ground: 'Green' Energy's Poisonous Legacy: Millions of Toxic Turbine Blades Destined for African Landfills[507]. Thousands of 45-70m blades (weighing between 10 to 25 tons) are being ground up and mixed with concrete used in the bases of other turbines erected later or simply dumped in landfill[508]. Which should worry locals: the plastics in the blades are highly toxic, and contain Bisphenol A, which is so dangerous to health that the European Union and Canada have banned it.[509]

Water Use for Wind

This is one of the few positives about wind turbines, as intermittent electricity from wind saved 2.5 billion gallons of water in California in 2014 by displacing water consumption at the state's thirsty fossil-fired power plants, playing a valuable role in alleviating the state's record drought.[510]

By displacing generation from these conventional power plants, U.S. wind electricity currently saves around 35 billion gallons of water per year, the equivalent of 120 gallons per person or 285 billion bottles of water.

In addition to directly offsetting freshwater consumption at thermal power plants, intermittent electricity from wind helps combat the impacts of drought by allowing grid operators to save hydroelectric electricity (in the form of water behind dams) until they need it to meet grid reliability needs. Sometimes grid operators use electricity from wind to store additional water behind dams where it can be used later to displace fossil fuel generation.

Life-cycle global warming emissions for Wind

The origins of the products for wind and solar are mined throughout the world, inclusive of more than 60 countries of Algeria, Arabia, Argentina, Armenia, Australia, Belgium, Bolivia, Brazil, Canada, Chile, China, Congo (Kinshasa), Cuba, Egypt, Finland, France, Germany, Greece, Guinea, Guyana, India, Indonesia, Iran, Ireland, Italy, Jamaica, Japan, Kazakhstan, Madagascar, Malaysia, Mexico, Mongolia, Morocco, Mozambique, New Caledonia, Oman, Pakistan, Papua New Guinea, Peru, Philippines, Poland, Republic of Korea, Russia, Saudi Arabia, Sierra Leone, Slovakia, South Africa, Spain, Suriname, Sweden, Thailand, Turkey, Ukraine, United Kingdom, United States, Uzbekistan, Venezuela, Vietnam, Western Sahara, and Zambia. [511]

A list of the sixteen components needed to build wind turbines are: aggregates and crushed stone (for concrete), bauxite (aluminum, clay, and shale (for cement), coal, cobalt (magnets), copper (wiring), gypsum (for cement), iron ore (steel), limestone, molybdenum (alloy in steel), rare earths (magnets; batteries), sand and gravel (for cement and concrete), and zinc (galvanizing). [512]

Recycling for Wind Turbines

The useful life of wind turbines is limited, generally from 15 to 20 years, but none of the decommissioning plans are public.[513] Mining projects, oil production sites, and nuclear generation sites are required to provide for decommissioning and restoration details down to the last dandelion. Would governments and greenies allow a decommissioned mine, oil, or nuclear site similar latitude given to renewable sites?[514]

Wind turbines may have an even worse disposal problem than solar panels. First, they are gigantic—a single blade can be longer than a wing on a Boeing 747 jumbo jet. Second, they are made of fiberglass, which must be cut by a diamond-studded saw to be carted away on fossil fuel driven giant trucks. And, as with solar panels, the only thing to do is to bury them, toxic materials, and all. This is done, as you can imagine, in enormous pits, creating yet another landfill problem.

There are not many options to recycle or trash turbine blades, and what options do exist are expensive.[515] Decommissioned blades are also notoriously difficult and expensive to transport. They can be anywhere from 100 to 300 feet long and need to be cut up onsite before getting trucked away on specialized equipment — which costs money — to the landfill.

Tens of thousands of aging blades are coming down from steel towers around the world and most have nowhere to go but landfills. In the U.S. alone, about 8,000 will be removed in each of the next four years. Europe, which has been dealing with the problem longer, has about 3,800 coming down annually through at least 2022, according to Bloomberg. It is going to get worse: Most were built more than a decade ago, when installations were less than a fifth the size of what they are now[516].

While most of a turbine can be recycled or find a second life on another wind farm, researchers estimate the U.S. will have more than 720,000 tons of blade material to dispose of over the next 20 years, a figure that does not include newer, taller, higher-capacity versions.

All mining and processing activities to get the iron ore and other metals that go into turbine manufacturing, transporting the huge blades to the sites, and decommissioning them, are all energy intensive activities that rely on fossil fuels, and the products from crude oil, and leave difficult wastes behind to dispose of during decommissioning.

We can be preached to forever about "clean electricity" messages and bedazzle farmers with the prospects of on-going revenue from renewables. However, the extensive mining worldwide for materials for millions of wind turbines and solar panels, and the decommissioning and restoration details, along with disruptive social changes that would be necessitated for societies to live with intermittent electricity, and without the thousands of products from petroleum derivatives, remain the dark side of the unspoken realities of renewables.[517]

The dark side of renewable wind electricity is that it is not clean, green, renewable, or sustainable.[518] They are horrifically destructive to vital ecological values that will last for generations to come.[519]

Government subsidies for Wind

To promote the growth of intermittent electricity sources, such as wind and solar, the federal government has given them special tax incentives. Chief among these are the production tax credit (PTC), which has been used primarily by wind generation and awards a substantial tax credit for every megawatt-hour (MWh) produced; and the investment tax credit (ITC), which is primarily used by solar electricity generators as a credit against construction costs. PTCs and ITCs can amount to more than one-third of the cost of building and operating wind and solar facilities. It is Time to End Subsidies for renewable electricity.[520]

The PTC for wind is now set to expire at the end of 2020, but intermittent electricity advocates are pushing to extend them again and the succeeded in the second stimulus bill for rescuing American coronavirus victims. The following are a few reasons renewable subsidies should not be extended.

- Subsidized renewables have distorted the electricity grid. Intermittent electricity from wind and solar do not provide the same value to the grid as conventional electricity sources. In addition to not operating on-demand, they provide little of the capacity value that is needed to maintain long-term reliability and cannot be counted on to provide the essential services the grid needs to maintain on-demand electricity. Instead, they rely on continuous backup from other electricity generators to provide the uninterruptable services they cannot, thus "imposing" those costs on other redundant generators and the grid. Though the wind and solar facilities do not pay these costs, ratepayers pay for the constant redundancy.
- Federal subsidies for renewables have failed to lower consumer electricity costs but succeeded in increasing costs to consumers. The ITC and PTC subsidies have lowered out-of-pocket costs for renewable project developers but have not led to similar savings for electricity ratepayers. This is evident when examining states that have enacted renewable portfolio standards (RPS) that require utilities to procure renewable electricity. A study from the Energy Policy Institute at the University of Chicago examined the impact RPS programs had on electricity rates across the country and concluded they led to higher electricity rates. Rates increased by 11 percent when the share of renewable generation increased by 1.8 percent, and by 17 percent when the renewable share increased by 4.2 percent according to the study, "These cost estimates ... likely reflect costs that renewables impose on the generation system, including those associated with their intermittency, higher transmission costs, and any stranded asset costs assigned to ratepayers." [521]
- Continued subsidies are no longer needed to support a fully developed renewables industry. After four decades of

federal subsidies and state mandates, the wind and solar industries are mature and able to compete on equal footing with conventional sources of electricity. As AWEA said, "the PTC has been successful ..." Or as we put it, enough is enough.

- A large portion of subsidies are sent overseas, i.e., "leaked" to foreign economies. Much of the wind and solar power deployed in the United States is owned by foreign firms, and the tax credits that these power projects generate are collected by international corporations. The best example is a Chinese wind farm currently being deployed in the heart of Texas.[522] Why? Who benefits? Both authors advocate for it to never move forward since it is environmentally unsound, leads to an unsecure electrical grid, and poses national security problems.

Wind subsidies are also inadvertently supporting the "leakage" to foreign countries for the mining and extraction of the minerals and metals required to support the wind farm installations, and the environmental degradation of land in those foreign locations.

SOLAR PANELS

Environmental degradation "leaked" to others for Solar.

Energy is more than electricity. We have had almost 200 years to develop clones or generics to replace the petroleum derivatives manufactured from crude oil that are used in thousands of products such as: medications, electronics, communications, tires, asphalt, fertilizers, military, and transportation equipment. The social needs of our materialistic societies cannot survive without continuous, uninterruptable, and reliable electricity from coal or natural gas generation backup. All those oil derivatives are manufactured out of crude oil. Our world is literally dripping in crude oil that's part of our daily lifestyles.

The mining operations required to build wind and solar facilities would involve removing and crushing hundreds of billions of tons of rock and ore, causing major habitat losses and widespread pollution. It would also create serious human health impacts, especially in countries that do not have modern equipment and health and safety protections.

Currently, more than 70 percent of the rare earth minerals used in renewable energy sources are mined in China or by companies under Chinese control, with much of China's production coming from areas north of Baotou, Inner Mongolia, though there are substantial reserves in other parts of the world.

Extensive mining is required to support the supply chains of the minerals and metals needed to build solar panels such as: Arsenic (gallium-arsenide semiconductor chips), Bauxite (aluminum), Boron Minerals, Cadmium (thin film solar cells), Coal (by-product coke is used to make steel), Copper (wiring; thin film solar cells), Gallium (solar cells), Indium (solar cells), Iron ore (steel), Molybdenum (photovoltaic cells), Lead (batteries), Phosphate rock (phosphorous), Selenium (solar cells), Silica (solar cells), Silver (solar cells), Tellurium (solar cells), and Titanium dioxide (solar panels).

Miners, young and old, labor for long hours under health, safety, and environmental conditions that would be intolerable in Western, industrialized countries. Filthy processing plants receive little or no regular maintenance, cleaning, or repair, which results in serious illnesses. The massive mining expansion that would be required to meet Green New Deal demands would further pollute lands and sicken human populations.

The Heartland Institute, a conservative think tank, highlights in their *Policy Brief* showing the incredible amount of mining required by the Green New Deal is anything but green. The terrible toll the Green New Deal would wreak on the environment and humanity is a reality its advocates must address.[523]

The recently released book Just GREEN Electricity – *Helping Citizens Understand a World without Fossil Fuel* explains that no one has the answer at this time as to how we can continue to make products and move things without fossil fuels, and the derivatives from petroleum, but most of all, the "aha moment" has arrived – electricity alone from wind and solar cannot replace the fuels and products that are manufactured from petroleum.[524]

The dark side of renewable solar electricity is that it is not clean, green, renewable, or sustainable.[525] Vital ecological values will be devastated for generations and billionaires such as Michael Bloomberg, Tom Steyer, and Bill Gates do not have answers for how to overcome this devastation that will last for generations to come.[526]

Longevity for Solar

No energy system, nor perpetual motion, in short, is actually "renewable or perpetual," since all machines require the continual mining and processing of millions of tons of primary materials, maintenance, and repairs, and the disposal of hardware that inevitably wears out.

As a general solar industry rule of thumb, solar panels last about 25-30 years. However, this does not mean that they stop producing electricity after 25 years – it just means that energy production has declined by what manufacturers consider to be a significant amount due to degradation of the panels.[527]

A 2012 study by the National Renewable Energy Laboratory (NREL) found that, on average, solar panel output falls by 0.8 percent each year.[528] This rate of decline is called solar panel degradation rate. Though this rate of decline metric will vary depending on which panel brand you buy, premium manufacturers like SunPower offer degradation rates as low as 0.3 percent.

Solar panel degradation rates are constantly improving as solar panel technology gets better over the years, and degradation rates below 1 percent are common throughout the industry. In the years since this 2012 study was conducted, more efficient technologies have been developed and many newer panels have just a 0.5 percent yearly decline in energy output.

What does panel degradation rate mean exactly? For the above example, a 0.8 percent degradation rate means that in the second year, your panels will operate at 99.2 percent of their original output; by the end of their 25-year "useful lifespan," they will still be operating at 80 percent. A slightly more durable panel with a degradation rate of 0.5 percent will likely produce around 87.5 percent as much electricity as it did when it was first installed. To determine the projected output of your solar panels after a certain number of years, you can simply multiply the degradation rate by the number of years you are interested in and subtract that number from 100 percent.

What Causes Degradation in Solar Panels?[529] There are many different factors that contribute to solar panel degradation, but many of them focus on something that is impossible to control - the weather. One common way microcracks form is through thermal cycling. When it is warm, just about everything expands. When it is cold, things contract. Solar panels are not immune to this, and it is that constant back and forth that put them under strain and create those little microcracks.

Similarly, strong winds can cause flexing of the panels - also known as a dynamic mechanical load. But if your solar system is installed by experts and you are using panels of good quality, this should not be a major contributor to degradation. Extreme cold and hot weather, humidity, and snow and ice also slowly contribute to solar panel degradation, as do solar system components with different voltage potentials.

Another question we get is "does sunlight cause degradation?" - the very thing that makes solar panels tick! Thankfully, most solar panel manufacturers create panels with UV blockers that protect the panels from most damage, but yes - the sun itself does contribute to degradation.

In fact, your panels degrade at one of the highest rates just hours after installation when they are first exposed to the sun and its UV rays. This is known as light-induced degradation (LID). Your panels can degrade 1 to 3 percent in this short amount of time. While solar panels lose a few percentages right away due to LID, the degradation rate slows down significantly for the remainder of their lifetime.

Land Use for Solar

Utility-scale solar electricity environmental considerations include land disturbance/land use impacts; potential impacts to specially designated areas; impacts to soil, water, and air resources; impacts to vegetation, wildlife, wildlife habitat, and sensitive species; visual, cultural, paleontological, socioeconomic, and environmental justice impacts, and potential impacts from hazardous materials.[530, 531]

All utility-scale solar electricity facilities require relatively large areas for solar radiation collection when used to generate intermittent electricity at utility-scale (defined for the Solar PEIS as facilities with a generation capacity of 20 MW or greater).[532] Solar facilities may interfere with existing land uses, such as grazing, wild horse and burro management, military uses, and minerals production. Solar facilities could impact the use of nearby specially designated areas such as wilderness areas, areas of critical environmental concern, or special recreation management areas. Proper siting decisions can help to avoid land disturbance and land use impacts.

Consider Tesla's most famous battery project, a 100 megawatt-hour lithium battery storage center in Australia. It provides enough backup power for 30,000 homes for 1 hour.[533] But wait - there are 9,000,000 homes in Australia, and 8,760 hours in a year. It would take 300 of those 100 megawatt-hour lithium battery storage centers to provide those 9 million Australian homes with a whopping 1 hour of electricity storage.

One of the largest lithium battery storage centers in the world in in Escondido, California. But it can only store enough power for about 20,000 homes for 4 hours.[534] But wait again -there are about 134,000,000 household in the United States.

Without large-scale ways to back up the intermittent solar generated electricity, California has had to block electricity coming from solar farms when it is extremely sunny, or even pay neighboring states to take it to avoid blowing out the California grid.[535]

Mark P. Mills treads the same path with his investigation into hypocritical claims that renewable intermittent electricity is all about saving the planet when wind and solar are simply devouring it.[536]

As policymakers have shifted focus from pandemic challenges to economic recovery, infrastructure plans are once more being actively discussed, including those relating to electricity. Green energy advocates are doubling down on pressure to continue, or even increase, the use of wind, solar, and electric cars. Left out of the discussion is any serious consideration of the broad environmental and supply-chain implications of renewable intermittent electricity.

Wildlife and Habitat for Solar

The construction of solar facilities on vast areas of land imposes clearing and grading, resulting in soil compaction, alteration of drainage channels and increased erosion.[537] Central tower systems require consuming water for cooling, which is a concern in arid settings, as an increase in water demand may strain available water resources as well as chemical spills from the facilities which may result in the contamination of groundwater or the ground surface.

The potential effects of the construction and the eventual decommissioning of solar electricity facilities include the direct mortality of wildlife; environmental impacts of fugitive dust and dust suppressants; destruction and modification of habitat, including the impacts of roads; and off-site impacts related to construction material acquisition, processing, and transportation.

The potential effects of the operation and maintenance of the facilities include habitat fragmentation and barriers to gene flow, increased noise, electromagnetic field generation, microclimate alteration, pollution, water consumption, and fire.

As with the development of any large-scale industrial facility, the construction of solar power plants can pose hazards to air quality. Such threats include the release of soil-carried pathogens and results in an increase in air particulate matter which has the effect of contaminating water reservoirs.

Large areas of public land are currently being permitted or evaluated for utility-scale solar energy development (USSED) in the southwestern United States, including areas with high biodiversity and protected species. However, peer-reviewed studies of the effects of USSED on wildlife are lacking.[538]

Facility design effects, the efficacy of site-selection criteria, and the cumulative effects of USSED on regional wildlife populations are unknown. Currently available peer-reviewed data are insufficient to allow a rigorous assessment of the impact of USSED on wildlife.

Mitigation measures for the impacts of large solar PV projects are complicated by the variety of bird and other species being affected. Some approaches being tested include turning off or replacing bright lighting at the sites with LED lights to avoid attracting insects and use of netting to exclude birds from the panel area. Other measures being explored include removing vegetation that could be attracting birds to the project area; creating patterns on the panels that attempt to visually deter birds; using predator and distress call recordings; shutting down projects during key migration times; and using roosting and perching prevention.[539]

Creation of habitat areas located outside large solar project areas, to draw birds away from the solar projects, has also been proposed.

Specific mitigation measures to reduce the number of streamers impact from concentrating solar projects have been more challenging. To date, regulatory measures promoted or imposed by USFWS or other agencies have been largely dependent on current experience and knowledge, which is not extensive and varies considerably depending on location.

The USFWS has under consideration several approaches, including the possibility of a permit program (perhaps like that for eagles) that may limit the number of injuries or fatalities allowed. Meanwhile, several interest groups have advocated suspending operations of these large concentrating solar projects until effective mitigation measures can be determined and proven.

One proposed concentrating solar project in California has already been withdrawn from consideration by the California Energy Commission, at least partially because of objections by local stakeholders about the impact on wildlife.

As the industry continues to monitor the collateral effects of solar electricity, mitigation solutions are being tested in project design, planning and avian avoidance measures. Remedies for existing and future solar projects will require increased collaboration between design engineers, environmental specialists, developers, and owners.

To provide a significant amount of intermittent electricity, solar farms require large tracts of land. Western states like California have deserts with abundant space and sunshine, but these areas, remote from where people concentrate, are also natural habitats that support wildlife.[540] For example, environmental reports underestimated the number of desert tortoises that would be displaced by the Ivanpah Solar Generating System in California's Mojave Desert. The same solar farm also came under scrutiny when an increasing number of bird deaths were reported on its premises. Many of their wings had been melted or burned off by heat from the solar farm's mirrors.

The impact that solar farms have on individual species can send ripple effects throughout entire ecosystems. For example, animals like burrowing owls in California's Mojave Desert rely on burrows dug by desert tortoises for shelter. When solar farms harm or remove species within a habitat, they also remove the valuable ecosystem services that they provide to the habitat. The habitat becomes less livable for plants and wildlife that have adapted to its specific conditions.

Public Health and Community for Solar

Left out of the often mistaken, never in doubt assertions of renewables' unalloyed goodness is the fact that the hardware used is hardly renewable.[541] It wears out and needs to be replaced. Then what? "The problem of solar panel disposal 'will explode with full force in two or three decades and wreck the environment' because it is a huge amount of waste and they are not easy to recycle," writes energy analyst Michael Shellenberger, quoting a Chinese recycling official.[542]

In his 2018 Forbes column headlined "If Solar Panels Are So Clean, Why Do They Produce So Much Toxic Waste?" Shellenberger also quotes a four-decade veteran of America's solar industry, who said "the reality is that there is a problem now, and it's only going to get larger, expanding as rapidly as the PV industry expanded 10 years ago". Researchers from the Institute for Photovoltaics in Stuttgart, Germany, found that "contrary to previous assumptions, pollutants such as lead or carcinogenic cadmium can be almost completely washed out of the fragments of solar modules over a period of several months, for example by rainwater."

More recently, Hazardous Waste Experts reported worn-out solar panels are "a potent source of hazardous waste," producing a "dilemma" that "is especially virulent in California, Oregon, and Washington, as those states started adopting solar electricity earliest in the game – suggesting that eco-virtue mightn't necessarily be its own reward".[543]

And just as solar and wind chew up immense tracts of real estate, so, too, will the retirement of solar electricity's constituent parts. Solar panels contain toxic metals like lead, which can damage the nervous system, as well as cadmium, a known carcinogen. Both are known to leach out of existing e-waste dumps into drinking water supplies.[544]

While nuclear waste is contained in heavy drums and regularly monitored, solar waste outside of Europe today ends up in the larger global stream of electronic waste.

Even Grist, a magazine that is surely read out loud in late-evening group sessions at Democratic Rep. Alexandria Ocasio-Cortez's congressional office, notes "solar panels are starting to die" and ponders "what will we do with the megatons of toxic trash".[545]

Solar panels are prone to violent storms, tornados, and hurricanes as panels get smashed like crockery at a Greek wedding, leaving 'green' power consumers in the lurch for weeks, if not months.

The millions of solar panels being dumped in landfills are a veritable toxic cocktail of gallium arsenide, tellurium, silver, crystalline silicon, lead, cadmium, and other heavy metals. At a time when renewable energy rent seekers and their political promoters are still raving about the benefits of so-called 'green' energy, here is a sobering account of where this insanity will end: a landfill near you and yours.

Of course, we have not even touched on the environmental

damage caused by renewables before they have even produced a single watt of energy. We will cover that later. For now, we will let stand as our argument the comment left by energy consultant Ronald Stein, who helpfully pointed out that much of the raw material used to build our "clean energy equipment comes from foreign countries that mine with no environmental regulations, which leads to unrecoverable environmental degradations".[546] Clean energy sure is a dirty business.[547]

Water Use for Solar

Solar farms need millions and millions of gallons of water to clean the mirrors and to generate power. Since most solar farms are built in the desert, we are talking about a precious resource already in short supply. "When push comes to shove, water could become the real throttle on intermittent solar electricity," according to Michael Webber, professor of mechanical engineering at the University of Texas at Austin.

Construction of solar facilities on large areas of land requires clearing and grading, and results in soil compaction, potential alteration of drainage channels, and increased runoff and erosion. Engineering methods can be used to mitigate these impacts.

Parabolic trough and central tower systems typically use conventional steam plants to generate electricity, which commonly consume water for cooling. In arid settings, any increase in water demand can strain available water resources.[548]

Concentrating solar thermal plants (CSP), like all thermal electric plants, require water for cooling. Water use depends on the plant design, plant location, and the type of cooling system.

CSP plants that use wet-recirculating technology with cooling towers withdraw between 600 and 650 gallons of water per megawatt-hour of electricity produced. CSP plants with once-through cooling technology have higher levels of water withdrawal, but lower total water consumption (because water is not lost as steam). Dry-cooling technology can reduce water use at CSP plants by approximately 90 percent. However, the tradeoffs to these water savings are higher costs and lower efficiencies. In addition, dry-cooling technology is significantly less effective at temperatures above 100 degrees Fahrenheit.

Many of the regions in the United States that have the highest potential for solar generating intermittent electricity also tend to be those with the driest climates, so careful consideration of these water tradeoffs is essential.

Life-cycle global warming emissions for Solar

Reality: all energy-producing machinery must be fabricated from materials extracted from the earth. No energy system, in short, is actually "renewable," since all machines require the continual mining and processing of millions of tons of primary materials and the disposal of hardware that inevitably wears out. Compared with hydrocarbons, green machines entail, on average, a 10-fold increase in the quantities of materials extracted and processed to intermittently produce the same amount of electricity.

This means that any significant expansion of today's modest level of green electricity—currently less than 4 percent of the country's total consumption (versus 56 percent from oil and gas)—will create an unprecedented increase in global mining for needed minerals and metals, radically exacerbating existing environmental and labor challenges in emerging markets in foreign countries where many mines are located, and dramatically increase U.S. imports and the vulnerability of America's electricity supply chain.

As recently as 1990, the U.S. was the world's number-one producer of minerals. Today, it is in seventh place. Even though the nation has vast mineral reserves worth trillions of dollars, America is now 100 percent dependent on imports for some 17 key minerals, and, for another 29, over half of domestic needs are imported.

Among the material realities of green energy:

- Building wind turbines and solar panels to generate electricity, as well as batteries to fuel electric vehicles, requires, on average, more than 10 times the quantity of materials, compared with building machines using hydrocarbons to deliver the same amount of energy to society.
- Oil, natural gas, and coal are needed to produce the concrete, steel, plastics, and purified minerals used to build green machines. The energy equivalent of 100 barrels of oil is used in the processes to fabricate a single battery that can store the equivalent of one barrel of oil.

A list of the seventeen components needed to build solar panels are: arsenic (gallium-arsenide semiconductor chips), bauxite (aluminum), boron minerals, cadmium (thin film solar cells), coal (by-product coke is used to make steel), copper (wiring; thin film solar cells), gallium (solar cells), indium (solar cells), iron ore (steel), molybdenum (photovoltaic cells), lead (batteries), phosphate rock (phosphorous), selenium (solar cells), silica (solar cells), silver (solar cells), tellurium (solar cells), and titanium dioxide (solar panels).

The origins of the products for wind and solar are mined throughout the world, inclusive of more than 60 countries of Algeria, Arabia, Argentina, Armenia, Australia, Belgium, Bolivia, Brazil, Canada, Chile, China, Congo (Kinshasa), Cuba, Egypt, Finland, France, Germany, Greece, Guinea, Guyana, India, Indonesia, Iran, Ireland, Italy, Jamaica, Japan, Kazakhstan, Madagascar, Malaysia, Mexico, Mongolia, Morocco, Mozambique, New Caledonia, Oman, Pakistan, Papua New Guinea, Peru, Philippines, Poland, Republic of Korea, Russia, Saudi Arabia, Sierra Leone, Slovakia, South Africa, Spain,

Suriname, Sweden, Thailand, Turkey, Ukraine, United Kingdom, United States, Uzbekistan, Venezuela, Vietnam, Western Sahara, and Zambia.

Recycling for Solar

The dark side of intermittent electricity from solar is that it is not clean, green, renewable, or sustainable.[549] They are also horrifically destructive to vital ecological values that will last for generations to come.[550]

All mining and processing activities to get the iron ore and other metals that go into solar panel manufacturing, transporting, and decommissioning them, are all energy intensive activities that rely on fossil fuels and the products from crude oil and leave difficult wastes behind to dispose of during decommissioning.[551]

In 2017, the United States installed 10.6 GW of new solar electricity. Using rough math (if every panel was 300 W), that is 35.3 million new solar panels installed last year. In about 30 years, a wave of 35.3 million panels may reach the end of their lifespans, not counting the hundreds of millions of panels that flooded the U.S. market in the last decade that may need to be disposed of sooner.[552]

The dark side of 'green energy' and its threat to the nation's environment is summarized by the question: What happens to old solar panels, windmills, and high-tech batteries? [553] A briefing paper released by the U.S. Environmental Protection Agency predicts these startling global numbers for countries by 2050 just for solar waste: [554]

- United States 11 million tons.
- Germany 4 million tons.
- China, 22 million tons.
- Japan 8 million tons.
- India 8 million tons.

Solar arrays have a life cycle of about 30 years, but the rapid adoption of solar in the United States and elsewhere has the problem of disposal creeping up in the rearview mirror — faster rather than later.

Currently the recycling of solar panels faces a big issue, specifically, there are not enough locations to recycle old solar panels, and there are not enough non-operational solar panels to make recycling them economically attractive. Recycling of solar panels is particularly important because the materials used to make the panels are rare precious metals, all of them being composed of silver, tellurium, or indium. Due to the limitability of recycling the panels, those recoverable metals may be going to waste which may result in resource scarcity issues in the future.[555]

Looking at silicon for example, one resource that is needed to make the majority of present-day photovoltaic cells and which there is currently an abundance of, however a silicon-based solar cell requires a lot of energy input in its manufacturing process, the source of that energy, which is often coal, determining how large the cell's carbon footprint is.

The lack of awareness regarding the manufacturing process of solar panels and to the issue of recycling these, as well as the absence of much external pressure are the causes of the insufficiency in driving significant change in the recycling of the materials used in solar panel manufacturing, a business that, from a power-generation standpoint, already has great environmental credibility.

Solar panels are an increasingly important source of intermittent power. They are also complex pieces of technology that become big, bulky sheets of electronic waste at the end of their lives — and right now, most of the world does not have a plan for dealing with that. But we will need to develop one soon, because the solar e-waste glut is coming. By 2050, the International Renewable Energy Agency projects that up to 78 million metric tons of solar panels will have reached the end of their life, and that the world will be generating about 6 million metric tons of new solar e-waste annually.[556]

The millions of solar panels and wind turbines being dumped in landfills are a veritable toxic cocktail of gallium arsenide, tellurium, silver, crystalline silicon, lead, cadmium, and other heavy metals. At a time when intermittent electricity rent seekers and their political promoters are still raving about the benefits of so-called 'green' energy, here is a sobering account of where this insanity will end: a landfill near you and yours.

CHAPTER FIVE

HUMANITY ATROCITIES

- ***Humanity abuses mining for the minerals and metals to support solar, wind, and EV batteries are occurring worldwide.***
- ***We get very excited when we "see" EV's powered by batteries and clean intermittent electricity being generated from wind and solar. What we do not see is the environmental degradation from the origins of the minerals and metals required to support and manufacture all that clean intermittent electricity.***

INTRODUCTION

We visually see wind turbines and solar panels and believe they are generating clean intermittent electricity, but we cannot see the worldwide environmental degradation and the staggering human atrocities that accompany the mining of exotic mineral and metals to support wind turbines, solar panels, EV batteries and battery energy storage systems for electrical grids "leaked" to other countries to achieve what we see just in our backyard.[557] Carbon leakage occurs when there is an increase in greenhouse gas emissions in one country as a result of an emissions reduction by a second country with a strict climate policy.

Our second book released in June 2020: **Just GREEN Electricity – *Helping Citizens Understand a World without Fossil Fuel*** explains that no one has the answer at this time as to how we can continue to make products and move things without fossil fuels, and the derivatives from petroleum. Most of all, the "aha moment" has arrived – electricity alone from wind and solar cannot replace the fuels and products that are manufactured from petroleum.[558]

America's green political interests do not want drilling, fracking, pipelines, nuclear, coal hydroelectric power plants – or the mining within the U.S. for the materials needed to manufacture wind turbines, solar panels, or EV batteries. They prefer to have that work done somewhere else, and just import the energy, cars, and consumer goods.

Rather than manufacture in the most environmentally regulated location in the world, the U.S., and European Union (EU) both prefer to "leak" emission generation to other locations that have significantly less environmental controls and virtually nonexistent labor laws.

As an example, California has long wanted a totally electric vehicle (EV) fleet, which they claim would be clean, ethical, climate-friendly, and sustainable. Of course, those labels hold up only so long as they look solely at activities and emissions within California state boundaries – and not where the mining, manufacturing and electricity generation take place. That kind of "life cycle" analysis would totally disrupt their claims.

Consider copper. A typical internal combustion engine uses about 50 pounds of this vital everyday metal, the International Copper Association says. A hybrid car requires almost 90 pounds; a plug-in EV needs 132 pounds; and a big electric bus can use up to 812 pounds of copper. California has more than 31 million registered vehicles, but if just half, or 15 million California cars were EVs, they would need almost 1 million tons of copper.[559]

But copper ores average just 0.5 percent metal by weight, notes energy analyst Mark Mills. That means 200 million tons of ore would have to be dug up, crushed, processed, and refined to get 1 million tons of copper. Almost every step in that process would require fossil fuels – and emit carbon dioxide and pollutants at the mining sites located outside of California.[560]

Just one electric car or backup-power battery weighs 1,000 pounds and requires extracting and processing some 500,000 pounds of various ores, Mills says. The true costs of "green" energy are staggering.

According to Cambridge University Emeritus Professor of Technology Michael Kelly, replacing all the United Kingdom's 32 million light duty vehicles with next-generation EVs would require huge quantities of materials to manufacture the EV batteries: [561]

- more than half the world's annual production of copper.
- twice its annual cobalt.
- three quarters of its yearly lithium carbonate output; and
- nearly its entire annual production of neodymium.

One can easily see that the world may not have enough minerals and metals for the EV batteries to support the EV growth projections when you consider that today:

- Combined worldwide car sales in 2019 were more than 65 million vehicles annually. [562]
- There are 1.2 billion vehicles on the world's roads with projections of 2 billion by 2035. [563]

Imagine replacing all the USA's nearly 300 million cars, SUVs, pickup trucks, buses, trucks, and other vehicles with electric versions under the Green New Deal – and then charging them daily. The millions of wind turbines, billions of solar panels, billions of backup-power batteries, thousands of miles of new transmission lines, grid upgrades and millions of fast charging stations all across America would also require copper, concrete, all these other metals and many more materials, in incomprehensible quantities.[564]

Alaska's Pebble Mine hosts critical Green New Deal ores such as 6.5 billion metric tons of measured and indicated averaging 0.4% (57 billion pounds) copper, 0.34 grams per metric ton (71 million ounces) gold, 240 parts per million (3.4 billion pounds) molybdenum, 1.7 g/t (345 million oz) silver and 0.41 ppm (2.6 million kg) rhenium. [565]

The Pebble Mine copper alone is nearly two times the world's 2019 output of that essential element. Permits were blocked for years for questionable reasons.[566] But the US Army Corps of Engineers recently found that mining would not have a "measurable effect" on sockeye salmon numbers in the Bristol Bay watershed and should be allowed to proceed, under tough U.S. pollution control, reclamation, wildlife protection, workplace safety, fair wage and child labor laws.[567]

Still, environmentalists intend to delay the Pebble Mine if possible – and block other U.S. exploration and mining projects with a regulatory minefield of labyrinthine local, state and federal rules" that will turn permitting into a two- to three-decades adventure in frustration. That is why most mining and processing is "leaked" to locations overseas, much of it in China and Mongolia or by Chinese companies in Africa, Asia, and Latin America, where none of these laws apply.[568]

Most of the world's rare earth ores are extracted near Baotou, Inner Mongolia by pumping acid into the ground, then processed using more acids and chemicals. Producing one ton of rare earth metals releases up to 420,000 cubic feet of toxic gases, 2,600 cubic feet of acidic wastewater, and a ton of radioactive waste.[569] The resulting black sludge is piped into a foul, lifeless lake. Numerous local people suffer from severe skin and respiratory diseases, children are born with soft bones, and cancer rates have soared.[570]

Lithium comes largely from Tibet and arid highlands of the Argentina-Bolivia-Chile "lithium triangle." Dead, toxic fish join carcasses of cows and yaks floating down Tibet's Liqi River, which has been poisoned by the Ganzizhou Rongda mine. Native people in the ABC triangle say lithium operations contaminate streams needed for humans, livestock, and irrigation, and leave mountains of discarded salt.[571]

The world's top producer of cobalt is the Democratic Republic of Congo, where some 40,000 children as young as four years of age toil with their parents for less than $2 a day up to 12 hours a day.[572] Many die in cave-ins, or more slowly from constant exposure to toxic, radioactive mud, dust, water, and air that puts dangerous levels of cobalt, lead, uranium, and other heavy metals into their bodies. The cobalt ore is sent to China for processing by the Chinese-owned Congo Dongfang International Mining Company.[573]

That is just to meet current raw material requirements for wind and solar intermittently generated electricity, and for battery energy storage systems (BESS) for electrical grids, and EV batteries. Try to picture the raw material demands, Third World mining and child labor conditions, and ecological destruction, under the humongous demands of the Green New Deal.[574]

Green advocates often say they support sustainable and ethical coffee, sneakers, handbags, and diamonds. No child labor sweat shops or unsafe conditions tolerated. But it is a different story with green energy and EVs for transportation.

In 2019, California Assembly Bill 735 proposed that the state certifies that "zero emission" electric vehicles sold were free of any materials or components that involve child labor. Democrats voted it down.[575] The matter is complicated, they "explained."

It would be too hard to enforce, cost too much and imperil state climate goals. And besides, lots of other industries also use child labor. (So, shut up about it.)

Recently, the U.S. House of Representatives had an opportunity to legislate a national certification that federally funded electric buses and charging stations would not include minerals mined with child labor. The Transportation Committee approved the amendment 43-19 (all 19 nay votes were Democrats). But Pete DeFazio (D-OR) quietly replaced the enforceable certification language with a meaningless statement that "it is the policy (not a requirement) of the United States" that funds "should not be used" for items involving child labor. Burying one's head in the sand is the perfect solution to avoid the realities of the dirt connected to green energy.[576]

DeFazio claimed certification is unnecessary because U.S. trade agreements prohibit child labor. But there is no agreement with the Congo, and China has shown no interest in ending child labor in its supply chains. Plus, the matter is complicated, hard to enforce and perilous for climate and Green New Deal goals.

It is easy for Nancy Pelosi and her colleagues to wear Kente cloth stoles in solidarity with Black Lives Matter. And for Sierra Club staff to criticize the organization's "history and culture of white supremacy" – what we call callous, deadly, and arguably racist eco-imperialism and carbon colonialism. We need real reform, and an end to the cancel culture that silences discussion about the horrors of what is going on in too many non-white areas of the United States and the world when it comes to the rare earth minerals and mined ore needed for solar panels, wind turbines, electric vehicles, and battery energy storage systems for electricals. These are the components that make up the so-called green transition, decarbonization plans for electrical grids, and mitigating or ending climate change within the Paris Climate Accord agreement.

ELECTRIC VEHICLE BATTERIES

The number of electric vehicles on the world's roads is rising fast. Latest figures show there are more than three million and sales are growing at close to 75 percent a year. But now doubts have been raised about the ethics of buying one, as the dirty secrets of electric vehicles are gaining recognition.[577, 578, 579]

Worldwide cars sales in 2019 were about 78 million. Of that total, less than 3 percent or 2.1 million new electric vehicles were sold worldwide.[580] China is the world's largest electric car market, accounting for 1.2 million - 56 percent of all electric vehicles sold in 2018.[581] China also accounts for 99 percent of sales of electric trucks, buses, motorcycles, and scooters. The U.S. came a distant second with 361,000 new electric cars sold in 2018, almost half of which were the new Tesla 3 model. In terms of market share, Norway leads the way - 49 percent of new cars sold were pure or hybrid electric.[582, 583]

Of the 17 million vehicles of all types sold just in the United States in 2019, the horrific EV sales data shows that only 325,000 electric and plug-in hybrid vehicles were sold in the U.S. in 2019, down from 349,000 in 2018. A dismal 2 percent for the entire nation. California remains the primary buyer of EV's with half of all EV's registered in America being in California, while the rest of America has shown little interest in the incentives and the increasing choices of models.[584] Are EV carmakers driving off a cliff?"

Today, there are 1.2 billion vehicles on the world's roads with projections of 2 billion by 2035.[585] By some estimates, the total number of vehicles worldwide could double to 2.5 billion by 2050. Registration of electric vehicles is projected to only be in the single digits, around 5 to 7 percent.

If projections come to reality by 2035, 5 to 7 percent of the 2 billion vehicles would equate to 125 million EV's on the world's roads, and potentially double that number if governments step up the pace of legislative change. The bad news is that would also represent more than 125 BILLION pounds of lithium-ion batteries that will need to be disposed of in the decades ahead.

The pressure to go green is increasing as bans on the sale of new fossil-fueled cars loom in Europe. Germany will stop the sale of all new petrol and diesel cars from 2030, Scotland from 2032, and France and the UK from 2040.[586]

Even California has jumped on the EV train with Governor Gavin Newsom (Democrat) issuing an Executive Order in 2020 to ban the sale of gas-powered vehicles by 2035 that will be devastating to the state's economy and both the state and world environment.[587]

The goal of zero-emission driving is still a long way off. Electric cars still only account for 2.6 percent of global new sales and despite Norway's stellar sales rate, electric cars still account for only one in eight of vehicles on the road in Oslo.[588]

Sales in other European countries are much lower. In Italy only 0.26 percent of new sales last year were electric cars, with buyers still preferring diesel over petrol. Even in Spain, which is a major car producer, only 0.5 percent of new sales were electric.

Factors holding back electric vehicle adoption.

1. High purchase price
2. Limited range on a single charge
3. Limited access to plug-in stations
4. Battery life and limited disposal options
5. Environmental degradations worldwide resulting from the mining for EV materials.
6. Humanity atrocities occurring worldwide supporting the mining for EV materials.

The actions to phaseout the sale of gas-powered vehicles will most likely result in parts of the world looking like Cuba in the decades ahead with hundreds of millions of vintage gas-powered vehicles being continuously re-registered each year, that are less fuel efficient, and bigger emission polluters, than state-of-the-art gas-powered newer vehicles that will be banned.

The number of cars manufactured with internal combustion engines (ICEs) is projected to be surpassed by battery electric vehicles before 2050 with more than 1 billion for ICE's and EV's as the BEV fleet grows.[589] Morgan Stanley projects that there may be 1 billion EV's on the road in 2050, as Electric vehicles should overtake traditional sales in just 20 years.[590] [591]

2019 may go down as the year the auto industry started putting some muscle into electric vehicle sales. Amidst a steady stream of pledges to deliver more EVs than ever over the next five years, Ford filmed an electric prototype of its F-150 pickup truck (a favorite gas guzzler among Canadians) towing an entire freight train in a CN railyard in Montreal. Not to be outdone, the forthcoming Tesla Cybertruck then hauled the F-150 uphill in a tongue-in-cheek tug-of-war.

The brawny marketing stunts carried a simple message: electric cars are not just for tree-hugging Leaf, Prius, and Bolt lovers anymore. The message is timely, with global leaders, including Prime Minister Justin Trudeau, committing to carbon pollution targets of "net zero" by 2050.

Tough new emissions standards are coming out of Europe, and a smattering of governments following Norway's early lead on banning gas-powered-car sales as soon as 2025. For most automakers that have cautiously dipped their toes in the EV market, the race to net zero is officially on.

So, which car companies are coming clean? Amnesty International challenges industry leaders to clean up their batteries.[592] At the Nordic Electric Vehicle (EV) Summit in Oslo, the organization is highlighting how lithium-ion batteries, which power electric cars and electronics, are linked to human rights abuses including child labor in the Democratic Republic of

Congo (DRC), and environmental risks which could undermine their green potential.

"Finding effective solutions to the climate crisis is an absolute imperative, and electric cars have an important role to play in this. But without radical changes, the batteries which power green vehicles will continue to be tainted by human rights abuses," said Kumi Naidoo, Amnesty International's Secretary General.

Other emerging threats to the human rights abuses are the environmental impacts of producing batteries. Most of the current manufacturing of lithium-ion batteries are concentrated in China, South Korea, and Japan, where electricity generation remains dependent on coal and other polluting sources of power.

This means that, while electric vehicles are essential for shifting away from fossil fuels and reducing greenhouse gas emissions, more needs to be done to reduce the carbon footprint within the mining and manufacturing phases in foreign countries.[593] Meanwhile, rising demand for minerals like cobalt, manganese and lithium has led to a surge in interest in deep-sea mining, which studies predict will have serious and irreversible impacts on biodiversity.

Amnesty International is also calling on companies to ensure that batteries are disposed of responsibly. There is already significant evidence showing that battery waste from electronics, which contains various hazardous materials, has been irresponsibly disposed of, contaminating soil, water, and air.

Electric vehicles are currently not as ethical as some retailers would like us to believe. Years of unregulated industry practices have led to detrimental human rights and environmental impacts, which governments and industry are not doing enough to tackle.

Amnesty International has documented serious human rights violations linked to the extraction of the minerals used in lithium-ion batteries, particularly in the DRC and the Indigenous peoples living near lithium mines in Argentina.

An Amnesty 2016 investigation found children and adults in southern DRC working in hand-dug cobalt mines facing serious health risks, neither protected by the government nor respected by companies that profit from their labor.[594] Amnesty's research has linked these mines to the supply chains of many of the world's leading electronics brands and electric vehicle companies.[595]

Those least responsible for climate change – indigenous communities and children – will pay the human atrocities price for the shift away from fossil fuels. The energy solutions of the future must not be based on the injustices being inflicted on developing countries. Amnesty International first thrust the dark side of mining for car and smartphone-battery minerals into the spotlight in 2016 with a damning investigation into the child labor-plagued cobalt mines of war-torn Congo which is home to 60 percent of global cobalt reserves.[596]

Amnesty International's work will focus on all three phases of the battery lifecycle:[597]

1. Extraction: Mapping supply chains of key minerals, calling for human rights impacts to be identified, prevented, and addressed, and calling for a prohibition on commercial deep-sea mining.
2. Manufacturing: Calling for carbon footprints to be properly disclosed, minimized, and offset; and for rights to and at work, including health, equality, and non-discrimination, to be legally protected and enforced.
3. Re-use and recovery: Calling for products to be designed and regulated so that their potential for re-use is optimized and waste is penalized; and illegal or dangerous exportation and dumping of batteries is prevented.

As a first step, companies should publicly disclose information about how human rights abuses and environmental risks are being identified, prevented, and addressed throughout the lithium-ion battery's lifecycle. With the climate taking a top focus for the U.S. and western-aligned nations, consumers have the right to demand that products marketed as the ethical choice really stand up to scrutiny.

Companies who overlook human rights concerns as they clean up their energy sources are presenting their customers with a false choice - people or planet. This approach is gravely flawed and will not deliver the sustainable changes we need to save humanity from the mining atrocities and human right abuses emanating from the clean energy transition. We are asking industry leaders to think hard about what kind of future they want to build?

Amnesty points to serious health risks to children and adult workers in cobalt mines in the Democratic Republic of Congo, documented in a report it issued. More than half the world's cobalt comes from southern DRC, much of it from artisanal mines that produce 20 percent of the country's output.[598]

Artisanal miners as young as seven years of age were seen by researchers who visited nine sites including deep mines dug by hand using basic tools. Miners, the youngest of whom were earning as little as $1 a day, reported suffering chronic lung disease from exposure to cobalt dust. Yes, almost half the world — over three billion people — live on less than $2.50 a day. At least 80 percent of humanity, or more than 6 billion, lives on less than $10 a day.[599]

Cobalt from these mines is sold to major producers. No country has laws requiring producers to report on the transparency of their supply chains.

Despite projections that the demand for cobalt will reach 200,0000 tons per year by 2020, no country legally requires companies to publicly report on their cobalt supply chains.[600] With more than half of the world's cobalt originating in southern DRC, Amnesty International says the chance that electric vehicle batteries are "tainted with child labor and other abuses" is unacceptably high.[601, 602]

Battery manufacturing now accounts for 60 percent of the 125,000 tons of cobalt mined globally each year.[603] But a move last year by the London Metal Exchange to ban the sale of tainted cobalt was opposed by a consortium of 14 non-governmental organizations, including Amnesty, on the grounds it would simply drive the trade underground.[604] They called for greater traceability of the mineral's sources.

The World Economic Forum's Global Battery Alliance notes two major challenges:

- First, raw materials needed for batteries are extracted at a high human and environmental toll. This includes, for example, child labor, health and safety hazards in informal work, poverty, and pollution.
- Second, a recycling challenge looms over the eleven million tons of spent lithium-ion batteries forecast to be discarded by 2030, with few systems in place to enable reuse and recycling in a circular economy for batteries."[605]

The Organisation for Economic Co-operation and Development is an intergovernmental economic organization (OECD) Forum on Responsible Mineral Supply Chains met in Paris in 2019, where members discussed how to demand companies begin to identify their cobalt sources. Apple, BMW, Daimler, Renault, and battery maker Samsung SDI have already agreed to publish their supply chain data.[606]

Amnesty says most manufacturing of lithium-ion batteries takes place in China, South Korea, and Japan, where electricity generation remains dependent on coal and other fossil fuels. They said makers should ethically disclose the carbon footprint of their products.[607]

Environmental and human rights advocates, along with

international heavyweights at the World Bank and World Economic Forum, say there is an elephant in the showroom. The EV revolution has been racking up a supply chains of trouble around the globe (including a recent lawsuit) related to an onslaught of often-contentious new mines opening to meet surging battery-metal demand, not to mention the coming tide of e-waste from old batteries.

The transport sector is currently the fastest-growing contributor to dirtier air in developing countries, according to the World Resources Institute, with road, rail, air, and marine transport accounting for 24 percent of global CO2 emissions in 2018. [608] Electrified transport – powered by low-carbon grids – could help clear deadly air pollution and cut millions of tons of greenhouse gas emissions per year. In the shift from burning planet-cooking fossil fuels to generating and storing electricity in batteries for our cars and, increasingly, our homes, one thing is certain: batteries are driving demand for more minerals and metals in the low-carbon transition.

Bloomberg New Energy Finance predicts that by decade's end the battery market will be worth $116 billion annually (not including investments in supply chains), up from $14.6 billion in 2017. [609] Trailblazing EV manufacturer Tesla and others have warned that underinvestment in the minerals and metals supply chain will lead to a shortage of nickel and other EV battery minerals down the road.

In its 2017 report The Growing Role of Minerals and Metals for a Low Carbon Future, the World Bank forecasted that global demand for low-carbon-economy minerals such as lithium, graphite, and nickel will skyrocket by 965 percent for lithium, 383 percent for graphite and 108 percent for nickel by 2050. But two years later, in 2019, as the World Bank noted that growing demand for minerals offers an "opportunity for mineral-rich developing countries to develop," it cautioned that "significant challenges will likely emerge if the climate-driven clean energy transition is not managed responsibly and sustainably."[610]

Then in late 2019, the issue hit the front pages again when a landmark lawsuit was launched against Tesla and a handful of tech giants on behalf of 14 Congolese families who say their children were seriously injured or killed working in cobalt mines earlier in the year.

With cobalt dubbed the "blood diamond of batteries,"[611] Tesla and others have been slashing their use of the controversial mineral and replacing it with nickel in a move that is said to prolong range per charge. Not that cutting and running from the Congo will help those mining in poverty, say activists, and without tough responsible mining standards in place, other EV minerals end up being called out for bad behavior, too.

The Washington Post reported in November 2019 that nickel mines in Indonesia are turning the oceans there red. The draining of water reserves for vast lithium mines in the salt plains of Latin America has been fingered for fueling water wars, social unrest and mine strikes in Chile, Bolivia, and Argentina. And a new frontier of destructive deep-sea mining for several "green" economy minerals has prompted the nation of Fiji, along with Greenpeace and others, to call for an immediate moratorium on the nascent practice.

At a MiningWatch conference in Ottawa in November 2019, the human and environmental implications of this new extractive rush were front and center. Representatives from Chile, Peru, Papua New Guinea, Congo, and northern Canada took the stage one by one, concerned about the green transition being used as justification for running roughshod over their ecosystems and human rights.[612]

The International Institute for Sustainable Development (IISD) has studied what it calls "green conflict minerals" (cobalt, nickel, lithium, rare earths, and aluminum). The problem with green economy minerals, says IISD analyst Clare Church, is that they are often found in countries with fragile governments, making their extraction prone to violence, conflict, corruption, and human rights abuses.[613]

The question, say the IISD and others, is whether green economy minerals – and the companies that source them – can help fuel (no pun intended) thriving, peaceful and sustainable development in communities with key mineral reserves – rather than exacerbating local unrest.

Perhaps because cars need such a large volume of minerals to provide for EV batteries that weigh in at roughly 1,000 pounds per car, compared to, say, a smartphone, and perhaps because EV owners tend to be a fairly conscientious bunch, EVs – and the companies that make them – are now driving demand for more ethical mineral sources.

One route to cleaner EVs involves boosting transparency. A few leading car companies – BMW, Daimler, and Renault, as well as Samsung and Apple – have started publishing supply chain data. While Tesla does not disclose cobalt suppliers, it does publish lists of its tungsten, tantalum, and tin suppliers, as mandated by California law regarding officially designated conflict minerals.

By 2025, the sale of light-trucks that includes SUVs, vans and pickups may make up 78 percent of sales, leaving sedans in the dust.[614] The larger SUV EV's require more power and are more costly than the smaller less experience EV sedans. Current and future SUV owners may be reluctant to downgrade to smaller EV sedans that are dominating the EV markets. The larger SUV EV's that require larger battery capacities and will put further pressure on the transparency of ethical means to support those batteries with the mineral and metal resources from developing countries.

Supplier disclosure is an important first step in shedding light on shadowy supply chains – something leading sneaker and clothing brands started doing years ago in response to sweatshop scandals.

Following the unveiling of Volvo's first fully electric car, the XC40 Recharge, the Swedish carmaker announced that it will begin using a blockchain platform which is essentially a decentralized digital ledger to trace its cobalt.[615] Volvo Canada's Matt Girgis tells Corporate Knights that while Volvo has long been marketed as the safest car, it is now trying to position itself as the safest car for the planet. Making sure its minerals and metals are "clear and safe from unethical issues," as its blockchain partner put it, is particularly pressing now that, as of 2020, all new Volvo models will be hybrids or plug-ins, with gas-only vehicles phased out by 2025.[616, 617]

It is a sign of the times that the world's largest cobalt miner, Glencore, announced in December that it too will start using blockchain for better traceability (days before it was named as the main supplier in the Congolese lawsuit).[618] Glencore and Fiat Chrysler are now the newest members of the Responsible Sourcing Blockchain Network, joining Volvo, VW, and Ford.

Not that blockchain alone will solve human rights or environmental violations. "Blockchain is a powerful tool for tracking," says Aimee Boulanger, executive director of the Initiative for Responsible Mining Assurance (IRMA), "but only if the information going in is quality" – that is, independently verified so that responsible practices are met throughout the supply chain.

In the race to ramp up EV production, a growing number of companies are concerned about how ethical EV's are and looking to lock down a steady and sustainable battery supply by wresting production away from coal-heavy China – which currently dominates global battery manufacturing – and bringing production home. Literally. [619]

VISION BOARD FOR ETHICAL EV BATTERIES

- Remove roadblocks to recycling at the design stage so batteries can be easily disassembled, reused, recycled, and aligned with the circular economy.

- Set national policy mandating EV battery recycling in all provinces, paid for by battery makers through "extended responsibility programs."
- Incentivize domestic battery-recycling facilities and give tax breaks to carmakers with the highest recycled content possible.
- For any minerals that cannot be sourced through recycling, ensure mines meet international environmental and human rights best-practice standards, such as Initiative for Responsible Mining Assurance (IRMA), and are audited by independent third parties.
- Push for stringent environmental and labor regulations for mines, both in Canada and abroad, and grant Canada's Ombudsperson for Responsible Enterprise (CORE) strong oversight powers to investigate and penalize companies that violate Canadian laws overseas.
- Provide grants that allow remote northern Canadian communities to shift from powering their communities with diesel to storing clean solar energy in refurbished EV car batteries.
- Accelerate national coal-power phase-out to ensure that low-carbon grids power EVs.
- The ultimate ethical battery is one that will be used by many. Funding mass expansion of electrified public transit will help ensure that the green transportation revolution is affordable and accessible to everyone.

In 20 years, will fair-certified cars with recycled-content logos be as commonplace as fair-trade coffee and Forest Stewardship Council–certified paper, scrutinized by third-party auditors and stamped with sustainable seals of approval? With so many batteries driving the clean energy transition, it is hard to see how we can have a sustainable electric future unless it is ethical to its core.

WIND TURBINES

The National Center for Biotechnology Information (NCBI) reports that industrial wind turbine noise has been identified for noise characteristics as plausible causes of annoyance and other health effects from amplitude modulation, audible low frequency noise, tonal noise, infrasound, and lack of nighttime abatement. Documented symptoms reported by individuals exposed to wind turbines, include:[620]

- Sleep disturbance
- Headaches
- Tinnitus
- Ear pressure
- Dizziness
- Vertigo
- Nausea
- Visual blurring
- Tachycardia
- Irritability
- Problems with concentration and memory
- Panic episodes associated with sensations of internal pulsation or quivering when awake or asleep.

During the past few years there have been case reports of adverse effects from wind turbine noise. A 2006 Académie Nationale de Médecine working group report notes that noise is the most frequent complaint. The noise is described as piercing, preoccupying, and continually surprising, as it is irregular in intensity. The noise includes grating and incongruous sounds that distract the attention or disturb rest. The spontaneous recurrence of these noises disturbs the sleep, suddenly awakening the subject when the wind rises and preventing the subject from going back to sleep. Wind turbines have been blamed for other problems experienced by people living nearby. These are less precise and less well described, and consist of subjective (headaches, fatigue, temporary feelings of dizziness, nausea) and sometimes objective (vomiting, insomnia, palpitations) manifestations. [621]

Reports of adverse health effects and reduced quality of life are also documented in Industrial Wind Turbines (IWTs) projects in Australia and New Zealand.[622]

Noise-induced annoyance is acknowledged to be an adverse health effect. Chronic severe noise annoyance should be classified as a serious health risk. According to the World Health Organization (WHO) guidelines for community noise, "the capacity of a noise to induce annoyance depends upon many of its physical characteristics, including its sound pressure level and spectral characteristics, as well as the variations of these properties over time." Industrial wind turbine noise is perceived to be more annoying than transportation noise or industrial noise at comparable sound pressure levels.[623]

The World Health Organization (WHO) has documented the irrefutable conclusion that the wind industry has been given a regulatory path to profits with an unfathomable license to hurt in the form of sleep deprivation (and associated disease) for a very long time.[624]

America's Rural Communities, like Arkwright, NY have gone on the offensive against big wind's turbine onslaught.[625] Andrew Cuomo, New York's wind power obsessed Governor is determined to carpet his state with thousands of wind turbines. Faced with the threatened onslaught, and already fed up with what they have had to suffer, the upstate New Yorkers have taken matters into their own hands and produced a documentary about noise from local wind turbines.

Ultimately the wind turbine installations in Falmouth,

Massachusetts are far worse than any experiment ever conducted in the United States as state and local officials always had first-hand knowledge of the damage to health and the requirement to take property with no compensation. State and local officials were following an agenda to achieve 2000 megawatts of commercial wind turbine power by the year 2020.

Unethical human experimentation in the United States performed on human test subjects in the United States is a fact of life. What happened in Falmouth is worse because the results of the turbine installations were well known in 1987, over forty years ago.

In 2017 we see the state and local officials using the Massachusetts court system to hide the human tragedy of creating a group of second-class citizens living under wind turbines in as many as twenty-one communities. Residents describe the noise as torture from lack of sleep causing health problems and unable to work near their property while the turbines operate.

History will show the Massachusetts wind turbine installations one of the most appalling atrocities in the past ten years and all those involved in the agenda need to be held legally responsible. No one has the right to take health and property rights without due process.

The Smoking Gun Letter to the Falmouth Public Works about the wind turbines generating 110 decibels of noise - as loud as a rock band.[626]

August 3, 2010

Mr. Gerald Potamis
Wastewater Superintendent
Town of Falmouth Public Works
59 Town Hall Square
Falmouth, MA 02540

RE: Falmouth WWTF Wind Energy Facility II
"Wind II", Falmouth, MA
Contract No. #3297

Dear Mr. Potamis,

Due to the sound concerns regarding the first wind turbine installed at the wastewater treatment facility, the manufacturer of the turbines, Vestas, is keen for the Town of Falmouth to understand the possible noise and other risks associated with the installation of the second wind turbine.

The Town has previously been provided with the Octave Band Data / Sound performance for the V82 turbine. This shows that the turbine normally operates at 103.2dB but the manufacturer has also stated that it may produce up to 110dB under certain circumstances. These measurements are based on IEC standards for sound measurement which is calculated at a height of 10m above of the base of the turbine.

We understand that a sound study is being performed to determine what, if any, impacts the second turbine will have to the nearest residences. Please be advised that should noise concerns arise with this turbine, the only option to mitigate normal operating sound from the V82 is to shut down the machine at certain wind speeds and directions. Naturally, this would detrimentally affect power production.

> *The manufacturer also needs confirmation that the Town of Falmouth understands they are fully responsible for the site selection of the turbine and bear all responsibilities to address any mitigation needs of the neighbors.*
>
> *Finally, the manufacturer has raised the possibility of ice throw concerns. Since Route 28 is relatively close to the turbine, precautions should be taken in weather that may cause icing.*
>
> *To date on this project, we have been unable to move forward with signing the contract with Vestas. The inability to release the turbine for shipment to the project site has caused significant [SIC] delays in our project schedule. To move forward the manufacturer requires your understanding and acknowledgement of these risks. We kindly request for this acknowledgement to be sent to us by August 4, 2010, as we have scheduled a coordination meeting with Vestas to discuss the project schedule and steps forward for completion of the project.*

Owing to the lack of adequately protective siting guidelines, people exposed to industrial wind turbines (IWTs) can be expected to present to their family physicians in increasing numbers. The documented symptoms are usually stress disorder–type diseases acting via indirect pathways and can represent serious harm to human health. Family physicians can effectively recognize the ailments and provide an empathetic response. In addition, their contributions to clinical studies are urgently needed to clarify the relationship between industrial wind turbine (IWT) exposure and human health and to inform regulations that will protect physical, mental, and social well-being.

For wind, the easiest thing to do is adopt a minimum safe distance between wind farms and human habitation and activities, i.e., setbacks. Scanning various attempts and incorporating oddball issues like spinning blades hurling chunks of ice as far as 1,750 feet, it appears as though a mile (more than 5,000 feet) or so is a reasonable set-back distance to use as a safety measure.[627]

Harnesses and personal protective gear for maintenance workers is also an easy fix and one that many other industries adopted to reduce worker death and harm.

Another subject tied to wind turbines is birds, but who cares about birds? In a December 2016 study released by U.S. Fish and Wildlife researchers on the number of eagle deaths by wind turbines ruffled some feathers in the industry, but the public remains tolerant, or maybe just not knowledgeable, about wind turbine sites participating in nationwide "incidental" killing of bald eagles annually without compensatory mitigation.[628]

In 2016, the former President Obamas' administration finalized a rule that allows wind-energy companies to operate high-speed turbines for up to 30 years — even if it means killing or injuring thousands of species protected under the Bald and Golden Eagle Protection Act and the Migratory Bird Treaty Act. [629] Under the rule, industrial wind may acquire an eagle "take" permit from the U.S. Fish and Wildlife Service (USFWS) that allows the site to participate in the nationwide killing of up to 4,200 bald eagles annually, under incidental "take" permits without compensatory mitigation.[630] It is shocking that industrial wind can legally obtain permits from the USFWS to kill those majestic bald eagles. I cry foul!

We wonder if the renewable industry is proud of those new renewable jobs being created also include those that need to clean up the mess from those creatures chopped up by the industrial wind generator blades and from those fried from the heat from the industrial solar panels?

Our politicians may be oblivious to the governments' acceptance of animal cruelty toward birds of prey, but the public is loudly rejecting this atrocity toward "taking", which is a nice word for killing, bald eagles.

What about humans being killed by wind farm accidents? Does this same philosophy about birds hold true for human deaths?[631] A paper from the Caithness Windfarm Information Forum 2013 titled "Wind Farm Accidents and Fatalities" was rather enlightening.[632]

- In England, there were 163 wind turbine accidents that killed 14 people in 2011.
- These are pretty low numbers. By contrast:
    - o In 2011 coal produced about 180 billion kWhrs in England with about 3,000 related deaths.
    - o Nuclear energy produced over 90 billion kWhrs in England with no deaths.
    - o In that same year, America produced about 800 billion kWhrs from nuclear with no deaths.

When you include fatalities of "other than energy" the world numbers get even worse...

- Currently, underdeveloped countries, mostly from energy starved countries, are experiencing about 11,000,000 child deaths every year of which more than 70 per cent are attributable to six causes: diarrhea, malaria, neonatal infection, pneumonia, preterm delivery, or lack of oxygen at birth.[633] About 29,000 children under the age of five – 21 each minute – die every day, mainly from preventable causes by the products from oil derivatives that underdeveloped countries have minimal or no access.[634]

- More than 40,000,000 abortions performed each year. [635]
- More than 8,000,000 world cancer deaths per year.[636]
- 5,000,000 tobacco related deaths per year, and current trends show that tobacco use will cause more than 8,000,000 deaths annually by 2030.[637]
- 3,800,000 deaths every year because of household exposure to smoke from dirty cook stoves and fuels.[638]
- 2,300,000 women and men around the world succumb to work-related accidents or diseases every year per.[639]
- 1,800,000 from the coronavirus in 2020. [640]
- 1,230,000 world traffic deaths_per year.[641]
- 270,000 pedestrians killed on roads each year.[642]
- 585,900 premature deaths caused by drugs per year.[643]

Since so many more people die from other causes, can we just forget about wind turbine fatalities? Like the eagles? Does any energy source kill a significant number of people? A post discussed human fatality by energy source titled "How Deadly Is Your Kilowatt?", and how coal is the biggest killer in U.S. energy at 15,000 deaths per trillion kWhrs produced, while nuclear is the least at zero. Wind energy kills a mere 100 people or so per trillion kWhrs, the majority from falls during maintenance activities (Toldedo Blade).[644] [645]

There is increasing concern that electricity generation from fossil fuels contributes to climate change and air pollution. In response to these concerns, governments around the world are encouraging the installation of intermittent electricity generation projects including Industrial Wind Turbines (IWT's).

But Whoa, Nelly! Concerned citizens, homeowners, patriots, and local businesses fighting to defend their rights and their families, sometimes referred to NIMBY's (Not-In-My-Backyard) have all rejected industrial wind turbines farms around the globe from Germany to Australia, California, New York, and Massachusetts are speaking loudly, and acting, to put a halt to the invasion of noisy wind farms in their backyards. Following numerous reports from Maryland to Canada to France on wind turbine noise, the NIMBY's are becoming energized (no pun intended).[646, 647, 648]

- The National Center for Biotechnology Information located in Bethesda, Maryland (NCBI) reported that industrial wind turbine noise amplitude modulation, audible low frequency noise, tonal noise, infrasound, and lack of nighttime abatement have been identified as plausible noise characteristics that could cause annoyance and other health effects. Documented symptoms reported by individuals exposed to wind turbines, include: Sleep disturbance, Headaches, Ear pressure, Dizziness, Vertigo, Nausea, Visual blurring, Irritability, Problems with concentration and memory, and Panic episodes associated with sensations of internal pulsation or quivering when awake or asleep.[649]
- The Green Energy Act (GEA) of Ontario, Canada may have been well-intended, but a recent analysis published by the Fraser Institute, Environmental and Economic Consequences of Ontario's Green Energy Act, demonstrates that the GEA has had disastrous impacts on Ontario's energy rates and is going to seriously threaten economic competitiveness for the manufacturing and mining sectors.[650, 651]
- The French Academy of Medicine Declare Wind Turbines Health Nuisance.[652] The planned extension of the terrestrial wind energy sector raises an increasing number of complaints from associations of local residents reporting functional disturbances achieving what is known as the "wind turbine syndrome". The report notes that noise is the most frequent complaint. The noise is described as piercing, preoccupying, and continually surprising, as it is irregular in intensity. The noise includes grating and

> incongruous sounds that distract the attention or disturb rest. The spontaneous recurrence of these noises disturbs the sleep, suddenly awakening the subject when the wind rises and preventing the subject from going back to sleep.

Back in 2014, Warren Buffett, the chairman and CEO of Berkshire Hathaway, and one of the world's richest people, famously said "We get a tax credit if we build a lot of wind farms. That's the only reason to build them."

One of Berkshire's companies, MidAmerican Energy, expected to collect a whopping $10 billion in tax credits for spending $12.9 billion on wind projects. Robert Bryce wrote *Buffett's company's push to add even more turbines in Iowa is being derailed by the bridges of Madison County.* [653]

Buffett's quest for more wind-energy tax credits is reaching its land-use limit. *On December 22, 2020, the province that has become renowned for its picturesque wood-covered bridges became the second Iowa county to ban wind projects. By a 2-1 margin, the Madison County Board of Supervisors passed an ordinance that prohibits installation of wind projects within 1.5 miles of non-participating landowners, limits the height of turbines to less than 500 feet, imposes strict noise limits, and eliminates property-tax breaks.* [654]

The significance of Madison County's rejection of Big Wind, I wrote, "goes beyond Buffet, Berkshire Hathaway, and Iowa. Since 2015, by my count, 291 government entities from California to Maine have rejected or restricted wind-energy projects." Indeed, land-use conflicts over renewables are increasing at the same time that the "Biden Administration, climate activists, academics, and powerful Washington lobby groups are pushing for massive increases in the deployment of renewables. Last summer, the Biden-Sanders Unity Task Force released called for installing half a billion solar panels and "60,000 made-in-America wind turbines" and doing so "within five years." [655]

Despite the political obsession for intermittent electricity generated from wind turbines, concerned citizens, homeowners, patriots, and local businesses fighting to defend their rights and their families are alive and well! Here is a sampling from around the globe that are stepping up to stop the installation of those monstrosities in their backyards:

- In Germany, thousands are being driven mad by practically incessant, turbine generated low-frequency noise and infra-sound, and are fighting back with a fury and energy which has rattled the wind industry. There are now hundreds of anti-wind industry groups across Germany.[656] And many of those groups and individuals have launched litigation against developers and government to either prevent wind farms from being built, or to seek substantial financial compensation for the loss of the use and enjoyment of their homes. In Germany, locals expressed their opposition in no uncertain terms – voting 25 to 1 against a giant wind project proposed for their patch of paradise.[657]
- In Australia and New Zealand, reports of adverse health effects and reduced quality of life are also documented in Industrial Wind Turbines (IWTs) projects in their countries.[658]
- In California, San Bernardino is the state's largest county, and "has banned the construction of large solar and wind farms of more than 1 million acres of private land." Residents do not want their rural desert community littered with renewables".[659]
- As previously mentioned in New York, Arkwright, NY have gone on the offensive against big wind's turbine onslaught.[660]
- In Massachusetts, the Town of Falmouth had documentation prior to the installations of its wind turbines that

> the megawatt wind turbines were too loud for residential locations.[661] The documentation includes emails, maps, a written letter warning and an admission by the Massachusetts Clean Energy Center in 2013 that "mistakes" were made in the original acoustic noise predictions. The first Falmouth wind turbine started operating in 2010. Ultimately the wind turbine installations in Falmouth are far worse than any experiment ever conducted in the United States as state and local officials always had first-hand knowledge of the damage to health and the requirement to take property with no compensation. State and local officials were following an agenda to achieve 2000 megawatts of commercial wind turbine power by the year 2020.

The list goes on and on at locations like New York, Wisconsin, and Scotland, etc.[662, 663, 664]

The state of Oaxaca in Mexico is home to at least 28 wind farm mega-projects, many of which have been contentious due to their association with indigenous land-rights violations, forced displacement, inadequate consultation, and violations by security forces. Due to intense opposition from local communities, wind farms have faced lawsuits and operational delays, posing significant financial risk to investors. Protestors have faced intimidation, death threats and violence from Mexican security forces, exposing the country's wind energy sector to heightened reputational risks.[665]

Major community resistance against wind farms, and indeed other solar and hydropower projects, has resulted in significant financial costs to renewable sector investors. The cancellation of a 61-megawatt wind farm in Kenya, Africa in February 2016, a result of widespread community opposition, included a court case over the project location and resulted in a shareholder investment loss of $66 million.[666]

If you need electricity to be happy and secure, and to live longer, how much are you willing to pay for it in dollars and death? What will the market bear for you to be happy and secure? 20 cents/kWhr? 50 cents/kWhr? And how many tax breaks are you willing to pay to make that happen - production tax credits, carbon credits, death tax credits or otherwise?

We are apparently so lazy we cannot save the hundreds of thousands of preventable deaths from smoking, auto accidents, obesity, and violent deaths each year, so I guess we should not care about a few thousand deaths from coal every year, and even less about a few dozen deaths from wind each year. It seems that it is just the cost of doing business.

Before too hastily pursuing those wind turbines for the generation of intermittent electricity, politicians should read the numerous published reports from Maryland to Canada to France about the effects of wind turbine noise, and listen to their concerned citizens, homeowners, patriots, and local businesses fighting to defend their rights and their families, have all rejected industrial wind turbines farm monstrosities in their backyards.[667, 668, 669]

SOLAR PANELS

Many of the tools needed to stop heating the planet already exist. The use of intermittent electricity generation resources is expanding in the West (China isn't following this path), but the production of electric vehicles, wind turbines, and solar cells needs to be scaled up.[670] To source all electricity from renewables by 2050—supposedly to limit global warming to 1.5 degrees Celsius—citizens will need 1 billion additional electric cars and more than a 30-fold increase in solar photovoltaic capacity.[671]

But as economies in the West address rising global emissions—another crisis is worsening elsewhere. Making all those vehicles, panels, and turbines requires resources such as copper, lithium, and cobalt—which, like fossil fuels, are extracted from the ground. But unlike fossil fuels, many raw materials for green energy come disproportionately from developing countries that are some of the worst ecological and human rights offenders that exist on earth.[672]

The human rights abuses and environmental degradation in places like the Congo do not bode well for the West's transition to sustainable electricity, which will stretch the demand for green energy materials. Some critics even cite such adverse effects of intermittent electricity production to argue against any transition to green energy, such as in the recent, widely criticized Michael Moore produced documentary *Planet of the Humans*—which also pushes the falsehood that manufacturing renewable technology consumes as many fossil fuels as burning them.[673]

Still, the question of how to source metals and minerals ethically remains a legitimate and urgent one. Policymakers, researchers, and mining industry leaders are now debating how to ensure that the transition to green electricity does not exacerbate environmental and social problems elsewhere.

One proposal is to improve the traceability of mining supply chains. At first glance, it makes sense: If manufacturers can achieve transparency from mine to final product, consumers will be able to make ethical purchases and put the "bad guys" out of business. Improved traceability has played an important role in pushing companies toward more ethical conduct. For example, some diamond retailers use blockchain technology as a tamper-proof transaction tracing system to ensure that conflict diamonds do not enter the legitimate supply chain.[674]

But applying this tracing model to something more complex, such as a battery, is more difficult. A battery consists of dozens of materials from factories around the world, each deriving resource from places with disparate environmental and social standards. "If you were to certify that a battery from an electric vehicle was free of child labor, it's not enough to focus on the facility where they manufactured the battery," Benjamin Sovacool, a professor of energy policy at the University of Sussex said. "You also have to certify that no children were involved when they collected the cobalt, alongside numerous other metals and minerals in supply chains that will change month-to-month based on availability, price, and stability."[675]

Renewable technologies create ethical issues at both ends of their life cycle. Benjamin Sovacool was part of a team of researchers who recently visited the two ends of technology supply chains: artisanal cobalt mining sites in Congo, where miners extract the metal using rudimentary tools or their hands, and electronic waste scrapyards in Ghana, a global cemetery for electronics such as solar panels. The team's findings reveal widespread child labor, the subjugation of ethnic minorities, toxic pollution, biodiversity loss, and gender inequality along the length of the supply chain.[676]

Much of the e-waste generated by the West is sent, sometimes illegally, to countries in Asia or Africa, where a small amount of it is mined for reusable materials and sold back to world markets. This economic opportunity often causes toxic pollution, fueling environmental and public health crises. Many workers at one of the largest e-waste processing sites in Ghana, Agbogbloshie, are children who help dismantle electronic goods, extract metals, and then burn the waste—producing smoke that envelops the surrounding communities.[677] Studies have found high levels of lead in the blood of those living near the processing sites.[678]

Although traceability has a role to play, there is no silver bullet that can solve these problems. There are already international treaties and conventions that aim to solve issues such as child labor, social inequality, environmental degradation, and biodiversity loss. But a recent United Nations review of 80 existing international initiatives that govern the mining sector concluded that they often fail to be enforced by national governments.[679]

Countries with weak governance, such as Congo—where mining companies from all over the world operate—are less likely to take part in such international treaties and practices. There are already rules for environmental protection and social responsibilities in Congo's mining code, but without adequate enforcement, it is left to mining companies to self-regulate. This structure allows unscrupulous mining companies to exacerbate socioeconomic inequality, violence, and conflict by sourcing resources from illegal miners, displacing communities, and polluting.

Such issues have galvanized a growing global community of anti-mining activists seeking corporate transparency, with some going as far as calling for the complete withdrawal of Western mining companies from developing countries. Others question the utility of this approach. "The risk is that responsible companies pull out and leave the field open to more unscrupulous players," said Tom Butler, the CEO of the International Council on Mining and Metals (ICMM), an international organization set up to improve sustainability in the industry.[680]

Corporate sustainability is not enough to address all the ethical issues in the mining supply chain, but some observers pin their hopes on it—fueled by a consumer appetite for sustainable materials.

Tom Butler of the International Council on Mining and Metals (ICMM) says that mining companies can sometimes fill the void left by weak governments, building schools, hospitals, roads, and water systems. Mining companies can reduce greenhouse gas emissions and prevent damage to the environment by making operations more resource efficient. They can also create benefits for people in local communities that extend beyond the lifetime of the mining operation, such as equipping locals with training and education that make them more likely to be hired in the future.

One of the largest mining companies operating in Congo, Switzerland-based Glencore, sources 96 percent of its staff locally across its global operations, and it paid $680 million in wages and benefits in Congo between 2015 and 2018. Similarly, the South Africa-based Gold Fields, which has operations in Ghana, hired 55 percent of its overall staff locally in 2019.[681, 682, 683]

For every job created in the mining sector, another 25 are created through the value chain indirectly via the procurement of goods and services or transport, according to Butler. "So, I say if you want to invest where it makes a difference, then a responsible mining company operating in the DRC is a good bet," he said. (Glencore has recently been accused of facilitating child labor in cobalt mining practices, although the company insists it does not tolerate child or forced labor).[684]

But there is a limit to what corporate social responsibility can achieve. The mining companies are often reluctant to shoulder too much of the responsibility of governments, as it risks undermining local efforts to strengthen communities. Individual companies operating responsibly cannot ensure that others follow the same standards. They also cannot solve the structural issues of poverty and the lack of domestic and international governance that underpin illegal mining, child labor, and the streams of e-waste from the West.

What is needed is collaboration between mining companies and governments and stronger policies. Though international treaties have limited impact in countries with weak governance, a unified policy that all countries in the mining supply chain agree on, could overcome that issue. "What we need is an international protocol that coordinates among existing environmental treaties," said Saleem Ali, a professor of energy and the environment at the University of Delaware.[685]

Benjamin Sovacool, Saleem Ali, and others have additional policy recommendations to lessen the burden of extractive industries on communities and the environment. Policymakers should ensure that mining companies are at least partially owned by locals, either by including requirements for minimum local ownership into national mining laws or by incentivizing mining companies through tax cuts or other financial benefits. They also recommend that countries consider mining impacts as part of their commitments to existing international treaties, such as the Paris climate accord.[686]

Benjamin Sovacool and his colleagues call this inequality the "decarbonization divide," which may widen as the world ramps up renewable electricity use. For example, solar generated intermittent electricity could meet the global demand for low-carbon electricity many times over. But the sheer size of solar panels, which often contain lead, cadmium, and other toxic metals, makes them one the largest global contributors of electronic waste. By 2050, which is the rough expiration date of solar panels manufactured today, the technology is estimated to produce 78 million metric tons of waste—some 80 percent more than the total annual waste from all combined technologies today.[687, 688]

- The overwhelming statistic mentioned above is worth repeating as it is almost incomprehensible: in 30 short years, the rough expiration date of solar panels manufactured today, the solar technology is estimated to produce 78 million metric tons of waste—some 80 percent more than the total annual waste from all combined technologies today!

A different set of solutions is needed to tackle the growing pile of e-waste at the end of the green electricity supply chains. Today, much solar technology is discarded after it expires rather than being reused due to the cost dynamics of solar markets. Maintaining affordability requires "better design, so that it is easier to repurpose, repower, or recycle components," Sovacool said.

Governments and major companies have difficult decisions to make, and the window of opportunity for addressing climate change is rapidly closing. While difficult, it is not impossible to convert the world to intermittent electricity without exacerbating existing inequalities. But the problem can only be solved through cooperation across the green electricity supply chain that incorporates technological and social strategies.

Intermittent electricity from wind and solar are set to lead efforts to temper climate change and expand access to electricity. But as the third energy transition gathers pace, a deeper examination of the industry's supply chains, and operational practices reveals a darker side when it comes to human rights.

Some of these challenges, such as those around the cobalt used in lithium-ion batteries, are well known. However, while the persistent presence of child labor in Congolese cobalt mines has left companies struggling to address supply-chain traceability, other human rights impact from renewable electricity may be flying under the radar.

Research shows that labor rights issues are present in renewable electricity supply chains, including in the manufacturing of solar panels and the cultivation of palm oil and sugarcane used in biofuels. Looking further down the value chain, the rights of vulnerable communities can also be put at risk by major renewable projects developed on their land.

So, while one side of the green revolution offers critical sustainability benefits, the image and reputation of energy operators, investors and end users can also be highly exposed to human rights issues that could cast a shadow over the sector's socially responsible credentials.

In theory, the intermittent electricity industry is in a strong position to address such concerns. Given its de facto status as a sustainable industry, shareholders are more likely to demand transparency and accountability beyond that mandated by laws such as the U.K. Modern Slavery Act.

As funding mechanisms, including green bonds, also increase in popularity, their requirements for socially responsible projects can drive best-practice efforts to reduce human rights violations within the sector. But this will remain challenging.

A case in point is a recent industry report from the nonprofit organization Business and Human Rights Resource Centre (BHRRC) that found almost half of all renewables companies surveyed did not have any human rights policies in place.[689]

With expansion moving forward in both developed and developing countries, addressing land-rights governance and supply-chain traceability is easier said than done. Identifying a clear picture of the human rights risks associated with renewable electricity production will remain a necessary step toward achieving truly sustainable energy processes.

With a history of excessive working hours and occupational health and safety (OHS) violations in the manufacturing sector, China's position as the top global producer of solar panels should prompt focused supply-chain due diligence in the sector. Although China has made significant improvements in environmental standards attached to PV cell manufacturing, OHS issues in its solar industry are not uncommon.

Further human rights risks may be looming if upstream inputs become a focal point. Industry supply chains for quartz, the raw material from which polysilicon for PV cells is produced, are somewhat opaque, posing a challenge for those seeking to understand upstream risks. Demand for solar panels is also driving exploration for high-purity quartz deposits, including in countries where labor standards are poorly enforced, such as Mauritania and Saudi Arabia.

The inherently hazardous nature of mining in poorly regulated countries, combined with quartz mining's links with the respiratory disease silicosis, significantly raises the potential for labor rights violations to occur.

Outside of labor rights issues, companies involved in land-intensive renewable energy projects, such as wind and solar farms, face exposure to risks related to land rights violations that can create delays and disruption.

China, India, Brazil, Turkey, Mexico, South Korea, and the U.S. are among the countries where significant expansion of the wind sector is expected in the next decade or so. Five out of these seven countries are assessed as "high" or "extreme risk" in indices covering indigenous peoples' rights, land rights and security force violations.[690]

Major projects in countries such as Kenya, Taiwan and Morocco have all been subject to scrutiny over these issues, but Mexico is the prime example of where we have seen these risks collide with negative impacts for operators and investors.

A failure to respect land and indigenous rights will result in prolonged community distrust toward renewable energy operators. While challenges in obtaining free, prior, and informed consent — the backbone of securing social license to operate — are significant, success stories have been recorded in high-risk countries such as Mexico.

Despite difficulties, Oaxaca, Mexico has also produced renewable energy projects that empower local communities, such as the Ixtepec wind project. Building on models such as this could form a roadmap for investors to reduce reputational and operational risks and benefit local populations.[691]

THE DARK SIDE OF CLEAN ELECTRICITY

The "Praise the Lord" (PTL) empire that preacher Jim Bakker built with wife Tammy crumbled thirty years ago. Today, it seems like we are being mesmerized again in the press and social media with rhetoric about dispensing with thousands of products from petroleum derivatives so we can save the world from human destruction by switching to intermittent electricity generation from wind and solar.[692]

Renewables by themselves may not be able to save the world

as they can only generate electricity, and intermittent electricity at best. The undisputable science is that renewables CANNOT manufacture any of the oil derivatives that are the basis of more than 6,000 products that are the basis of societies and economies around the world.

Everyone knows that electricity is used extensively in residential, commercial, transportation, and the military, to power motors and lite the lights; but more importantly it is the 6,000 products that get manufactured from crude oil that are used to make those motors, lights, and electronics. Noticeable by their absence, from turbines and solar panels, are those crude oil chemicals that wind and solar are currently incapable of providing.[693]

We have had almost 200 years to develop clones or generics to replace the products we get from crude oil such as: medications, electronics, communications, tires, asphalt, fertilizers, military, and transportation equipment. The social needs of our materialistic societies are most likely going to remain for all those chemicals that get manufactured out of crude oil, that makes everything that's part of our daily lifestyles and economies, and for continuous, uninterruptable, and reliable electricity from coal or natural gas generation backup.[694]

Electricity is one of those products that came AFTER the discovery of oil. All the mineral products and metals needed to make wind turbines and solar panels rely on worldwide mining and transportation equipment that are made with the products from fossil fuels and powered by the fuels manufactured from crude oil.[695]

A single electric-car battery weighs about 1,000 pounds. Fabricating one requires digging up, moving, and processing more 500,000 pounds of raw materials somewhere on the planet.

Never discussed by the Green New Deal or Paris Accord sponsors are the questionable and non-transparent labor conditions and loose, or non-existing, environmental regulations at the mining sites around the world for the products and metals required for renewables. To meet the goals to go "green" for just electricity will most likely cause a rare earth emergency as those "green" goals require a massive worldwide increase in mining for lithium, cobalt, copper, iron, aluminum, and numerous other raw materials such as.[696, 697]

- A list of the sixteen components needed to build wind turbines are: Aggregates and Crushed Stone (for concrete), Bauxite (aluminum, Clay, and Shale (for cement), Coal, Cobalt (magnets), Copper (wiring), Gypsum (for cement), Iron ore (steel), Limestone, Molybdenum (alloy in steel), Rare Earths (magnets; batteries), Sand and Gravel (for cement and concrete), and Zinc (galvanizing).[698]
- A list of the seventeen components needed to build solar panels are: Arsenic (gallium-arsenide semiconductor chips), Bauxite (aluminum), Boron Minerals, Cadmium (thin film solar cells), Coal (by-product coke is used to make steel), Copper (wiring; thin film solar cells), Gallium (solar cells), Indium (solar cells), Iron ore (steel), Molybdenum (photovoltaic cells), Lead (batteries), Phosphate rock (phosphorous), Selenium (solar cells), Silica (solar cells), Silver (solar cells), Tellurium (solar cells), and Titanium dioxide (solar panels).[699]
- The origins of the products for wind and solar are mined throughout the world, inclusive of more than 60 countries of Algeria, Arabia, Argentina, Armenia, Australia, Belgium, Bolivia, Brazil, Canada, Chile, China, Congo (Kinshasa), Cuba, Egypt, Finland, France, Germany, Greece, Guinea, Guyana, India, Indonesia, Iran, Ireland, Italy, Jamaica, Japan, Kazakhstan, Madagascar, Malaysia, Mexico, Mongolia, Morocco, Mozambique, New Caledonia, Oman, Pakistan, Papua New Guinea, Peru, Philippines, Poland, Republic of Korea, Russia, Saudi Arabia, Sierra Leone, Slovakia, South Africa,

Spain, Suriname, Sweden, Thailand, Turkey, Ukraine, United Kingdom, United States, Uzbekistan, Venezuela, Vietnam, Western Sahara, and Zambia.[700]

The signatories to the Green New Deal and Paris Accord, to sunset the fossil fuels industry for a world surviving on intermittent electricity would also sunset its own renewable industry that is supposed to be the salvation for the world, as there would be no components to build the turbines and panels![701]

All mining and processing activities to get the iron ore and other metals that go into turbine manufacturing, transporting the huge blade beasts to the sites, and decommissioning them, are all energy intensive activities that rely on fossil fuels and the products from crude oil and leave difficult wastes behind to dispose of during decommissioning.

We can be preached to forever about "clean electricity" messages and bedazzle farmers with the prospects of on-going revenue from renewables. However, the extensive mining worldwide for materials for millions of wind turbines and solar panels, and the decommissioning and restoration details, and the social changes that would be necessitated for societies to live without the thousands of products from petroleum derivatives remain the dark side of the unspoken realities of renewables.

When we read the WSJ article "The Best-Laid Energy Plans" about the Government planning and subsidies that were supposedly intended to make America the world's green-electricity superpower, create millions of jobs, and supercharge the economy, it recalls the most terrifying nine words in the English language:" I'm from the government, and I'm here to help."[702, 703]

In pursuit of a way to store the daytime intermittent electricity from solar panels, for use when the sun is not shining, the reality is closer to the financial failure at Crescent Dunes, a Nevada solar-energy plant that went bust after receiving a $737 million federal loan guarantee. No worries, it is only taxpayer money.

Crescent Dunes was the first concentrated solar power system that generated solar power by using mirrors or lenses to concentrate a large area of sunlight onto a receiver plant with a central receiver tower and advanced molten salt energy storage technology.[704, 705, 706] Hand the energy economy over to the government in the name of climate change, and there will be countless more Crescent Dunes fiascoes.

Basically, electricity can power the motors, lights, and electronics, but it cannot make the motors, lights, and electronics! Electricity could not exist without fossil fuels as all the parts for wind and solar renewables are made with fossil fuels. Electricity and fossil fuels are interwoven together; renewable intermittent electricity falls under this category. Without fossil fuels, electricity doesn't exist, and without electricity products made from petroleum derivatives in factories have a difficult time being made without baseload electrical generation.

Renewables need crude oil and electricity for the turbine blades and solar panels in the manufacturing process. But they aren't manufactured without electricity in the factories that assemble renewables and overwhelmingly that electricity comes from oil, coal, natural gas and nuclear.

Here are many of the things that electricity can and cannot do for civilization:

- Provide electricity to run the motors of vehicles, heating, air conditioners, but it cannot make the motors.
- Provide electricity for lighting, but it cannot make the lightbulb.
- Provide electricity for electronics, but it cannot make the electronics.
- Provide electricity for the medical infrastructure, but it cannot make any of medical products.

Energy storage could revolutionize industries in the next 10 years, but despite the preaching's about these renewable saviors, it is becoming obvious that due to their intermittency and unreliability, and their inability to replace any of the chemicals from crude oil that account for the products in our daily lives, societies and economies around the world may not be too thrilled about the needed social changes to live on just intermittent electricity.[707]

Lets' be clear about what that means. First, it is not renewable energy, it is only renewable electricity, and more accurately its only intermittent electricity. Renewables have been the primary driver for the high costs of electricity for residents of Germany, Australia, and California. Second and most important is, electricity alone is unable to support militaries, aviation, and merchant ships, and all the transportation infrastructure that support commerce around the world.

Germany tried to step up as a leader on climate change, by phasing out continuous uninterruptable electricity from nuclear and fossil fuels and pioneered a system of subsidies for intermittent electricity from wind and solar that sparked a global boom in manufacturing those technologies. Today, Germany is failing to meet its climate goals of reducing carbon-dioxide emissions even after spending over $580 billion by 2025 to overhaul its energy systems. Germany's emissions miss should be a "wake-up call" for governments everywhere.[708]

Power prices in Germany are now among the highest in Europe and globally among developed nations. Today, German households pay almost 50 percent more for electricity than they did in 2006. Much of that increase in electricity cost is the Renewable Surcharge that has increased over the same period by 770 percent.[709]

Hopefully, before committing to an all-electric world, with intermittent electricity, we can achieve the technical challenges of discovering a green replacement for the thousands of products based on fossil fuels being provided to every known earth-based infrastructure, and society will accept the consequences of altering their lifestyles that will result from less services and more personal input to accommodate losing the advances fossil fuels have afforded them over the last few centuries.

Rare earth metals, hard-to-find materials, with unfamiliar names such as lanthanum, neodymium, and europium, are used in wind and solar energy projects, but dwindling supplies could hinder a roll-out of low carbon technologies and slow China›s shift away from coal power. These compounds, which are highly toxic when mined and processed, also take a heavy environmental degradation toll on soil and water, posing a conundrum for policymakers in China, the world's biggest producer and consumer of rare earths.[710, 711, 712]

In 2012 the Chinese government named the city of Ganzhou, in the southeastern province of Jiangxi, a "rare earths kingdom"; even though at that time its rare earth reserves were already almost depleted. According to a rare earths white paper issued by the State Council News Office in 2012, the reserves to extraction ratio for rare earth elements in southern China was 15.[713] In other words, if mining continued at the existing rate, those reserves rich in Medium and Heavy Rare Earth Elements (MHREEs) would only last for another 15 years. [714] Will there be a shortage of rare earths after 2035?

Despite rapidly depleting reserves Ganzhou still accounts for more than half of all rare earths produced in China.[715]

Three years later and 6,000 miles away in Paris, 190 countries signed the historic Paris Climate Agreement, including plans to introduce a greater share of wind and solar intermittent electricity in a "decarbonized" future. But few of the delegates gathered in Paris seemed to realize how important one small south-central Chinese city would be to achieving this target; as almost all the clean, smart, and low-carbon technologies are reliant on rare earths.

This prompts the questions: do we have enough rare earths to build the clean and smart electrical future we are imagining?

Can China, supplier of 90 percent of the global rare earths over the last 20 years, meet expected growth in demand; and what are the environmental consequences?[716]

A visit to the mines and industrial parks of Ganzhou gives no sense of a glorious "kingdom." It is a scene of devastation: crude, open-air mines and smelters, and rough, muddy attempts at restoring the landscape. Water in and around the mining area is severely polluted. According to *China Environmental News,* the water supply for 30,000 people in the county of Longnan alone has been affected by rare earth mining, with 40,000 mu (6,589 acres) of farmland seeing reduced yields or complete harvest failure.

Over a decade of excessive extraction has left the surface water in the mining area, China's biggest source of ion-absorption rare earths, with ammonia and total nitrogen levels far above safe standards, while groundwater is nowhere near up to minimum drinking water standards.

In April 2012 a cross-ministry investigation headed up by the Ministry of Industry and Information Technology (MIIT) found 302 abandoned rare earth mining sites in Ganzhou, with 97.34 square kilometers affected. It would take 70 years just to deal with the 190 million tons of mining waste left behind.[717]

China's government says the country "meets 90 percent of the world's demand for rare earths but has only 23 percent of global reserves." In the early 1990s China overtook the U.S. to become the world's biggest producer and exporter of rare earths and since then has virtually become a monopoly supplier, with some rare earth products produced only in China. If you trace them back to the source most fluorescent lamps, offshore wind turbines, electric and hybrid cars, smartphones, and personal electronic devices have, thanks to the rare earths used in their components, "Chinese DNA".[718]

According to the U.S. Geological Survey, at one-point China

was accounting for 98 percent of global rare earth output. In 2015 that figure still reached 85 percent.[719]

Although on the surface China plays a leading role on the rare earths market, there is an unknown side to the apparent hegemony. The main importers who benefit from China's rare earths production, such as the U.S., Korea, and Japan, as well as the manufacturers and brands who use rare earths in their products, often tap into a substantial black market in rare earths. Every year tens of thousands of tons of rare earth ores are illegally mined and traded, leaving China through the black market.

Those lower down the supply chain turn a blind eye to this, and international cooperation on law enforcement is minimal. With no international traceability system, such as that for conflict minerals, companies have no way of monitoring supply chains and we cannot know if the electric cars we drive or the smartphones we use contain illegally mined and smuggled rare earths.

The huge profits to be made means Ganzhou is plagued by corruption and illegal mining. The China Rare Earth Industry Association estimates that in 2013 the actual supply of rare earths in southern China was over 50,000 tons, and over 40,000 tons in 2014. However, the Ministry of Land and Resources only permitted output of 17,900 tons per year for that period. That means the black market may be two to three times the size of the legitimate market.[720]

And rare earth mining, whether legal or not, entails shocking environmental costs. Research has found that producing one ton of rare earth ore (in terms of rare earth oxides) produces 200 cubic meters of acidic wastewater. The production of the rare earths needed to meet China's demand for wind turbines up to 2050 (in a scenario of radical wind power expansion) will result in the release of 80 million cubic meters of wastewater – enough to fill Hangzhou's West Lake eight times over. Not to mention the emissions from the rest of the product lifecycle, smelting, separation, processing, transportation.[721]

Businesses, policymakers, and consumers all need to think again: what actions can we take to ensure we meet our low-carbon goals in a way which is friendlier to both the environment and the climate? It is after all, both contradictory and unjust to sacrifice public health and the environment in a resource-producing area for the sake of low-carbon development.

To this date, 189 Parties have ratified of 197 Parties, the Paris Agreement, buttressing attempts at "decarbonization". That means now is the time to look again at the link between China's rare earth resources and the clean, low-carbon and smart technologies relying on those. Over the last 20 years the environment has paid the price for China's economic successes.[722]

China's rare earth reserves are much depleted; environmental costs in the trillions of yuan have not been factored into market prices; and a rampant black market in rare earths, both at home and abroad, has exacerbated environmental damage and the loss of resources.[723]

This has left the Chinese government with no option but to cover huge environmental remediation costs, while those living near rare earth mines are directly or indirectly suffering environmental and health problems.

Another issue is that China is no longer simply a supplier and exporter or rare earths as domestic demand for these resources has increased sharply. China is the main driver of global investment in wind power. In a scenario for radical expansion of wind power produced by the National Development and Reform Commission's Energy Research Institute, China could see installed wind power capacity of 2,000,000 megawatts (2 terawatts) by 2050. A typical 2-megawatt turbine contains 341-363 kilograms of the rare earth neodymium and about 59 kilograms of dysprosium.[724]

The quantities of rare earths needed just to allow for wind power growth are astounding, and this is before the increased rare earth demand arising from the "China Manufacturing

2025" plan, which aims to prioritize development in electric vehicles, marine engineering equipment and astronautic and aeronautic manufacturing, are considered.[725]

China may not even be able to meet domestic demand, never mind increasing demand from other nations. According to U.N. Conference on Trade and Development (UNCTAD) estimates, global demand for rare earths will be between 200,000 and 240,000 tons annually by 2020, with 70 percent of that demand coming from China. Even if China makes full use of its entire mining quota, there is still a gap of 35,000 to 63,000 tons between new annual output and expected growth in demand. How will that gap be met?

Looking at rare earths throws up other unanswered questions about our low-carbon future. How will all that wastewater be handled? Will there be new drinking water safety issues? Will the costs of better technology and management, intended to reduce emissions, be reflected in rare earth prices?

Back in 2014 the Chinese government declared a "war on pollution", which was followed up by "history's toughest" environmental protection law and standards for the rare earth industry on emissions and the use of water and energy. This means compliance costs for the industry are bound to rise; low-cost rare earth mining and processing are a thing of the past in China. EU and U.S. research bodies have pointed out that there will be a shortage of light rare earths in the short and mid-term, while the shortage of medium and heavy rare earths will be in the mid and long-term. The combination of increased costs and shortages mean price rises are inevitable.[726]

The world must ask if its low-carbon future may be limited by these "industrial vitamins."

It takes a special kind of delusion to continue believing that intermittent electricity from wind and the sun will power Berlin, Los Angeles, or San Francisco.[727] Placating environmentalists, their lobbyists, and beneficiaries of their campaign contributions needs to be understood before they destroy California and the U.S. by replacing fossil fuels (mainly coal, natural gas, and petroleum) and nuclear with just intermittent electricity from solar panels and wind turbines. Are this what California and America wants for their energy policies, environmental health, and economic future? Let us hope America does not follow Germany into ruin.[728] Or the remainder of the European Union and western-aligned nations such as Japan, South Korea, and Australia.

The dark side of renewable wind, solar, EV batteries, and biofuel energy is that they are not clean, green, renewable, or sustainable. [729] They are horrifically destructive to vital ecological values that will last for generations to come.[730]

CHAPTER SIX

CHINA AND INDIA

- ***The two largest countries in the world desperately need affordable, reliable, scalable, abundant, and flexible sources for electricity.***
- ***With more than 35 percent of the world's 8 billion population, they have amassed the most coal fired power plants and are building more at a record pace.***

INTRODUCTION

Over 2.5 billion people of the 8 billion on earth are in China and India. We are in the beginnings of the Asian hemisphere version of the Industrial revolution. When the first industrial revolution took place it affected a couple hundred million – now we are witnessing billions vying for energy and electricity to power factories, transportation, militaries, and economies that will lead the 21st century into prosperity or environmental annihilation.

An astonishing number when you realize that the west, which consists of approximately a billion people from the post-World War II and Cold War alliance of the United States (U.S.), Europe (the European Union or EU), and Asian allies Japan, South Korea, Taiwan, Australia; and increasingly the Philippines and Vietnam since both are under direct attack from China over annexation of the South China Sea.[731]

Intermittent electricity from renewables (defined as solar panels, wind turbines, and destructive biomass for electricity) are now in vogue throughout the halls of the United Nations (UN), and the west.[732] Ask most engineering students in accredited western universities and they want to work on supposedly free, limitless electricity the sun and wind provide. A Green New Deal for Europe and America is being promoted to deliver results against anachronistic fossil fuels and scary nuclear energy, but something being free, and renewable is a misnomer.

What is more unrealistic is believing China and India are doing away with coal-fired power plants, natural gas-fired power plants, or nuclear. They are certainly not eliminating the over 6,000 products from oil derivatives. China and India's future is coal, natural gas, oil, petroleum, and nuclear. Now with Africa beginning to grow at a record pace that is being fueled overwhelmingly by coal a competition for coal, natural gas, oil, petroleum and products from a barrel of crude oil is now taking place like never before.[733] China and India are never moving away from fossil fuels this century unless a better alternative is created. Renewables are only meant to placate gullible western environmentalists, foundations, academia, and governments. China effectively uses this tactic; particularly, their latest pledge in late 2020 of being carbon neutral by 2060.[734]

Why would China and India rid themselves of fossil fuels or nuclear when both governments could show their citizens the earth is actually cooling and not warming.[735] 400 papers in 2020 deeply questioned the efficacy of the notion that man's over use of fossil fuels is uncontrollably warming the earth leading to climate change.[736] Then add a "120 years of climate scares," and it's predictably easy to understand why China and India are going to use more instead of less fossil fuels for the rest of the century.[737] If you the reader, want to deeply explore the issue of anthropogenic global warming (man-caused warming) please go to the source at the end of this sentence to check out, and read hundreds of books questioning the viability and factual relevance if man is ruining the earth over using fossil fuels.[738]

Mankind's activities of the burning of fossil fuels, massive deforestations, the replacing of grassy surfaces with asphalt and concrete, the "Urban Heat Island Effect" and more are creating extensive harmful pollution and leading to the additional warming of our planet. Yes, we believe we should be "going green" whenever and wherever possible. However, some of the long-term warming and cooling of global temperatures trends from 2500 B.C. projected to 2040 A.D. may be the result of climatic cycles, solar activity, sea-surface temperature patterns and more.

Randy Mann, Meteorologist at Harris-Mann Climatology prepared the following graph that illustrates the warming and cooling cycles that have occurred over time. Most of the 78 major temperature swings in the last 4,500 years occurred when fossil fuels were not part of the equation.[739]

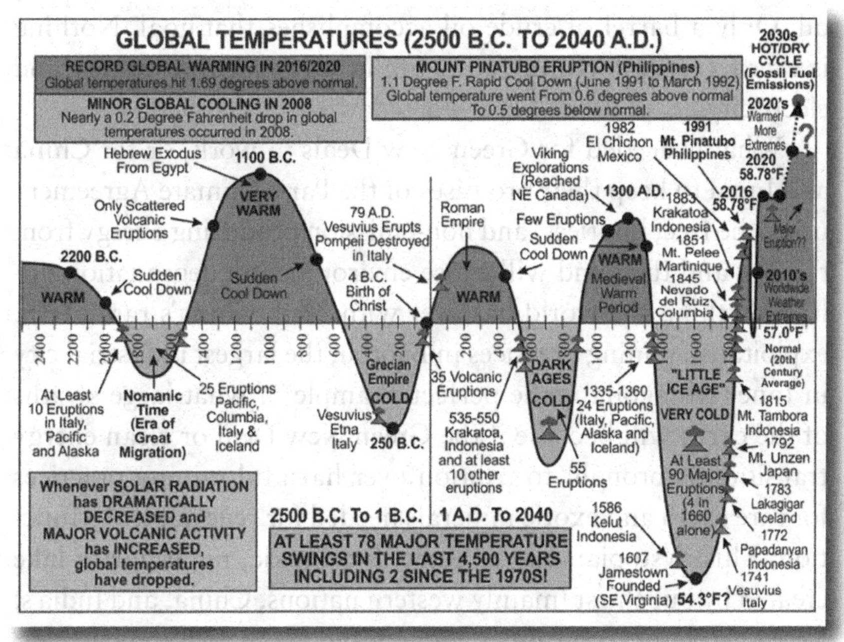

Figure 6-1 Global Temperature Trends From 2500 B.C. To 2040 A.D

Obviously, billions of people between two countries offers constraints never seen in recorded human history. Nature only has so many resources so why not use the sun and the wind? If only it were that simple. Economics references opportunity costs. For intermittent electricity from the wind and solar to work it must be abundant, affordable, scalable, reliable, and flexible. Otherwise, electricity from renewables is not free, and the positive opportunity cost vanishes. Then there is the issue that wind and solar cannot manufacture the petroleum derivatives that produce today's widely used products like makeup, plastics, and COVID-19 vaccines.

Machines are used in exploration and production (E&P) of coal, oil, and natural gas. For renewables they need rare, deep earth minerals and metals such as cobalt, copper, and lithium to produce electricity. Renewables cannot produce petroleum derivatives that are the basis of more than 6,000 products made from oil. Only a barrel of crude oil accomplishes that goal. Nothing is ever free. Opportunity costs are in every facet of energy and electricity.

What is needed for Green New Deals to work, or for China and India to keep their promises of the Paris Climate Agreement using the free, limitless, and non-emission producing energy from the sun and the wind will cause environmental degradation unlike anything the world has ever witnessed. China's rare earth, exploitative mining practices in Baotou, the largest industrial city in Inner Mongolia is the perfect example of what large swaths of the earth will become if the Green New Deal or clean energy transition is brought to fruition over harmful mining practices for rare earth and exotic minerals needed for renewables to function. This dystopian city is home to a "toxic, nightmarish lake created by our thirst (mainly western nations, China, and India's) for smartphones, consumer gadgets and green tech (renewables, EV batteries and battery energy storage systems for electrical grids)."[740]

Solar panels and wind turbines for billions of people is "the invisible (environment-destroying) elephant in the room," no one wants to discuss.[741] The gargantuan amount of nature that will be destroyed if China and India attempt decarbonization or a green transition is unfathomable with virtually non-existent environmental regulations and labor laws in place.

Begin with the transportation sector's push for electric vehicles (EVs) and China is at the forefront for the supply chain for EV battery materials. China leads the world for rare earth minerals that are used in the manufacturing process for EV batteries.[742] India wants the same EV opportunities for their growing country. The physical components bumping against nature's realities is shocking since EVs are not eco-friendly vehicles as they are necessitating worldwide environmental degradation from the mining for the minerals and metals to support EV battery construction. Using America as an example: there are approximately one million EVs on the road (only 0.5% of current U.S. passenger vehicles) after billions in U.S. taxpayer dollars subsidizing this small amount.[743]

What about the batteries and the amount of goods and materials to make one battery for over 2 billion people in China and India? A single EV battery typically "weighs 1,000 pounds, requires digging up roughly 500,000 of materials."[744] That's over 10-times more than what is used by a typical petroleum-based vehicle over the entire life of the vehicle. Rare earth minerals and metals such as neodymium, copper, nickel, and lithium are needed in an EV. An EV uses substantially more copper than a normal vehicle and global demand is expected to rise over 1,500 percent in the next two to four decades.[745] All rare earth minerals for EVs comprise a suite of "critical energy minerals that will drive a 200 percent to 8,000 percent increase in demand."[746]

More on EV rare earth mineral extraction amounts and the destruction to the environment in the conclusion of this chapter and in Chapter Four: Environmental Degradation.

China along with the Democratic Republic Congo where cobalt is mined, and the lithium triangle in South America use fossil fuels to mine, transport and ship these minerals are being destroyed in this bloodlust for rare earth minerals needed for solar panels, wind turbines, EV car batteries and battery energy storage systems for electrical grids. Oil demand will rise using EVs and renewables, and in the case of EVs they are expected to only be 10 percent of global vehicles by 2040.[747] This is with heavy government mandates in place. Markets for EVs typically drop when subsidies and tax advantages are taken away.[748]

Solar panels and wind turbines (renewables) use just as many rare earth minerals and metals. On average renewables "use 10 times more primary materials to produce the same energy output compared to hydrocarbon machines (fossil fuels and nuclear energy).[749] Ten times, which means globally, mining will grow rapidly, and all of it powered by coal, oil, petroleum, and hopefully natural gas and nuclear if the west is lucky since both forms of energy lead to lower emissions.

The west will have to look to China for every rare earth mineral and metal for a Green New Deal to meet decarbonization policies or the goals for carbon-free societies. China supplies over 90 percent of the minerals needed for EVs, solar panels, wind turbines, and battery energy storage systems for electrical grids to electrify entire nations and continents.[750] At this time the U.S. is 100 percent dependent on imports from China and others that are some of the worst ecological and human rights offenders that exist on earth for 17 key minerals and metals and another 14 out of 28 minerals that are needed for a clean energy economy to materialize.

China also controls over 90 percent of the battery industry's cobalt refining.[751] This is needed for raw ore, otherwise it is worthless. Russia continues growing their nickel industry – renewables and battery energy storage systems do not function without nickel. American and western sources are nowhere to be found on this list or India as well. The west's baseless worries over emissions should be concerning for this reason: energy-related emissions in the U.S. declined 3 percent in 2019, after rising by 3 percent the previous year.[752] U.S. electricity emissions are down an astonishing 33 percent from their peak in 2007 due to the fact U.S. electricity is switching to cleaner natural gas over coal.[753]

Except for California which is shuttering most of its in-state natural gas power generation, is contrary to the nationwide rise in natural gas as a fuel for electric power generation in the U.S. that has gone from 14 percent of all power in 2000, to 31 percent at the end of 2019. But if the U.S. moves to rid itself of fossil fuels in favor of renewables their electrical grids will crash when the sun does not shine or wind does not blow, which is why the entire electrical generation system in place within the European Union is on the verge of collapse.[754]

Electrical grids were never meant for intermittent sources of electricity; simply put, industrial solar and wind turbine farms do not work as currently proposed and envisioned.[755] Under current technological constraints battery energy storage systems for electrical grids cannot store enough electricity when the sun and wind are intermittent.[756] Storage systems for electrical grids are currently used as grid-stabilizers instead of storage systems.

This rise in natural gas takes place, because the U.S. is an advanced western nation, and can afford the multi-billion-dollar costs it takes to explore, produce, transport, liquify in terminals and employ natural gas for electrical generation. The U.S., EU, and other western-aligned nations have outsourced, or "leaked" mining to nations with minimal or non-existing environmental regulations and hostile to freedom, human rights, and democratic institutions over CO2 emission worries, which are not tied to factual reality.

Add this hostile action when China enacted legislation that goes into law in 2021 "allowing banning exports of strategic minerals to companies and nations that China considers a national security threat."[757] Say goodbye to renewables, battery energy storage systems, EVs, and other industries and technologies that need rare earth minerals without acquiescing to China. India will resist their nation being overtaken by China. The west might, but India will not.

Will not China and India, two countries that have cities that are some of the dirtiest in the world, be concerned about environmental atrocities associated with renewables, battery energy storage systems, and EVs? Likely not – since both countries routinely are considered the highest CO2 emitters. Remember, China and India are entering their Industrial Revolution with infinitely more people than the U.S. or Europe in the 1850s.[758]

Should the west attempt to emission-shame China and India into an all-renewable future of intermittent electricity based on the theory the sun and wind are free and do not cause pollution or respiratory illnesses? Not so fast – the Australia Institute for Sustainable Futures says the rush for exploding growth in rare earth minerals and metals will catapult miners into "some remote wilderness areas that have maintained high biodiversity because they haven't yet been disturbed."[759]

China and India each have forests, parks, and remote biodiverse areas in their respective countries. Should these areas be bulldozed for rare earth minerals and metals when biodiverse areas are exactly what both countries need to combat their skyrocketing emissions?

The current United Nations (UN) Human Rights Council has North Korea, Syria, and Cuba on its committee, and they review abuses against countries. You can laugh now, but it is likely after they are done bashing Israel, they will report child labor abuses from mines in the Congo where 70 percent of the world's cobalt originates.[760] Other disgusting human rights abuses and unfair labor practices can be reported in connection to mining for copper and nickel, which is used for EVs, solar panels, wind turbines, and battery energy storage systems for electrical grids.[761]

Billionaires and supporters of renewables, EVs, and battery energy storage systems for electrical grids in China, India, the

U.S., and the EU, all pledge "ethical sourcing," which is not an uncommon saying. [762] What is uncommon is the gargantuan amount of rare earth minerals and human rights abuses taking place and set to increase from energy policies such as the American and EU Green New Deal.

China and India will never concern themselves with ethical sourcing or human rights when the two countries combined have over 2.5 billion people to feed, clothe, provide electricity for, heat and cool their countries, and use the over 6,000 products that come from a barrel of crude oil. If they can use green products – of course they will – but do not pretend they will attempt anything like the 2010 Dodd-Frank bill from the U.S. Congress requiring reporting and sanctioning of anyone using conflict minerals and metals.

The world's largest economy – the U.S. – is beholden to China for components for solar panels, wind turbines, EV batteries, and battery energy storage systems for electrical grids. Again, it cannot be stated enough – the U.S. imports 90 percent of its solar panels. President Joe Biden's 2021 Climate Plan is fully supportive of America continuing its dependency on China, even if it puts America's economy at a national security risk.

For wind turbines, it is 80 percent of the parts that are imported.[763] China and other Asian companies are the global leaders in the so-called green transition. This is the wholesale destruction of the U.S. and western model of energy independence and the best chance the west has of checking China's ascendant rise without firing a shot. Eliminating trillions in hydrocarbon economic activity means China wins, and global environmental goals and human rights are negated.

India then becomes the only check on China's hegemonic intentions to conquer Asia and subjugate freedom-loving nations – and all this over renewables. Let us say renewables could electrify entire economies with intermittent electricity – there is still the humongous increase in the amount of land it takes to deploy solar panels and wind turbines at any sort of effective rate over coal-fired, natural gas-fired or nuclear power plants for electricity.[764] Continue to remember that renewables are only for intermittent electricity and cannot replace or create the over 6,000 products that come from a barrel of crude oil.

Without subsidies, mandates, and onerous regulations, renewables would die-off. China and India are laughing at the west over their suicidal energy policies.[765] Shouldn't China and India follow the International Energy Agency's (IEA) advice and wholeheartedly adopt renewables to stop catastrophic climate change? Read the fine print in their latest 2020 forecast revealing only 3 percent of the world's electricity is derived from solar panels and wind turbines.[766] China and India will never adopt renewables as their primary electricity source under this 3 percent scenario as their billions of people need abundant, reliable, affordable, scalable, and flexible hydrocarbons to meet their electricity needs.

The wealthiest nations, which are primarily western-based account for less than 10 percent of the world's population and the UN with the IEA's backing can announce green policy intentions, decarbonization plans, elimination of fossil fuel-based vehicles for EVs, carbon-free societies, and Green New Deals, but it will not happen when over 2.5 billion people need energy, electricity, transportation, and a better way of life. The IEA predicts natural gas and petroleum will rise to pre-coronavirus levels once COVID-19 vaccines are distributed. Higher fossil fuel usage will continue for at least two to four decades ahead.

Mark P. Mills, the noted energy scholar from the U.S. based Manhattan Institute summarizes the IEA report best on the subject of rare earth minerals and metals needed for battery energy storage systems (BESS') and renewables by stating:

> "That same IEA report (says) that reliable supplies of the critical minerals and metals...are vital for

> reaching green goals. History may see that as the understatement of the decade."[767]

If China is a menace from self-inflicted environmental degradation from mining and to global health, peace, and prosperity under the current Communist Party regime, is India any better? India keeps one eye on China, one to the west for geopolitical, realist balancing options against their enormously aggressive neighbor (China), and both feet firmly planted in their own interests. In a short amount of time (approximately 8 years after this section was written in late November 2020) India will be the largest country in the world.[768] So what does this truly mean?

INDIA SAYS ONE THING AND DOES ANOTHER

India and China make up 2 out of 7 people in the world, and India is also using more coal than ever before to deliver electricity to their growing population and economy.[769] The IEA's Renewables 2020 report stated worldwide wind and solar capacity for intermittent electricity will double over the next five years, going from 3 percent to 6 percent. The report also says by 2023 solar and wind generating capacity will overtake natural gas, and then surpass coal in 2024. Renewables will then reach 18 percent of electrical generation globally by 2025.[770] Seems to be pie-in-the-sky projections from the IEA, because what India witnesses is:

> "China put 38.4 gigawatts (GW) of new coal-fired power capacity into operation in 2020, according to new international research, more than three times the amount built elsewhere around the world and potentially undermining its short-term climate goals."

Keeping with the theme of this book and the environmental degradation that takes place when renewables are implemented, how is India expected to achieve a large majority of their electricity from wind turbines and solar farms, while lowering emissions? Do they sit back and watch China continue to dominate global manufacturing supply and value chains powered by coal-fired power plants? While India attempts to counter with the wind and the sun to run their economy and military against their powerful, aggressive neighbor?

India also understands mining and transportation are mainly derived from fossil fuels, and both sectors are what bring rare earth minerals and metals to market for India, China, and the world to reach the 18 percent threshold by 2025 the IEA predicts. Do they frack for oil and natural gas while building millions of acres of industrial solar and wind farms? What about the Not-In-My-Backyard (NIMBY) protesters of solar and wind farms in their backyards and the environmental protesters with their hatred of fracking?[771]

Prime Minister Narendra Modi is a Hindu populist who is generally wary of China, but open to renewables and concerned about world opinion. Therefore, Mr. Modi and his New Delhi based government embraces the IEA projections. Modi is a smart politician when his government claims they are adhering to the Paris Climate Accords (PCA) and do the exact opposite through their humongous usage of the dirtiest forms for generation of electricity - coal-fired power plants.[772]

Who isn't for sensibly using the freedom of the sun and wind for electricity? India, however, needs basic sanitation, proper landfill management, newer transportation systems while somehow minimizing the negative ramifications of wholesale mining needed for renewables in their billion-plus citizenry country. India could and should plant millions of trees to help clean up their environment, but "close to 8 billion people live on the Earth and they feed 80 per cent of their hunger for energy with hydrocarbons or fossil fuels."[773]

India is growing, and another 3 billion people will likely be added in the next fifty years within the borders of India, China, Africa, and Asia.[774] Considering the growing need for state-of-the-art medical facilities, new air conditioners, electronics, cars, airplanes, and all the components of a modern society will increase the demand for electricity, petroleum, natural gas, coal, and nuclear dramatically. India is aware of all of this. But India's history will be difficult to overcome. Especially since the best-known hero of India's founding – Mohandas Gandhi said these incredibly insensitive, racist, and insidious statements during World War II. Concerning India's imminent doom at the hands of the Japanese:

> "Leave India in God's hands, in modern parlance, to anarchy, and that anarchy may lead to internecine warfare for a time, or to unrestricted dacoities [gang robberies.] From these a true India will arise in place of the false one we see."[775]

Gandhi's advice to Britons during the horrors of the London Blitz was:

> "Invite Hitler and Mussolini to take what they want of the countries you call your possessions. Let them take possession of your beautiful island with its many beautiful buildings. You will give all this, but neither your minds nor your souls."[776]

To Ethiopians struggling to defend themselves against Mussolini's Italian hordes this came out of Gandhi's mouth:

> "Allow themselves (the Ethiopians) to be slaughtered by the Italians since, after all Mussolini didn't want a desert."[777]

The Jews in Germany were told this by Gandhi after the Kristallnacht (Night of broken glass against Jews; a pogrom) atrocity where he (Gandhi) wanted Jews to approach their holocaust with his approach of:

> "Non-violent action what has today become a degrading man-hunt can be turned into a calm and determined stand offered by unarmed men and women possessing strength of suffering given to them by Jehovah, as that would convert the SS to an appreciation of human dignity."[778]

In May 1940 Gandhi told a friend,

> "I do not consider Hitler to be as bad as depicted. He is showing an ability that is amazing and seems to be gaining his victories without much bloodshed.[779]

Finally, Gandhi praised: "Hitler's dedication to bravery and devotion to the Fatherland (Germany)...Nor do we believe that you are the monster described by your opponents."[780] Ironically, Hitler's words to Lord Halifax on Britain in 1937 when they met at Berchtesgaden was – "SHOOT GANDHI!"[781]

Why should India, or you, the reader care about appalling comments from a western hero emblazoned on mugs and shirts in college bookstores, Oscar-winning films, and an Indian martyr? When the past is not recognized, and atoned for, the future stays dreadfully the same. India has a caste system, widespread rape of women, overt racism embedded in their society, and religious persecution against people groups who do not follow Hindu belief systems coupled with nationwide chauvinism against women and young girls.

Abundant energy resources and atoning for past sins is a good start for India to begin leaving the Dark Ages and joining the world community. Otherwise, India will continue having open defecation and urination in their streets, discarded children living in landfills for food and shelter, and continue receiving United Nations studies proclaiming their country as the filthiest in the world.[782]

India can and should do better. If they do not, the Chinese are waiting at the doorstep to crush their country and subjugate them like what the Japanese did to the Chinese in the 1930's and World War II. Will India atone for their past sins? Not likely, as the country is trying everything it can to keep up with exploding demand for electricity by any means necessary. The truth is:

> "Coal-fired electricity generation accounted for 70 percent of India's power output in 2019, according to India' Central Electricity Authority. India's coal-fired electricity generation rose 9.4 percent in the first half of September."[783]

India is the third-largest energy and electricity consumer in the world after China and the U.S.[784] This figure is climbing and eventually India will overtake the U.S. and China over their dynamic economic growth, climbing population, and modernization plans under Prime Minister Modi, whose party – the Bhartiya Janata Party (BJP) was re-elected as the majority party in May 2019 for another five-year term.

While COVID-19 has temporarily halted India's growth, and this section argues for confession of past sins, energy security is the country's top priority.[785] Next steps are overall infrastructure development from roads to sewage treatment and liberally opening the country's economy are what will keep India from collapsing to contain their exploding population.

What will not accomplish these goals are relying on wind turbines and solar panels for intermittent electricity.[786] For a country this size, and growing, it is unrealistic to believe renewables can accomplish anything India requires. Primary energy and electricity consumption almost tripled:

> "Between 1990 and 2018, reaching an estimated 916 million tons of oil equivalent. Other renewable fuel sources make up a small portion of primary energy consumption."[787]

Only coal, oil, petroleum, natural gas, and nuclear can provide abundant, reliable, affordable, scalable, and flexible hydrocarbons creating a continuous uninterruptable electricity generation portfolio India desperately needs. Hopefully, India attempts to insert more clean and abundant natural gas for the country's electricity consumption since their country "was the third-largest consumer of crude oil and petroleum products after the United States and China in 2019."[788]

India's demand for electricity is consistently leapfrogging all nations over population growth and a sclerotic yet burgeoning economy. Crude oil use reached over 4.9 million barrels per day (mbp) in 2019 whereas production within India could only supply roughly 1 mbp of oil and petroleum derivatives.[789] Here is what India cannot do for decades ahead if they do not want to end up in crippling energy poverty.

INDIA CANNOT FOLLOW THE WEST

The U.S. is closing four zero-emission, hydro-electric dams in fiscal year 2021.

Four dams on the Klamath River that flows 257 miles through

Oregon and northern California in the United States, emptying into the Pacific Ocean, three in California, and one in southern Oregon have been identified for removal at an approximate cost of $450 million. Iron Gate, Copco #1, Copco #2, and J.C. Boyle Dams make up the Klamath River Hydroelectric Project. Removing all four will be the largest dam eradication project in the world opening hundreds of miles of salmon and steelhead spawning habitat.[790]

PacifiCorp's 169-megawatt (MW) Klamath Hydroelectric Project (FERC No. 2082) is in a predominantly rural area in southern Oregon. This project consistently generates approximately 716 gigawatt-hours of emission-free electricity – enough power to supply the electricity needs of approximately 70,000 households.[791] Only nuclear power can deliver this sort of large-scale, consistent electricity without emissions.[792]

Good for the U.S. – the wealthiest country in the world and history of mankind – who is also blessed with some of the largest natural gas supplies in the world within the Bakken, Eagle Ford and particularly the Marcellus Shale in Pennsylvania and lower state New York can tear down efficient and clean dams. India does not have that option with the country's terrain. Monsoons brutalize the country regularly. But India does not have the option to utilize dams that are a life-saving feature that gives clean, emission-free electricity and can be used for flood control.

This is exactly the type of infrastructure project India needs, and it gives electricity that does not continue polluting their cities and countryside. In no way, shape, fashion, or form does India need to discontinue using dams or following the wealthy United States into green virtue-signaling territory by dismantling hydro-electric dams. Nor should India ever consider following convoluted climate regulations from the likes of Columbia University; or any western university, think tank, non-governmental organization (NGO)/non-profit that advocates western environmentalism.

Appreciating the hard work Columbia put into climate policy papers is admirable, but nothing worth following for India's billion and a half people, which is more than the U.S., European Union (EU), and western-aligned allies in Asia combined.[793]

The counter to building and employing dams for India could come from the premise it is unusually sunny and hot most days of the year. So why not deploy industrial solar electricity farms on a wide-scale basis? Because new reports have shown solar to be less reliable than previously thought in sunnier parts of the world.[794]

PV Solar is useful in countries such as Spain and Morocco, which have limited and expensive access to natural gas then solar makes sense. Concentrating solar is costly to build, operate, unreliable, intermittent, and the electricity is overly expensive.[795] None of this makes sense for India.

Then use wind if solar does not work for India's electricity needs. Conventional wisdom says the wind is abundant, and offshore wind for India's thousands of coastlines should make sense. Except it does not when you consider it takes "45 tons of rebar and 481m3 of concrete to build one wind turbine, and its carbon footprint is 241.85 of CO2 emissions."[796] India is already a dirty, polluted, high-emission country. Why use a form of electricity generation that leads to environmental degradation?

Still, solar and wind are reaping trillions in investments; with taxpayers globally footing the bill and investment management firms/institutional investors jumping onto the so-called clean energy transition.[797] The west may jump onto the faddish-renewable-craze, but India plans to double oil and petroleum refining capacity in the next 5-years.[798] Arch-rival China plans to become the world's largest refiner on the back of India's announcement.[799]

Any emission or pollution reducing actions the west takes are overtaken by India and China. If you believe in and want action on climate change then start with these two countries and move to Africa. Nothing matters environmentally whatever virtuous environmental action the west takes as long as China, India, and Africa are spewing emissions faster than the west can eliminate all forms of fossil fuels.

Encouraging India, China, and Africa (best shot is somewhat Democratic India) to accept U.S. LNG over European objections who are more concerned about methane emissions than they are about Russian dominance via their trillions in natural gas reserves.[800] Sensitivity about fracked gas should never enter India's energy portfolio. China is a bigger concern, and the Europeans through U.S.-backed NATO can bluster environmental credentials while using environmentally unsound Russian natural gas.

Hydrogen is the answer for India. Use hydrogen. Clean, abundant, reliable what is not there to like about hydrogen for electricity; and it can be combined with solar. Germany is the leader for this new forward-thinking technology. Except there are problems with this newfound technology and energy philosophy. Read our German Chapter Two to understand how Germany's switch to wind and solar for electricity has been a disaster according to global consulting firm McKinsey & Co.

As part of Germany's COVID-19 stimulus package, German lawmakers have:

> "Earmarked 9 billion euros ($10 billion) for the expansion of hydrogen capacity at home and abroad in a bid to meet emissions targets. The stimulus package also sets fixed prices for Germany's renewable energy surcharge, the EEG, under which power consumers help fund the country's solar and wind power expansion and which accounts for about a fifth of their electricity bill."[801]

India does not have their national security needs taken care of by a NATO-based organization in Asia backed by U.S. nuclear weapons the way Germany does. India also does not have hundreds of billions of dollars to waste on solar panels, wind turbines, and hydrogen for over a billion and a half people desperate for petroleum, oil-based product derivatives and reliable electricity. Hydrogen is a horrible waste of India's money and treasure for an energy source that is decades away from being a reliable source of electricity or usable for transportation.[802]

Hydrogen cannot work at this time based upon the notion of the energy return on investment (EROI); the ratio between the energy delivered by a particular fuel – in this case hydrogen – whose source of energy to society and the energy invested in the capture and delivery of that energy.[803] EROI are why fossil fuels (oil, petroleum, natural gas, coal) and nuclear generated electricity are the only forms of electricity at this time that meet EROI requirements and the basics of energy needing to be abundant, reliable, scalable, affordable and flexible. Hydrogen is not close to meeting those requirements for the U.S. with 350 million people and certainly not for India's billion-plus citizens.[804]

The counter to hydrogen, wind turbines, and solar panels not working are they can be backed up by battery energy storage systems for electrical grids. The above paragraphs are not meant to be a book unto themselves or a chapter, but an overview of where India stands in relation to providing life-giving electricity to over a billion people. We in the west have this panacea idea where the sun, wind, hydrogen, and anything exotic sounding over fossil fuels will power the world when the exact opposite is the case.[805]

That is the case with batteries. For a battery to work it must store hours, days, weeks, months, or years of electricity available at a moment's notice. Our second book – *Just Green Electricity* – extensively details why this type of technology is not currently available. Grids only work if they have reliable and continuous uninterruptable electricity; and even this reliable form of electricity has three different redundancies built into the systematic grid.[806]

The wind and sun are volatile, and battery storage for electrical grids "still needs to demonstrate that it eventually can become cost-effective and reduce its significant ecological footprint," says Dr. Sebastian Luning.[807] The lithium-ion battery industry is the main component in batteries for electrical grids and electric vehicles.[808] This growing sector of the global economy "is expected to grow from 100 gigawatt hours of annual production in 2017."[809] Latest figures projected out to 2027 will see an increase of over 800 gigawatt hours – mainly from China's push for electric vehicles and India's push for renewables based upon their all-of-the-above approach for electricity.[810]

To reach this enormous lithium figure as an example a Tesla Model S needs roughly 12 kilograms of lithium to run properly, and a battery energy storage for electrical grids requires more strip-mining for lithium than mining and transportation companies are currently performing.[811] Land, water, people displacement and overall misery is what will take place at the lithium triangle in South America and regions in China and the U.S. where lithium is mined.

Promising outcomes and delivering results for over a billion people is the crux of why batteries could be decades away. The six types that are the most promising and have achievable breakthroughs towards a renewable society are: rechargeable batteries, pumped hydro energy storage, power-to-gas, and power-to-liquid, compressed, thermal, and flywheels. None currently are anywhere close to having the technology to store enough electricity for one American city much less India, China, or Africa. Only fossil fuels and nuclear meet that requirement.

What causes hydrogen, batteries, solar panels, and wind turbines to take on a Christ-like savior quality for emissions? McKinsey and Company is the world's largest consulting firm with revenues over ten billion dollars a year. But they are promoting human suffering and scorched-earth environmentalism through their McKinsey Global Institute.[812]

They recently published a report titled, *McKinsey on Climate Change.*[813] An excerpt should explain why countries like India may say they are worried about climate change when they could care less. They blatantly avoid the need for transparency to the world of environmental degradation and humanity atrocities of the supply chain for the material to support the "green" movement. This is palaver-based research for unserious times. From the report:

> "The changing climate is poised to create a wide array of economic, business, and social risks over the next three decades. Leaders should start integrating climate risk into their decision making now."[814]

When you scare people who run businesses with questionable climate change tactics, they look to firms like McKinsey who stands to make billions scaring them to death over changing weather patterns. The overpriced consultants at McKinsey have zero answers for how to give over 4 billion people dispersed throughout India, China, and Africa the products from petroleum derivatives, and electricity.

They only know how to scare gullible westerners more afraid of bad Twitter reviews than thoughtful, serious, far-reaching energy polices for billions in India and elsewhere desperately needing reliable electricity without environmental degradation that renewables, battery energy storage systems for electrical grids, and EV batteries, offer over their rare earth mineral requirements.

India should never follow anything that has to do with renewables, battery energy storage systems for electrical grids, hydrogen or climate change driven energy policies for decades ahead until the technology catches up with the demand India requires. India should go ahead and drill for fossil fuels, use nuclear energy whenever possible, and never follow the U.S. Democratic

Party's policies of eliminating exploration and production on federal lands.[815]

India will only bring suffering, misery, environmental catastrophe, lower human rights, and enrich billionaires and multi-national corporations over the average Indian citizen and worker if New Delhi follows western environmental models over sensible energy and electricity polices that uses coal, natural gas, oil, petroleum, and nuclear over renewables.

THE CHINA SYNDROME

In our first book – *Energy Made Easy* – Chapter 3 details how World War I & II were won with energy from oil, petroleum, and coal. The risk involving China with the Chinese Communist Party (CCP) in power will be their continued savagery for energy and electricity to dominate Asia and eventually the world.[816] This is an unprecedented threat that is larger than the combined size and resources of Nazi Germany, Imperial Japan, and the Roman Empire. None of these regimes and ideological forces had the size, scope of resources, economic heft, and ability to menace the world the way China does for this century.

Further explanation is warranted to understand the enormity of the problem with Chinese government policies, how these inflame India, and what that means for energy when almost 3 billion people (China and India population figures combined) could go to war at any moment.[817]

The People's Republic of China has a population of roughly 1.4 billion, is governed by the Chinese Communist Party (CCP) with 90 million members strong and has another 300 million elites who are strongly invested in the regime's legitimacy and success.[818] Access to abundant and reliable energy and electricity is the lynchpin of this growing dictatorship. Even during the coronavirus epidemic China still imports record amounts of oil.[819]

State-owned enterprises are privately-run businesses tied into the idea of "socialism with Chinese characteristics."[820] This provides trillions in monies for the Chinese Communist Party to spread Chinese characteristics globally and domestically. China's CCP and business model working together only works if oil is abundantly stored and used.

This is in direct conflict with the United Nations and western model of freedom of speech, thought, and religion for the individual. Nothing works in China, India, or elsewhere without energy and electricity. China understands this concept better than any nation on earth.

Inefficiency should seep into the Chinese economy since they incorporate government-controlled capitalism, Marxism, Maoist, and Leninist schools of thought and macroeconomics – but in truth it works. To assist state-run apparatus China has an incredibly large global intelligence network through its Ministry of State Security – and all of it does not work without abundant energy in all forms.

A Neo-mercantilism economy and government does everything in its power to steal in the areas of industrial espionage, cyber warfare, and economic coercion. Since 2010 China has stolen over $6 trillion worth of just U.S. intellectual property from Silicon Valley, Hollywood, and medical labs based out of New England in the northeaster portion of the U.S.[821] Backing the Chinese Communist Party (CCP) and Ministry of State Security in its nefarious dealings is the People's Republic of China (PRC) military with over 2 million men, and the world's largest navy.

The lengths to which the Chinese Communist Party (CCP) will go should be understood by reading the famous book published in 1999 by two Chinese Colonels titled, *Unrestricted Warfare*. Warfare, however, only works if it is backed by coal, natural gas, oil, petroleum, and nuclear; otherwise, it is a paper tiger.

Understanding a rival's use of energy and electricity it is

important to analyze how these life-giving components are used for good or evil. China uses their power and energy policies for unrivaled wickedness. The treatment of its own people has the Chinese Communist Party (CCP) killing through war, starvation, or execution at approximately 100 million. Mass murder on this scale took place by the Red Guards during the Cultural Revolution (1966-1976) and Mao's Great Leap Forward (1958-1962). The Cultural Revolution resulted in conservative estimates of over 70 million Chinese citizens butchered by their ideological government. Add to this malevolence – China had a One-Child Policy in place until recently that ordered forced abortions: estimates are 500 million Chinese babies were murdered by the Communists running China resulting in 600 million mainly defenseless Chinese killed by the Communists.[822]

Global media outlets were more fixated on the previous Orange Man in the U.S. White House than reporting on concentration camps for 1-2 million Chinese Muslim Uyghurs. Where Chinese uses their energy powers against the Muslim Uyghur's results in:

> "Abduction of Uyghur women for sexual use by Chinese soldiers, how the government harvests the organs of the Uyghur population for sale both in China and abroad. This latest (Chinese) atrocity has become a multi-billion-dollar industry.[823]

Japan never had the scope or scale of killing the way China does to its own citizens. To put this level of killing into context please read Dr. Victor Davis Hanson's *The Second World Wars* to understand the Japanese were viciously brutal during their occupation of Japan during the 1930s and World War II; but they did not even kill 10 percent of the Chinese the CCP has done to their fellow neighbors and friends.

This duplicitous behavior does not end with slaughtering

Chinese citizens but continues with Chinese government lies about committing "to peak its carbon dioxide emissions before 2030 and to be carbon neutral before 2060."[824] Pledges are never the same as enforceable actions, because China is building new coal-fired power plants for electricity that have a 50-60-year usage life. Currently, China is building – not planning to build – but building 250 gigawatts (GW) of coal-fired power plants for generating capacity.[825]

A coal boom is how this should be spoken from the mountaintops when the numbers are more sobering than poor attempts at sardonic humor. This 250 GW of new Chinese coal-fired electrical generation capacity are more than the U.S. has in its entire coal generating fleet.[826] Once these 250 GW are built, "China will have more coal-fired capacity than the entire generating fleet in the United States from all power sources.[827]

Building sprees of coal-fired power does not stop with this 250 GW. In 2018 China added 32 GW and in 2019 the Chinese Communist Party deployed another 44 GW with another 100 GW under construction and 152 GW in the planning stage.[828] This is done to stimulate their domestic economy without any environmental considerations, nor humanity atrocity considerations. Pre and during the COVID-crisis from January 1-June 15, 2020 China permitted 17.0 GW of new coal-fired capacity for construction.[829] Keeping with their green ambitions to control the world's minerals and metals and the west's suicidal energy policies – China does this to power over 7.9 GW of long-distance transmission lines from renewables on their west coast to city-centers in the east.

All this use of coal for electricity makes it hard to believe China will be carbon neutral in 40 years like they told the September 22nd, 2020 United Nations General Assembly. Say this out loud: the country that killed over 600 million of its own citizens will tear down new coal-fired plants that employs hundreds of workers, supports countless mining operations, and allows China to power their quest for global hegemony. Not likely.

On top of this scenario China "has substantial domestic coal reserves, 142 billion metric tons as of the end of 2019, 13 percent of the world total, and the fourth largest in the world."[830] Coal is a reliably secure domestic energy source for China's overwhelming electricity needs. Coal consumption by Chinese industry and domestic use "increased by 2.3 percent and accounted for 57.6 percent of its energy use and 51.7 percent of the world's total coal use."[831]

Emissions are the global catchphrase for cleaner environmental health, but China is doing the exact opposite to control respiratory illnesses and Chinese health when roughly two-thirds (approximately 43.8 GW of the 68.3 GW constructed) and newly commissioned coal-fired power plant generating capacity was constructed in China.[832] From 2000 to currently (this section written in November and December 2020) and outward to 50-60 years from now China's coal fleet will grow exponentially.

Damn the Paris Climate Accords, treaties, agreements, protocols, and basically any type of environmental framework that will be tossed aside to ensure China has more energy than anyone on earth. A total of 1,040 gigawatts – essentially half of the world's total – has been accorded use inside China's borders.[833] 80 countries in the world use coal for electricity, over 36 percent of global electricity came from coal in 2019, and this is the highest share in decades while having the greatest share of electricity than any other fossil fuel, nuclear, or renewable source. Coal is king, and China is its loyal servant since estimates have country electrical generation at 65 percent capacity from coal in 2019.[834]

The average age of these coal-fired power plants is 14 years, that is very young for a coal fleet globally.[835] China will likely never be carbon-neutral in this century. As long as China is governed by the communist party, they will continue saying they are for clean energy, but using the dirtiest forms of coal, exploiting their advantage in rare earth minerals and metals that are needed for renewables to work, and continuing to kill their own people in record numbers when it suits their purposes, whether for clean energy mining, or national and international security benefits.

THE U.S. POSITION IN ASIA AND WHAT THAT DOES FOR CLEAN ENERGY EXPLOITATION

During the U.S. Presidency of Barak Obama, he "pledged to concentrate on efforts to advance security, prosperity and human dignity across the Asia Pacific."[836] A noble pledge, and the U.S. is the only country to buttress against, and possibly halt, the horrific forward-march of the communist hordes coming from Beijing. Much less attempt to control the sulfur oxide spewing calamity emanating from Chinese coal-fired power plants.

Only the U.S. can stand up to China in the areas of energy, foreign policy, economics, national security, and global health. The EU, NATO, Australia, Japan, South Korea, the Philippines, Vietnam – even India – all need U.S. military might, its diplomatic soft power, and continued usage of the U.S. dollar as the principal global currency for financial transactions to slow Chinese gains.

The Obama administration seeing these variables aligned against U.S. interests invoked a new policy known as "the pivot."[837] A realignment from the Middle East and Europe to Asia. Obama believed his Nobel Peace Prize, charming personal narrative, and overwhelmingly positive adoration from U.S. and western media outlets could shift America without incurring inertia. Energy from the fracking boom, which allowed America to leave the Great Recession of 2007-09 gave the former President room to pivot and reshape the Asian narrative.[838]

China took note, continued to use coal and all forms of energy while blindsiding Obama and the unwieldy American foreign policy establishment. Obama had accomplishments such as diplomatic breakthroughs with Myanmar.[839] He also had better relations with Southeast Asian countries, which showed the pivot was working. Vietnam is now a U.S. ally, a major diplomatic achievement. Notably to counter China, the U.S. supported Japan over China during a territorial dispute that has its origins from World War II.[840]

Through all of this, China continued building and exploiting their rare earth mineral and metal advantage, dumping solar panels on the world market (particularly to the U.S.), and positioning itself to rid the world of all U.S. post-World War II, liberal-led order advantages. Instead of transitioning to a clean energy society using natural gas over coal the Chinese decimated their environment with every form of energy – never taking the time to understand or mitigate the downside and costs to their country.

President Trump, like Obama, also tried to pivot to Asia using energy as a soft power tool by extending the olive branch of friendship and business to India. Neither President could extract the U.S. from Afghanistan or Iraq (ISIS is defeated but radical Islam still thrives) to fully concentrate on China. Obama, unlike Trump, was caught flat-footed over China's use of weaponizing energy to expand in the South China Sea and use infrastructure as a debt-trap via the Belt and Road initiative.[841]

Both Presidents Obama and Trump were unable to negotiate the Trans-Pacific Partnership free trade deal. Obama's U.S. Democrats were against the deal since it left Labor and Unions on the sidelines, and Trump withdrew from the negotiations leaving the enormously influential Asian free-trade agreement called the Regional Comprehensive Economic Partnership (RCEP) between China and 14 other nations signed in late November 2020.[842] China is now on equal footing with the U.S. in Asia. What once was U.S. territory to roam at-will is now a struggle between Washington and Beijing with abundant energy at the forefront. Trying to contain or even mitigate exploitation for clean energy is an afterthought under this new scenario.

According to the outspoken former permanent secretary of Singapore's ministry of foreign affairs, Bilahari Kausikan said: "The pivot to Asia was a good idea, but it was never properly implemented."[843] Asian-led China presents challenges the soft power of energy can ascertain for solutions over military options. The obstacles are numerous. Besides an abysmal record on human rights, China is challenging India's territorial claims over a shared border.[844] Threats against Taiwan are a daily occurrence, no matter who is the U.S. President.[845] Menacing behavior in the South China Sea is now Chinese norm.[846]

This sweeping vision of power from President Xi will be powered by energy, electricity its daily engine, and all forms of renewables with humanity and environmental degradation cast aside for communist domination.[847] The west better wake-up, because this dredge on freedom and the human spirit will make Nazi ideology seem like a starter course for global war that will dwarf World War II in death and destruction.

Joe Biden, Tom Steyer, Michael Bloomberg, Bill Gates, Mark Zuckerberg, and the men who run Google, and George Soros, all bow to China, and look the other way at the exploitations and relentless Chinese destruction of life, liberty, property, and the pursuit of happiness. Since each one of these white billionaires greatly profits off renewables, mining rare earth minerals and metals, and electric vehicles watch them continue to make billions off Chinese markets.

The green energy transition lines their pockets from the have-nots to the haves; this is today's modern U.S. Democratic Party.[848] A party of billionaires looking to line their pockets using climate change for their benefit. Greed is the motivation. Environmental health and environmental degradation to the world's landscape are in the rearview mirror.[849]

CONCLUSION

Leaving a post-COVID decimated economic wasteland should be the highest priority for wealthier nations that represent less than 10 percent of the world's population, the U.S., EU, NATO militaries, western-led Asian allies, Canada, and Australia, because it will be for China and India. Requiring industry and business to use fossil fuels and nuclear since they meet the basic requirements of energy will rescue post-COVID economies, when intermittent electricity from renewables do not have the ability to be abundant, affordable, reliable, scalable, and flexible even with fossil fuels backing them up on electrical grids should be a priority.[850] Only reliable and affordable electricity and access to petroleum and the over 6,000 products made from oil derivatives will allow India to no longer be an environmental wasteland while allowing freedom-loving nations to counter the butchers-of-Beijing.

Reason, logic, good sense, and abandoning renewables is the place to start. China is power hungry for energy, so is India, whereas the west is relying on the fluctuations in weather under the scenario of using the intermittent electricity from wind and sun to drive their economies.[851] Holding energy reason to delusional ideology is something China and India will never beholden themselves to for their power or energy policies. For proof of this assertion, ask yourself, why are China and India "exempt" from the Paris Accord emissions reduction targets until 2030, while the west must pay dearly for our failure to meet targeted reductions?

So far, only the U.S. has met the emission reductions by using natural gas-fired power plants over coal-fired for domestic electrical generation needs.[852]

A rabbit hole of massive subsidies for billionaires for intermittent electricity from wind and solar is the western plan moving ahead; for China and India it is coal-fired power plants, which are the most reliable generation fleet. Suicidal attempts to wreck their countries following economic socialism and social justice is laughable to Xi and Modi (leaders of China and India).

Clean energy is exploitative and will only bring environmental degradation and human misery – if under current technological constraints it is widely deployed and used for transportation and electricity.[853] Pointedly, China and India will be world leaders in electric vehicle use, sales, and technology. Particularly, the Chinese, but the facts show a devastating environmental impact if just EVs replace internal combustion engine (ICE) vehicles.

Using Britain as an example since current Prime Minister Boris Johnson (this section written in early December 2020) announced on November 17, 2020 a complete ban on all new gasoline-and-diesel-powered vehicles taking effect in 2030.[854] Professor Richard Herrington and his colleagues at the Natural History Museum in London researched what it would take for the UK to achieve these goals.

According to Cambridge University Emeritus Professor of Technology Michael Kelly, replacing all the United Kingdom's 32 million light duty vehicles with next-generation EVs would require huge quantities of materials to manufacture the EV batteries:[855]

- more than half the world's annual production of copper.
- twice its annual cobalt.
- three quarters of its yearly lithium carbonate output; and
- nearly its entire annual production of neodymium.

One can easily see that the world may not have enough minerals and metals for the EV batteries needed to support EV growth projections when you consider that today:

- Combined worldwide car sales in 2019 were more than 65 million vehicles annually.[856]

- There are 1.2 billion vehicles on the world's roads with projections of 2 billion by 2035.[857]

The U.S has approximately 276 million registered vehicles – and millions of unregistered vehicles. This number is about 9 times as many vehicles as the UK. Assume Professor Herrington's numbers are correct, then electrifying the entire U.S. fleet needs:

> "18 times the world's current cobalt production, about nine times global neodymium output, nearly seven times global lithium production, and about four times world copper production."[858]

Basic physics also dictate countries such as China and India will increase their use of fossil fuels to power their transportation sector since energy density is one of the main factors for all energy usage. While batteries for EVs are improving they still have mountains of work ahead of them before they reach the same energy density that an ICE vehicle achieves. Gravimetric energy density, or the amount of energy contained per kilogram means EVs cannot rival the ICE vehicle when:

> "Gasoline and diesel contain about 80 times more energy per unit of weight than the best lithium-ion batteries." (around 90 percent of all EV and batteries for electrical grids are lithium-ion based)[859]

EV lithium-ion batteries could increase two-fold exponentially, and still a petroleum-based gasoline and diesel vehicle is over 40 times better and more efficient than an EV battery. Our 2nd book *Just Green Electricity* details the bigger problem for EVs to overcome is the current configuration of gasoline/petrol filling stations and electrical grids needing to be completely overhauled, replaced, or entirely new grids built to handle the multiple gigawatts of new electricity needed to charge hundreds of millions of EVs at the same time or during staggered periods of time.

Then, if EVs are adopted wholeheartedly, what is to say China and India will follow this course and do away with oil's value for lubrication, and the over 6,000 products ranging from cosmetics, plastics, pharmaceuticals, bowling balls, toothbrushes, and on and on and on...oil is not going anywhere. For the west or China and India, it is unwise, untruthful, or an agenda is in place to think otherwise.

China and India claim they are following the Paris Climate Accords, but this chapter has definitively shown that it is naïve posturing on the part of western environmentalists, the UN, and global media outlets wanting a narrative to take shape that simply is not true. In actuality, and worth repeating – the U.S. is the only modern country with a growing economy to meet the Paris Climate Accords (PCA), because:

> "In 2019 (using natural gas from shale deposits) helped the U.S. achieve the largest absolute decline of energy-related carbon dioxide emissions of any country in the world, surpassing all signatories of the Paris Climate Accord."[860]

Only the skyrocketing growth in natural gas for electrical generation can mitigate the current fleet of Chinese and Indian coal-fired power plants, which are young, growing, and more are coming down the energy conveyor-belt for decades ahead.[861] The only way to decline China and India's growing emissions is convincing them (no small task) to use natural gas for their electrical generation. Let it be a bridge fuel to begin the herculean mission of weaning both countries away from overwhelmingly using coal-fired power plants.

However, both countries will need all the natural gas they

can import from the U.S., Middle East, and Russia to meet this demand. To think otherwise is like believing both countries (China and India) will be carbon-free, decarbonize, or meet any of the requirements for a clean energy transition. Believing either country will meet peak carbon dioxide emissions in the next 50 years is hard to believe while using the dirtiest forms of coal-fired power plants for their gargantuan energy to electricity needs.

Making promises at international gatherings, informal discussions with other western, global leaders, or interviewing with CNN to gather western investment capital is entirely different than actually delivering oil-based products, petroleum, jet fuel, diesel for the transportation sector, and electricity for over 2.5 billion people (the approximate combined population of China and India).

Stupidly, the U.S., EU, even NATO will take promises with no contractual requirements attached or enforcement mechanisms in place – nothing more than verbal promises – to justify similar actions and commitments that destroy their economies and hinder national security. China and India have fashioned a narrative that runs counter to their actions, because western billionaires line their pockets with a gullible voting public who believes the west is warming the earth uncontrollably and only industrial solar panel and wind turbine farms backed-up by battery energy storage systems for electrical gids can keep humanity from cooking itself to death, when the exact opposite is true.

The average westerner and global citizen yearning for energy should realize over:

> "80 percent (comes from) fossil fuels. In the UN's most likely scenario for the 21 century, this will change little over the next 80 years."[862]

This story about China and India and the western environmental model is never told, because it does not fit an activists' media-driven narrative about clean energy. If the truth were told, then more than likely Donald Trump is still the U.S. President over Joe Biden.[863]

China and India – want to underplay the U.S. reduction in CO2 emissions – allowing them to operate on the global stage the best way to achieve significant reductions is "attend fancy confabs (UN meetings), draw-up (meaningless) treaties (the Paris Climate Accords) and hold self-congratulatory press conferences to focus on process instead of outcomes."[864]

Global emission will continue rising because of China, India, and eventually Africa, the remainder of this century. Only natural gas used for electricity can stem this tide along with zero-emission nuclear electricity. The U.S has proven that growing economies and environmental health do not have to be autonomous from each other. Instead of unstable, dangerous, intermittent, environmentally degrading while adding to death, misery and human suffering, China and India should be encouraged from the U.S., UN, EU, NATO, and Asian trading partners "to invest in groundbreaking energy technologies though innovative policies, including robust support for (fossil fuel and nuclear-based) R&D projects."[865]

Without this type of technical innovation, the world is doomed to overwhelming emissions and continued electricity duplicity and filthy cities from China and India. Each population will find they lack business competitiveness, lack upward mobility, and little hope of owning affordable homes, starting entry-level careers, or making decent incomes. Being pro-business, pro-growth, pro-environment, and pro-ethics does not mean China and India drift into economic malaise, it means they join the world community responsibly for the first time in the modern era.

CHAPTER SEVEN

FINANCIAL AND ENVIRONMENTAL RACIAL BIASING

- ***Energy and environmental policies impose economic stresses for energy on billions of low-income and minority communities that can least afford it.***
- ***The financial racial biasing of climate and environmental policies against the poorest residents of the world is unconscionable.***

FINANCIAL BIASING

Poverty is the worst environmental problem. A well-fed person has many problems, a hungry person has but one.

In many developing countries, smoke from indoor cooking and heating has the biggest impact on public health. The World Health Organization (WHO) estimates 2 million people die each year from breathing indoor smoke from burning wood, dung, and other farm waste. Today, this is the world's most serious health issue that stems from pollution.[866]

China and India, with 40 percent of the world's population, are only now entering the automobile and air conditioning era in a significant way, where commercial energy consumption increases rapidly. They are prolific in their use of coal for continuous uninterruptable electricity.

There are 62,500 power plants around the world operating

today, representing all types, including more than 10,000 coal fired power plants. [867]

With limited income and poverty status, China, and India with 2.7 billion people, continue to pursue coal as their primary energy source as it is abundant, reliable, and affordable. There are 62,500 power plants around the world operating today, all types, generating electricity for the world's inhabitants.[868] Of that total, more than 2,449 are coal-fired power plants. Over half (1,363) of the world's coal power plants (2,449) are in China and India whose populations of mostly poor peoples is roughly 2.7 billion. Together they are in the process of building 284 new ones of the 546 total.[869] [870] They are putting their money and backs into their most abundant, affordable, and available source of energy – coal.[871]

China and India are not the only ones pursuing the most abundant, affordable, and available source of energy – coal, but there are 1,600 new coal plants planned or under construction in 62 countries. The new plants would expand the world's coal-fired power capacity by 43 percent. [872]

The term capacity factor is used to describe the amount of electricity actually produced compared to the potential if the generator were operating at 100 percent of its capacity, 24 hours a day, 365 days a year. Large base-load power plants, such as coal, nuclear, and hydroelectric, typically have capacity factors of 70 to 90 percent or higher as they run continuously, except for repairs and refueling in the case of nuclear power plants.

The average capacity factor of solar is 10 to 25 percent. That is the main reason why, at the standard cost of electricity it will take more years than their projected longevity to pay for these systems. Not even solar panels last forever. It does not take a genius to realize solar power is a waste of good money on the grid. [873]

One can only conclude that wind electricity and solar electricity are investment bubbles that will eventually burst. Only very rich countries that think they have money to burn can afford these technologies. To expect that countries in Africa will adopt them without huge subsidies from rich countries is far-fetched. It appears equally far-fetched that rich countries will provide such subsidies. In many ways these very expensive technologies for intermittent electricity are destroying wealth as they drain public and private investment away from more affordable and reliable electrical generating systems. It seems this lesson will be learned the hard way.

Climate change in the U.S. will hurt poor people the most, according to a bombshell federal report.[874] Heart and lung disease, heat stroke and bacterial infections are just a few of the health consequences associated with climate change. Low-income populations "typically have less access to information, resources, institutions, and other factors to prepare for and avoid the health risks of climate change," the report says, leaving them especially vulnerable. Lack of health insurance among the poor will also intensify the risks of illnesses caused by climate change.

Today, almost half the world -- over three billion people — live on less than $2.50 a day. At least 80% of humanity, representing more than 6 billion on this earth, lives on less than $10 a day. More than one billion people earn less that one dollar per day, and they face daily risks and hardships that determine their very survival. More than 80 percent of the world's population lives in countries where income differentials are widening, especially in America. [875]

Just a few of the richer developed countries are on the green bandwagon, while more than 6 billion live in abject poverty. For those living in the richer countries, their residents are forced to incur some of the highest costs for electricity that the poorer still developing countries cannot afford. [876]

Richer countries	Population	World Share
United States	331,002,651	4.25 percent
Germany	83,783,942	1.07 percent
United Kingdom (UK)	67,886,011	0.87 percent
Australia	25,499,884	0.33 percent
Totals	508,172,488	7.32 percent

Despite all the hype and the media coverage about electrifying the world with intermittent electricity, controlled by less than 8 percent of the world population located in the richer countries, the U.S. Energy Information Administration (EIA) has not wavered from their continuous forecasts that the other 90 percent of the world's population will continue its demands from industries that were virtually nonexistent before fossil fuels entered the horizon a few centuries ago, such as: agriculture, medical, plastics, communication, electronics, military equipment, airlines, merchant ships, and construction equipment.

Poverty is a global problem that affects citizens around the world. The development community, including government agencies, banks, and non-governmental organizations, seeks to improve the livelihoods of impoverished citizens through poverty reduction strategies that address the root causes of poverty and its crippling effect on people trapped in adverse situations. But after years of implementing programs to solve these issues, poverty remains a multi-dimensional problem with many faces as poverty is a global problem that affects citizens around the world. [877]

Efforts are being made to recognize the importance of natural resources for poor countries and poor households. Most poor countries and most poor communities are dependent on natural resources. Natural capital is a much greater share of wealth for the poor than for the rich. Poor countries generally have small service and industrial sectors and poor people have limited access to financial and produced capital. Thus, natural resources must be recognized as a major—if not the major—asset of the poor. This understanding should be integrated into country strategies and programs, including through Poverty Reduction Strategy Papers (PRSPs), approaches to the Millennium Development Goals (MDGs), US AID Mission strategies, decentralization policies, and other economic and political instruments; governments, Non-governmental organizations, and other institutions must give careful consideration when developing such programs. Given the importance of natural capital in the economic portfolio of the poor, both economic growth and poverty reduction programs must focus on improving the poor's access to and use of natural resources.

Incredibly, global poverty affects the poorest 40 percent of the world's population, or more than three billion, and they account for just 5 percent of global income. The richest 20 percent accounts for three-quarters of world income. Water problems affect half of humanity.[878]

Climate change remains one of the most serious threats to the integrity of life on earth. But we still need the world to have compassion for the trade-off to eliminate fossil fuels too quickly as it will allow the continuation of 11 million children in the world dying every year from preventable causes of diarrhea, malaria, neonatal infection, pneumonia, preterm delivery, or lack of oxygen at birth as many developing countries have no or minimal access to the thousands of products from oil derivatives enjoyed by the wealthy and healthy countries.[879]

While it is possible to survive without electricity, it is not possible to achieve a high quality of life. Whether it is literacy, education, health care, public transit, industry, or entertainment, electricity forms the foundation of a civilized life.

The U.N. trade body, United Nations Conference on Trade and Development (UNCTAD,) has already issued a report_breaking down some of the unintended negative consequences of the green shift to foreign countries, which include ecological degradation as well as human rights abuses.[880]

Additionally, Amnesty International_has documented children and adults mining cobalt in narrow man-made tunnels, and the exposure to the dangerous gases emitted during the procurement of these rare minerals, not to mention the destruction of the local ecosystems when the waste water and other unusable ores are let loose onto the environments they have no choice but to live in because their wages are so infinitesimally small, it causes me the take a step back and examine my moral obligations to humanity. [881]

In the best interests of humanity, we should require transparency of the worldwide environmental laws and labor laws that are being taken advantage of for their nonexistence, to support the exotic mining in foreign countries for the minerals and metals needed for wind turbines, solar panels, and EV batteries.

Wealth and urbanization will stabilize the human population. Agriculture should be mechanized throughout the developing world. Disease and malnutrition can be largely eliminated by the application of modern technology. Health care, sanitation, literacy, and affordable electrification that is continuous and uninterruptible should be provided to everyone.

America has only about four percent of the world's population (330 million vs. 8 billion). Elected officials and environmental activists should know that oil and gas is not just an American business with its few refineries, but an international industry with more than 700 refineries worldwide that service the demands of the 8 billion living on earth. [882]

Even the slightest increase in energy use by the poorest two-thirds of humanity will overwhelm any conservation savings we can accomplish in the developed world. This is not to suggest conservation is not worthwhile. Wherever we can economically increase energy efficiency, in our vehicles, homes, and appliances, we should do so. But at some point, you cannot diet your way out of starvation. Conservation cannot conserve what is not produced.

ENVIRONMENTAL BIASING

During the 1980s the concept of "Environmental Justice" (EJ) gained prominence in the United States with the publication of studies by the General Accounting Office (1983) and the United Church of Christ (1987), suggesting that predominantly minority and low-income communities were being exposed to disproportionately higher levels of environmental hazards. Since that time, the literature has evolved from examining simple correlations between community characteristics and exposure to environmental harm to relatively sophisticated analyses that seek to control other factors that may affect exposure to environmental hazards. Today, poor and minority populations face more hazards. [883]

We are all inundated with ads for EV's and the need for more and more wind and solar to generate intermittent electricity as many of our elected representatives and environmentalists believe that electricity can replace fossil fuels, but they fail to comprehend the limitation of just electricity. But before we discuss Financial Racial Biasing in the U.S., let us look first at the world challenges, since America only represents 4% of population and has 135 refineries that did not exist before 1900 that are providing products and fuels.

When most people decide they agree with all your reasonable ideas the only way you can remain confrontational and antiestablishment is to adopt ever more extreme positions, eventually abandoning science and logic altogether in favor of zero tolerance policies.

To a considerable extent the environmental movement was hijacked by political and social activists who learned to use green language to cloak agendas that has more to do with anticapitalism and anti-globalization that with science or ecology.

In urban areas, which produce 80 percent of greenhouse gas emissions in North America, the poor "live in neighborhoods with the greatest exposure to climate and extreme weather events," the report says. This includes living near pollution sites and in housing developments without sufficient insulation or air conditioning. Additionally, disruptions to infrastructure during natural disasters can have an outsized impact on city residents who rely on public transportation.

Rural areas often have agriculture-dependent economies, so the livelihoods of low-income residents are more vulnerable to changing environmental conditions.

Many rural households also suffer from energy poverty, the report states, meaning they "are not able to adequately heat or provide other required energy services in their homes at affordable cost." As average temperatures continue to rise, people who cannot affordably cool their houses will continue to feel financial strains.

Racism, bias, and economic inequality impair the financial well-being of millions. The fact that 44 million Black Americans, taxpayers, and earners are being met with resistance and denial as they ask white people to consider economic and health care obstacles that Black people confront daily is a clear example. To simply dismiss the concerns of an entire race or ethnicity when you have not walked one day in their shoes is privilege. To ignore the plight of millions because you "don't see it" in your life is white privilege. The point that you deny as truth anything you do not see, face, or negatively experience daily is blatant privilege. Not having to worry about bias and racism at every job, in every labor market, school, shop, street and neighborhood you encounter is privilege. [884]

Over the past decade, the governments of El Salvador and Costa Rica have taken bold steps to implement de facto moratoria on financially lucrative but environmentally destructive mining. Both did so to protect the environment, despite the prospect of historic profits and in the face of retaliatory measures from transnational gold mining firms. NIMBY's (Not-In-My-Back Yard) are alive and well and speaking out in El Salvador and Costa Rica. [885]

The next section of this co-authored chapter is by Todd Royal.

THE POLITICS OF CLEAN ENERGY

Nothing strips the soul of dignity and basic humanity the way poverty has for the course of human history. Poverty is the worst environmental problem since the poorer a state, nation, or continent, leads to higher levels of pollution, emissions, and with health and disease eradication rates at sub-par levels.[886]

Most of the developing and certainly failed states such as the Democratic Republic of Congo, and Haiti, deal with respiratory illness from increased emissions since smoke from indoor cooking and heating mainly comes from wood. This leads to the biggest impact on the environment and overall societal health. Poor, failing countries produce environmental carnage.[887]

The World Health Organization (WHO) "estimates 2 million people die each year from breathing indoor smoke from burning wood, dung (animal and human feces), and other farm waste."[888] The argument can be made this is the world's most serious health issue stemming from elevated pollution levels. Even more so than emissions from power plants.

According to a December 9, 2020 report by the International Energy Agency (IEA) and the Nuclear Energy Agency (NEA) titled, *Projected Costs of Generating Electricity* gave these findings:

> "By 2025 the economics of low-carbon generation technologies are poised to disrupt conventional fossil fuel generation so dramatically, onshore wind could have the lowest levelized cost

> of electricity (LCOE) on average, (and good news for zero-emission electricity) nuclear power could emerge as the dispatchable low-carbon technology with the lowest expected costs."[889]

Great news for millions burning cow dung, and dying of respiratory illnesses in countries all over the world, right? Unfortunately, no. Substituting one form of death for another more potentially damaging catastrophe is the worst trade-off imaginable.

Even the technologically advanced, first-world British government cannot make their commitment to wind power work.[890] During a November 2020 spectacle titled, "Wind Week!" the wind produced hardly any electricity, and they had to rely overwhelmingly on their coal-fired power plants to keep their electrical grid from blackouts.[891]

Bloomberg News, which is all-in on global warming and renewables being humanity's saving grace has confirmed British reliance on wind turbines for electricity is on the verge of destroying the British electrical grid.[892] Good for Michael Bloomberg's company confirming these things do not work. The Germans are no better for their reliance on intermittent, dramatically expensive industrial solar and wind farms for electricity.

Our German Chapter Two extensively covers this issue, but it is worth reminding that Germany has the highest electricity prices in the world, and in Europe causing over 300,000 German homes to no longer be able to afford basic electricity.[893] These homes are now literally going into German forests and cutting down wood for fuel. What will become of poor, developing countries if for financial reasons solar and wind are thrust on them?

Mining for the rare earth minerals and metals needed for the green energy transition (solar panels, wind turbines, electric vehicles, battery energy storage systems) such as lithium, cobalt, and neodymium will see hills, mountaintops, and entire forests ruined, chopped in half and cleared with chemicals and polluted water sources: "And for what? So eco-minded urbanites in Paris, Berlin, and New York City can feel good about driving around in zero-emission cars."[894]

Worse yet are the amount of rare earth minerals and metals needed for solar panels, wind turbines, and battery energy storage systems for electrical grids to work. This destruction is far worse than electric vehicles being adopted over a petroleum-based internal combustion engine.

Why – because on a basic economic level, industrial solar and wind farms (onshore or offshore) are always more expensive to produce electricity than coal-fired, natural gas-fired, or nuclear electricity power plants.[895] From direct and indirect subsidy costs to government-mandated usage no one would rely on the wind and the sun for intermittent electricity. The costs to lives and pocketbooks is staggering when the only one's profiting are politicians and the billionaires supporting these energy schemes.[896]

All this financial nonsense is predicated on the man-made global warming issue that ranges from escalating climate temperatures to dangerously high sea-levels. Neither of these perspectives is remotely true, and certainly mankind is nowhere near disappearing or being engulfed by oceans spilling over and drowning billions.[897] These misnomers are how U.S. Democrats, European Greens, western environmental organizations, and their billionaire supporters will make trillions off scare-tactics and flat-out lies.

Data from the U.S. National Oceanic Atmospheric Administration (NOAA) confirms the U.S. is neither overly warm nor cool, but somewhere in the middle. Here are the average mean temperatures in the continental U.S. from 1998-2019:

To refute the first view, we turn to data generated by the National Oceanic and Atmospheric Administration (NOAA) for the relevant years under discussion.

The table below reports the average mean temperature in the continental U.S. for the years 1998 through 2019:

1998	54.6	degrees
1999	54.5	degrees
2000	54.0	degrees
2001	54.3	degrees
2002	53.9	degrees
2003	53.7	degrees
2004	53.5	degrees
2005	54.0	degrees
2006	54.9	degrees
2007	54.2	degrees
2008	53.0	degrees
2009	53.1	degrees
2010	53.8	degrees
2011	53.8	degrees
2012	55.3	degrees
2013	52.4	degrees
2014	52.6	degrees
2015	54.4	degrees
2016	54.9	degrees
2017	54.6	degrees
2018	53.5	degrees
2019	52.7	degrees

Source: *National Climate Report – Annual 2019.*[898]

If, however, sea levels are rising something should be done, right? Sure, but what? Does moving to solar panels and wind turbines which requires trillions of tons of environmental destruction for rare earth minerals and metals over 700 times the land use required for industrial solar and wind intermittent electricity farms ensure sea levels will stop rising? Of course not, but it will add billions to the financial statements of men like Bill Gates, Michael Bloomberg, and Tom Steyer: all white male billionaires.

Scientific consensus based on satellite laser altimeter readings since 1993 have shown the rate of increase in overall sea levels at approximately 0.12 inches per year.[899] This is disputed when "empirical evidence of any substantial increases taken from land-based measurements has been ambiguous."[900] A controversial professor from Stockholm University, Professor Niles-Axel Morner is loathed in climate circles for arguing based on his study of sea levels in the Fiji Islands – that "there are no traces of any present rise in sea levels; on the contrary, full stability."[901]

His views are peer-reviewed backstopped by empirical evidence to substantiate and give credence to his climate-nihilism studies regarding his non-sea level rise hypothesis. What should be argued, the world has enormous problems in the areas of electrification, disease, famine, and hunger in undeveloped countries such as most of Africa, South America, Central Asia, and large swaths of Asia – with limited resources and political will to address the plight of billions of wretched souls who desperately need help.

Wasting financial resources on renewables and global warming is not the solution. Only white billionaires are enriched off the backs of black, brown, yellow, and reddish-skinned people; instead, building electric grids is the worthiest achievement mankind could accomplish this century. That is the politics of positivity and thriving, which should be embraced.

Electrifying the world by building electrical grids so every person on the planet had access to reliable electricity would rival the first-time Astronauts landed on the moon, and when vaccines were introduced leading to increased global health and human prosperity. But the politics of western environmentalism and its movement, which began in the 1960s from achieving these worthy goals is under attack from an insidious strain of protectionism coming from China.

A new Chinese national security law gained strength in January 2021, which allows the Chinese Communist Party (CCP) to limit rare earth mineral exports on the grounds of national security. Any country or entity the Communists disapprove of for any reason will no longer be allowed to receive rare earth mineral exports from the Chinese government. When China controls between 80-90 percent of all rare earth minerals including lithium and cobalt this is a major national security problem for the countries that need those materials from China to support their electrification needs, and a major problem for the finances behind the clean energy transition and the transition being stopped dead in its tracks.[902]

Moreover, it is no secret China understands whoever controls the sale and distribution of rare earth minerals and metals controls the world's military, transportation, and commerce. China wants to be that one country; and is willing to put its renminbi currency behind the deception of intermittent electricity from Industrial wind and industrial solar being the way to save the world from itself.

Without rare earth minerals wind turbines, solar panels, electric vehicle batteries, and batteries for the electrical grid cannot function. Any country that attempts to decarbonize, meets the requirements of a carbon-free society, or implement climate plans will have the "rug (pulled) out from under them whenever China wants." All this for what – climate change! – what if the impact is uneven or unknown?[903] If that is the case, then China is the new global superpower over western-acquiesce to ideological climate change.

CHINA AND THEIR ENVIRONMENTAL LACKEYS STAND AGAINST PROGRESS

The incoming Biden administration's climate policies are corrupted by the Chinese Communist Party (CCP).[904] The former head of U.S. President Barack Obama's U.S. Environmental Protection Agency (EPA), Administrator Gina McCarthy who currently runs the uber-liberal, pro-global warming, pro-renewables, viciously anti-fossil fuels environmental agency, The Natural Resources Defense Council (NRDC) is a CCP "tool."[905] That is not hyperbole, unfortunately it is true.

The NRDC has been accepting millions from China for over twenty years while supposedly advising them on environmental policies.[906] The NRDC is vehemently anti-coal in the U.S., all western-aligned nations, and the developing world, but nothing comes from the NRDC against China's overwhelming use of coal-fired power plants for electricity. More on this issue later in this chapter. Hypocrisy is not a strong enough condemnation against faux-environmental organizations like the NRDC.

A report released by a little-known non-partisan educational think tank – The Global Warming Policy Foundation (GWPF) – has unequivocally asserted the entire western environmental movement is being controlled by the Chinese Communist Party (CCP). The new report damningly states, "western environmentalists are being controlled and used by the communist regime in China."[907]

Beijing has co-opted all western environmentalists through money and coercion. Author of the report, Patricia Adams writes:

> "Like all western non-governmental organizations, green groups are only allowed to operate in China so long as they bite their tongues and toe the party line (Chinese Communist Party). But Beijing is also able to influence their behavior through funding bodies like Energy Foundation China, a US-based body that distributes money from American billionaire foundations."[908]

Mouthpieces for China's authoritarian President Xi Jinping

are what western greens have become. Western environmental organizations were once the authoritative voice of reason for lowering emissions and pollution. Now money has corrupted their mantles.

Western billionaires and environmentalists who back them and receive funding in return have made a deal with the devil when it comes to China. They praise Chinese commitments to fighting climate change with renewables without saying a word about China's massive and continuous use of coal-fired power plants for electricity.[909]

The Greens never say a word when China does not honor commitments to the Paris Climate Accords (PCA) or fair trade through the World Trade Organization (WTO). China abuses the world, and certainly has no intention of lowering their carbon footprint while western environmentalists stay silent.

U.S. and European multinational companies are also complicit in this deception since these firms are "riddled with members of the Chinese Communist Party (CCP)."[910] Boeing, Qualcomm and Pfizer are only three of what is termed a "mass infiltration of American companies."[911] The extent to which incoming U.S. President Joe Biden and his family are infiltrated and influenced will be known in the coming months and years. National security and foreign policy will likely be forever changed for the worse.

China plays every U.S. company and environmentalist under their sway for the useful idiots they are when corporate America and multinational firm's trip over themselves to prove their climate change mitigating credentials; and how all of them use renewables and are moving towards a carbon-free marketplace for their products. Meanwhile, billions suffer without electricity, but western environmental billionaires grow richer and sell out western freedom and human rights for the ability to sell their goods to over a billion domestic Chinese. Useful idiots indeed.[912]

Western environmentalists are fools believing the Chinese Communist Party (CCP) ever cares about democracy, open societies, the value of human life, or mitigating pollution in their country and global emissions. There are a few western leaders who understand the damage China is doing to global environmental health, and how they are the greatest threat to freedom since Nazi Germany.

Garnett Genuis, Vice-Chair of the Special Committee on Canada-China Relations of the Canadian House of Commons said China is incapable of reform/environmental health:

> "A government that is genocidal and totalitarian is one that cannot be trusted. And I think it is becoming harder and harder to argue the contrary, in this current environment. Today, there is just no excuse for having your head that far in the sand and yet, you know, we still have too many politicians that do."[913]

This environmental destruction and squashing of human rights begin with the head of the Chinese Communist Party (CCP) led by President X Jingping. Financial incentives for white, mainly western-based billionaire's clouds everything about China and causes deep mis-assessments as to the nature and evil intent of the Chinese regime.

According to Paul Evans, Professor at the School of Public Policy and Global Affairs of the University of British Columbia:

> "China has only become more of a bad actor. Under Xi Jinping, China is more repressive domestically and along its periphery than at any time since Mao Zedong."[914]

Another China-watcher, Bonnie Glaser puts the Chinese communists under a harsh light and the foolishness that western environmentalists and their billionaire enablers have undertaken believing China shares western environmental values. Miss Glaser prudently states:

> "China does not respect the rule of law (environmental or human rights). It does not share liberal democratic values, and it does not protect human rights. It is seeking to alter the international system (and environmental protocols and treaties) in ways that are favorable to China and detrimental to western interests. China's tool of domestic governance, and its detention of over one million (Muslim) Uighurs, its censorship of expression, and its social credit system should not be a model for the rest of the world."[915]

The western world, really most of the world does not share these values. But for western environmentalists and the billionaires and multinational corporations that profit off this malfeasance these are their values. China's strength and march towards domination benefits each of these recipient's personally, professionally, and their sleazy financial income statements and balance sheets.

GREEN ENERGY ECONOMICS

Back to the Germans who have a specific word for when the wind does not blow, and the sun does not shine: *dunkelflaute*. German winters can be largely gloomy seasons, which shutter the sun and wind's ability to produce electricity. No output from these industrial power plants, or a fraction of what they are projected to produce.

Belgium vies with Germany for the highest electricity prices in the world and Europe since it overly relies on the wind turbines for electricity. The phenomenon for Belgium was critical in January 2017:

> "When it (the Belgium nation) experienced a whopping nine-day calm and dull spell. Despite having just 9 per cent of its capacity from renewable sources, the country's network had to scramble to supply sufficient electricity to avoid disruption."[916]

This stagnant weather is generally a European occurrence, but renewable heavy systems still have major problems.

California is the 5th largest economy in the world and has some of the best research universities in the world with Stanford, Caltech, Berkeley, U.C.L.A., and Silicon Valley. Home to Google, Facebook, Apple, and a host of other companies, which is now the highest valued sector in the world.

California is now so inept the state cannot keep its electrical grid from rolling blackouts when the weather is hot and windless.[917] Managing solar and wind intermittency cannot be solved by the greatest minds at Caltech or Silicon Valley. Weather-dependent, low-carbon electricity is nowhere close to working, and the challenge will continue causing human misery, suffering at global levels, and overly expensive electricity that will cause increased forest clearing for wood to heat homes and cook food.

That is the true economics of the green transition: human misery and loss of freedom.

Opposite problems can then occur when too much sun is shining, or gusting winds cause an abundance of electricity. Electrical grids cannot handle renewables the same way they are unable to handle the demand load that will be placed on them if governments and politicians get rid of all cars that are using gasoline and replace them with electric vehicles (EVs).[918]

Grids needs reliable, constant electricity 24/7/365, otherwise power plants must shut down from overloading the complexity of an electrical grid. Our second book *Just Green Electricity* details this process how grids work and why fossil fuels and nuclear is the only option available currently for electrical grid reliability from consistent electricity generating sources.

If Denmark, Germany, Great Britain, and the U.S. cannot figure out renewables, blackouts, and out-of-control electricity prices, why do these governments persist in bringing disaster onto their citizens and economies. It is a money play – pure and simple. Right now, western environmentalists who sunk millions into electing Joe Biden to the U.S. Presidency are lining up at the federal and global trough for the trillions that energy and climate polices affect.[919]

None of these people care about selling out hard won freedoms to Chinese communists, grid blackouts, or solving any problems that makes life better for people, because they only care about money and power.

Please do not think these groups or people can be saved, reasoned with, or will seek a middle ground. Climate change is their ideology with renewables backed by battery energy storage systems for electrical grids and EVs their sacred text just like *The Bible* is for Christians. The days of looking to eliminate western levels of elevated smog, or lakes, rivers, and oceans so polluted they would light on fire are over. Paul Driessen, Bjorn Lomberg, Michael Shellenberger, Patrick Moore, Dr. Roger Pielke – each one of these folks are sensible environmentalists, but those days are long past.

Each of these men are castigated and maligned for views outside of today's western environmental movement's mainstream that only embraces killing off fossil fuels, ridding society of zero-emission nuclear generated electricity, and only believing intermittent electricity from renewables will solve their questionable premise of man-made global warming.

These people, foundations, non-governmental organizations,

and organizations are Nazi-like in their fervor to destroy abundant and reliable fossil fuels while being supported by the Chinese Communist Party (CCP). Being progressive now means you only support killing economies, discarding millions of jobs, litigating out of existence life-saving pipelines, and eliminating the over 6,000 products that come from a barrel of crude oil.

Ironically, these would be the same people to line up for the COVID-19 vaccine without ever realizing the vaccine or mitigating measures against the coronavirus do not exist without crude oil. They are literally injecting crude oil into their bodies and masking their faces with products from Big Oil companies.[920]

What do these people want? Besides power, control, rigged markets, and money is embodied in pressure Joe Biden is now under to:

> "Declare climate change an emergency on day one (of his Presidency), uphold a campaign pledge to end fossil fuel leasing on federal lands, block big oil pipelines like Keystone XL, and direct 40 percent of climate related investments to environmental justice communities (no one has defined or given specific policy proposals for what this actually means)."[921]

As Rupert Darwall's book, *Green Tyranny* eloquently makes the case that "climate alarm is more about power and control, and less about the climate or the environment."[922]

This is the same reasoning reiterated by Michael Shellenberger, Patrick Moore, Paul Driessen, Alex Epstein, Edward Ring, and Bjorn Lomberg. All respected environmentalists and folks who agree environmentalism is now about destroying the environment over the use of renewables and battery storage systems while wrecking the global economy in their favor.

Just Germany's use of wind for electricity has resulted in

over 1.35 million tons of toxic wind turbine blades that cannot be recycled, disposed of properly, or eliminated without burying them in the ground.[923] It gets worse.

Children for decades have loved *Grimm's Fairy Tales*. Even this is under assault by the profit-seeking advocates of the green transition who seek renewables to solve environmental issues. One of Germany's most idyllic, fairy-tale like forests; the *Reinhardswalk* located in the hilly region west of Gottingen is:

> "Slated to be industrialized for "green energy." The 20 million square meters of 1000-year old "fairy tale" forest to be designated as an industrial wind park zone. The *Reinhardswalk* is known as the "treasure house of European forests" or "Grimm's fairy tale forest."[924]

Incredibly, a natural, undisturbed habitat, which environmentalists such as Michael Shellenberger fight for, and have dedicated their life towards preserving will fall victim "to the greed of crony policymakers and industry lobbyists."

Here are some of the most influential environmental groups cramming this agenda down our collective throats to a gullible and uninformed public: 350.org's Fossil Free Media, Environmental Defense Action Fund, League of Conservation Voters, the Sierra Club, EDF Action, Sunrise PAC, Youth Climate activist group Sunrise Movement, Greenpeace, Earthjustice, National Wildlife Federation, California Environmental Justice Alliance, Center for Biological Diversity, Natural Resources Defense Council, Pacific Institute, and Friends of the River along with foundations such as the Ford Foundation, Rockefeller Foundation, and Gates Foundation.[925]

These are western, mainly U.S. based environmental organizations, but with extensive global reach. From the halls of the UN to every significant capital city in the world. Focusing on the

U.S. is obviously important since it is the largest economy in the world, and still ensures global security through the liberal-based order that has been in place since World War II.

Here are a few of the individuals who are using climate change to win elections, garner taxpayer monies and crushing human development and ingenuity by not delivering electricity to the over 4 billion people globally without reliable service or none-at-all.[926] The people in the next paragraph are not interested in solving problems since it does not fit the climate change/ renewables belief system.

Number one: Joe Biden and his energy and climate appointments. Gina McCarty, the new Biden "Domestic Climate Coordinator," where everything in the U.S. federal government, foreign policy, national security, and global relations via the UN Security Council makes all policy through the lens of fanatical climate change and lowering emissions.[927]

Without any guarantee emissions will ever be lowered pursuing these climate and energy policies.[928] Ms. McCarthy has admitted under U.S. Congressional testimony that she has no idea if emissions, global health, poverty, or pollution can be lowered, mitigated, or alleviated focusing on climate change or lower emissions.[929] Ms. McCarthy famously mishandled the Flint, Michigan water crisis when she was head of the U.S. Environmental Protection Agency (EPA), but still she persists with climate change from this new perch in the Biden administration while killing off human progress.

Current Michigan Governor Jennifer Granholm (section written in December 2020) is a champion for renewable electricity, and the projects she advocated for have costs taxpayers over $1 billion. None of the projects ever delivered electricity as promised. She will be the next U.S. Secretary of Energy. New EPA administrator, Michael Regan is Secretary of the North Carolina Department of Environmental Quality and wants "environmental justice" instead of a clean environment and the balance of environmental health with economic growth.[930]

Congressman Deb Haaland of New Mexico will be the new U.S. Secretary of the Interior and she is opposed to energy development of U.S. federal lands, which have done more for global security and New Mexico's economic resurgence more than anything in previous memory.[931] New Mexico would continue being a dirt-poor wasteland within the U.S. without fracking on federal lands to that state's benefit.[932]

U.S. foreign policy will again be led by former U.S. Senator and former U.S. Secretary of State John Kerry who will be a "Special Envoy" on climate change with a seat on the U.S. National Security Council. Mr. Kerry along with Ms. McCarty, Ms. Granholm, and Ms. Haaland means the most powerful country on earth will "shift climate and energy policies in a radical, ominous direction."[933] This list is only the tip of this insidious profit and control-driven iceberg.

But still it is climate, climate, climate for these people the way Nazi's were obsessed about Jews. Lust for power. Lust for money. Lust for control. Adolf Hitler appeared on the cover of *Time Magazine* on multiple occasions. Most famously, on January 2, 1939, when he was named their Man of the Year.[934] Moral of the story: the mainstream media is not always reliable. Whether environmentalists or Jew-hating maniacs who wanted to burn down the world for their own personal gain. Sounds familiar, huh?

These groups, governments and their billionaire corporate powers use the levers of sovereignties under the guise of climate change. It is why the Paris Climate Accords was created, which will literally do nothing to change temperatures up or down.[935] Nor will they alleviate emissions since the accords do not change the behavior of China and India. If anything – climate change – is enabling Chinese power through Green-aligned non-governmental organizations.[936]

Why was it created then? To ruin the U.S. Nations that were

aghast when the Trump administration backed out of the Paris Climate Accords. Trump was right and they were wrong because the Paris Climate Accords has nothing to do with securing environmental health, alleviating pollution, or lowering emissions.

The United Nations *Emissions Gap Report 2020* highlighted the U.S. is the only country to have met the Paris Climate Accord emissions reductions over their abundant use of natural gas-fired power plants instead of coal-fired power plants for electricity.[937] The U.S. had a red-hot economy in 2019 before the coronavirus pandemic and still was the only industrialized country to hit the emission-lowering benchmarks. The U.N. said, "America is already cutting so much carbon it doesn't need the Paris Climate Accord."[938]

The U.S. should keep doing what it is doing, and use more natural gas while exporting to China, India, and Africa to help them from using so much of the dirtiest forms of coal for electricity. Then what should an individual, village, town, city, county, state, country, or continent think about climate change? Jobs and human thriving should be the answer. That is how to mitigate changing weather and climate patterns while still growing jobs, economies, education opportunities, and the ability for affordable and reliable electricity and access to petroleum derivative products fueling developed and developing nations daily lives.

The pushback against this idea of human thriving is fierce. Edward Ring, Senior Fellow at the California Policy center states it best – what he terms the "The Silicon Valley" mentality that will be globally exported under the Joe Biden/Kamala Harris administration:

> "(This mentality) has a monolithic opinion on issues that strike to the heart of how the United States and the rest of the world will develop over the next few decades." Their wealth and power (now include billionaires who promote renewables

> and climate change for their gain) is matched by intolerance for dissenting points of view. But if they can stifle the aspiration of humanity, enforcing rationing (energy, land use, agriculture, transportation), scarcity, micromanagement, technology-driven surveillance (like China), and billions for the bureaucrats and litigators, instead of for the bulldozers and builders, their legacy will be one of destruction and decline."[939]

Where anyone with skin color other than a white billionaire should understand these policies also promote abortion that is wiping out African Americans in the U.S.[940] Where U.S. Democratic Party aspirations thrive, abortion is widely pursued and used for birth control. What type of energy policies put in place can determine life or death for millions of unborn children.

What would Senator Jeff Merkel (D-Oregon) say about climate polices that hurt the individual? Senator Merkel is all-in on climate change and stopping fossil fuels and nuclear to achieve undefined climate goals.[941] Doubting Senator Merkel's intentions is not the issue – asking what climate policies produce is likely the largest energy issue facing the world within the next U.S. administration. Other global leaders take a different, positive view of thriving and energy production.[942]

One thing is clear: environmentalists and all the monies, people, foundations, and non-governmental organizations that support them are not going away or backing down; they feel emboldened. When COVID-19 U.S. stimulus relief funds are used for green tax credits to protect industrial wind and solar electricity farms it should be understood this is about greed and destruction over mitigating the worse effects of climate change through prudent energy policies.[943]

These advocates of destruction are frauds, hucksters, charlatans...I continue to say Nazi's. So where do we go from here moving forward under global leadership that wants taxpayer monies for climate change and renewables/EVs over fossil fuels, nuclear, and the over 6,000 products that comes from a barrel of oil that also fuels the entire global and mining transportation structure? It is murky at best to figure out how energy and human thriving intersect.

CONCLUSION: WHERE DO WE FINANCIALLY AND SOCIETALLY GO FROM HERE?

Toyota president, Akio Toyoda, says electric vehicles are "over-hyped."[944] He continues excoriating EVs and the idea they control emissions:

> "The more EVs we build, the worse carbon dioxide gets. Converting entirely to EVs could cost hundreds of billions of dollars and make cars unaffordable for average people. Why would anyone concerned about CO2 emissions ever pay more for a car that increases them?"[945]

The reason why – because money and power go into the pockets of white billionaires. That is why.

If you truly care about the environment, then you advocate and heavily use natural gas for electricity the way the Trump administration did from 2016-2020. Therefore, emissions fell in the U.S. while the economy and jobs grew without exchanging environmental health for prosperity. The "EV dream" is simply a dream without factual evidence backing the claims the same way renewables are the answer for man-made global warming.[946] Neither are the answer.

It is ultimately about control and destroying economies for financial gain. Whether a global investment firm or Michael Bloomberg's media empire. Destroy the economy and gain the world. Here is a chilling example of how it will take place. Joe Biden has stated, "he will prohibit new oil and gas permitting on (U.S.) public lands and waters."[947]

After examining the economic pain inflicted by this ban the Wyoming Energy Authority concluded this would cost "$670 billion in lost GDP and wipe out more than 72,000 jobs annually (from 2020-2024) while offering zero guarantee of reduced CO2 emissions."[948] This ban will likely shift oil, coal, and natural gas exploration and production to other regions and areas such as Mexico Canada, the Middle East, and Russia, which have less environmental controls and mitigation measures in place.

Focusing on the U.S. immediately highlights what the environmental left does to humanity and prioritizing emissions reduction at all costs without giving credit to natural gas or nuclear to achieve their goals. The U.S. achieved energy independence for the first time since 1957 while continuing to lead the world in emission reductions.

The U.S. Environmental Protection Agency (EPA) reported in November 2020 the same findings the UN concluded: "U.S. emissions from large facilities fell by nearly 5 percent from 2018-19."[949] Biden and the environmentalists and billionaires who control him and funded his path to the U.S. Presidency want him to begin achieving net-zero emissions by 2050.

It cannot be stated enough or written about over and over in this chapter and book: this is an Eco-Activist collusion at the highest levels of government, academia, media, foundations, non-governmental organizations, and multinational corporations. A backlash could possibly be building against this coalition since they are pushing the U.S. and world to eliminate natural gas in favor of electricity for cooking, heating, and daily needs.[950] This will force people to use more expensive electric stoves, ovens, and laundry machines at the expense of the poor and most vulnerable globally.

Only aim to be achieved will be sending billions of taxpayer monies to favored "green groups" and well-connected "green energy interests.[951] Whether EVs, banning natural gas, renewables being the cure-all, demonizing fossil fuels, scaring westerners out of using nuclear, or controlling your life, it all comes down to billions, even trillions, frittered away into nothing.

Not one bit of environmental health will be achieved by these people in the U.S., European Union, UN, or any western organization pushing this agenda down people's throats, because of China. China is the biggest global and environmental threat on the planet. When President Trump pulled the U.S. out of the Paris Climate Accords (PCA) between January 2017-May 2019 the U.S.:

> "Shuttered 50 coal-fired power plants, with 51 more shutdowns announced, bringing the total shutdowns to 289 (330 once announced shutdowns also take place) since 2010, leaving under 200 (coal-fired power plants) still operating."

Currently, China has 2,363 active coal-fired power plants, and building another 1,171 domestically.[952] Plus, developing countries are building hundreds more in Africa, Asia, and Pakistan. All three recipients are desperate for affordable, reliable, and abundant electricity, which coal provides.

China tries to domestically stop planetary destruction by using modern-pollutant scrubbing technology on over 80 percent of its coal-fired power plants.[953] But no scrubbers are used on any Chinese-built coal-fired power plants in Africa, Asia, Pakistan, or other Chinese energy largesse beneficiaries.

Carbon dioxide emissions will likely increase from these coal-fired power plants China is building without scrubbers. The Paris Climate Accords does not address this issue from China. They will continue getting a pass from western governments, environmental organizations, foundations, non-governmental organizations, the UN, and all who stand to gain financially from green technology and the move to address global warming/climate change above all else.

Harvard University China specialist Edward Cunningham says:

> "China is building, planning or financing more than 300 coal plants, in places as widespread as Turkey, Egypt, Vietnam, Indonesia, Bangladesh, the Philippines, India, South Korea, Japan, and South Africa."[954]

Even Germany, the supposed global leader, and example to follow for how to transition an economy to only using solar panels, wind turbines, and biomass for electricity is now building and reopening coal-fired power plants.[955] This is done to backup intermittent and unreliable renewables.

Here is what China understands and the global community – mainly the U.S. and UN – will have to contend with now:

> "China knows the world will need oil, natural gas and coal for decades to come. It sees "green" as the color of money. Communist Party leaders seek global military and economic power. Global control of electricity generation, raw materials extraction (exploration and production) and manufacturing of wind turbines, solar panels and battery modules (electrical grid scale battery energy storage systems) they will sell to address the West's obsession with the "man-made climate crisis" and "renewable, sustainable" energy (to electricity)."

A Green New Deal for the U.S., EU, or anyone attempting the transition to a "renewable, sustainable" path for the electrical grid for electricity will up their reliance on China with devastating consequences for global security and sustained peace.[956]

The guardians of the environment stay silent as eco-manslaughter takes place in China to the world's detriment. Now the power of governments such as the U.S., Europe, parts of Asia, the World Bank, multinational banks, and global investment management firms will steer all things towards climate change.[957] Policy and investments will be barred from hydrocarbons, fossil fuels, and nuclear.

Corporations jump in to show their green bona-fides such as the outdoor clothing manufacturer North Face. North Face is particularly repugnant since they bill themselves as being environmentally conscious, an eco-friendly company, but:

> ", however, its business is predicated on selling clothes, shoes and outdoor gear made of petroleum-based synthetic materials and then shipping those products on fuel-guzzling ships and airplanes."[958]

What if this entire energy transition could be one big bubble without considering the over 6,000 products from a barrel of crude oil, reliable electricity, and how billions are living in abject poverty?[959]

Are these billionaires and the politicians they support really going to keep people away from reliable electricity over climate change? Yes, they will if it supports their financial agenda. What if rising emissions are a natural occurrence as they have been previously?[960] Previous periods of climate history show entirely different outcomes such as the "Holocene climate and glaciers dynamic" times.[961]

So, then what of the economic activity and products we consume every day that produce carbon emissions? Add into the mix all types of travel, transportation, heating, and air conditioning, and we are looking at harming or eliminating goods and services globally. For what? Climate change? Renewables? Battery energy storage systems for electrical grids? EVs? What?

Finite resource are not infinite resources with infinite amounts of money, time, and energy attached to them. Battery energy storage systems for electrical grids we are now told is how to hook up zero-emissions solar panels and wind turbines to electrical grids, because these units can store electricity for an infinite amount of time. Is this true, because the example below is the murky, dark side of the green transition and power and profits behind this movement.

Let us examine the financial impact on just New York City attempting to use battery energy storage systems (BESS') for intermittent renewables attached to their electrical grid.

If New York City went 100 percent wind and solar without any coal-fired, gas-fired, or nuclear power plants as backups, the city approximately peaks at 32,000 MW a day to keep their electrical grid working 24/7/365.[962] Summer months are hard on New York City when they experience high pressure systems for 5-7 days at a time.

Wind turbines need sustained winds between 10-30 mph, but high-pressure systems do not produce enough wind, but could theoretically produce enough sun from solar panels to the battery energy storage systems (BESS).

Reliability is a factor since grids need constant energy, otherwise blackouts and systems malfunctions take place. The math for this to happen is:

> "32,000 MW x 168 hours (number of hours in a week) = 5,376,000 MWh of stored juice (electricity) needed to make it. Normal reliability for the grid needs another 20 percent (capacity)."[963]

This figure does not account for transitioning to electric

vehicles over current transportation technology. New York is bringing online the world's largest 400 MWh battery energy storage systems for an electrical grid in 2021 within their Con Edison electrical grid configuration.

For New York City to keep the grid running they need 5,376,000 divided by 400, which equals 13,440 more 400 MWh battery energy storage systems units to go. How much would it cost to meet the 13,440 figure? The U.S. Energy Information Administration (EIA) reports the average utility scale battery energy storage system costs approximately $1.5 million a MWh of storage capacity. For New York City's new battery energy storage systems in 2021 the cost is likely around $600 million to the taxpayer and ratepayer.[964]

So, the approximate cost considering finite taxpayer and ratepayer resources to reliably provide battery storage on their electrical grid using only the wind and the sun to scale would be "just over $8,000,000,000,000 or Eight Trillion Dollar."[965]

This is only for one city in the U.S. What is the cost for the U.S.? Europe? China? India? The math is staggering. Not to mention the amount of raw earth minerals needing to be mined, environmental degradation that would take place, and human atrocities on a scale not seen since World War II.

Utilities make out handsomely from these battery energy storage systems (BESS'), all over the U.S., and globally. They understand the math and the limitations, but politicians promise wind and solar are cheaper than coal, natural gas, and nuclear when that simply is not the case. It is a hoax. A scam. Most voters, the media, or the average person trying to live their life have no idea how rich utilities and their backers are getting off building renewables and batteries to back them up. New York City is in for a rude awakening.

The maddening part is none of the trillions wasted on climate change, global warming, solar panels, wind turbines, hydrogen, electric vehicles, decarbonization, or utility-scale battery energy storage systems will provide roads, power plants, electrical grids, flood control, sanitation, water management systems, medications, disease eradication, affordable family housing, schools from Pre-K to PhD's, energy production, modern machinery, militaries, or transportation. It all comes overwhelmingly from oil, petroleum, natural gas, coal, and nuclear.

A green ramrod is what lays ahead, and this will play out in every part of the global economy so billionaires can become wealthier at the expense of poverty and citizens. Rejecting renewables or disputing man-made, ideological global warming does not make a person a NIMBY or denier, it makes them a realist.[966] Environmentalists are lobbying for their causes and countless litigators are imposing scarcity and fraud onto entire countries and continents.

Serfdom is the future until this is recognized, the financial gain had by the Overlords of environmentalism is discarded in favor of affordability, abundance, and human life is valued. There are acceptable tradeoffs between "environmental protection and economic growth."[967]

Unless the Green tyrants are stopped from reigning supreme the future looks bleak. One does not have to choose climate change over lifting billions out of poverty by building electrical grids the same way the environment does not have to be destroyed for economic prosperity. One does not beget the other. But as the environmental green-movement goes, so goes the western world with China, Russia and even India waiting in the wings to establish a new order no longer based on freedom for the individual. God help us if that takes place.

This chapter, and this book chooses abundance and the prosperity that energy and electricity bring to every human anywhere in the world over environmental degradation and human atrocities.

CHAPTER EIGHT

SUMMARY OF THE 2020 ELECTIONS IMPACT ON THE ECONOMY AND ENVIRONMENT

Our reflection on the 2020 elections is disappointment for those that voted for the Democratic platform. The platform loves intermittent electricity from wind and solar that will doom millions into higher electricity prices and into energy poverty throughout the entire economy that hurts working families and the poor struggling to survive.

The Democratic platform loves California and wants to clone California's policies and regulations for the other 49 states. We know what the Democratic platform has done to California, now with the some of the highest costs for electricity and fuels in the country, and the highest poverty and homeless rates in America. Taxes are skyrocketing with no relief in sight. As we write, the California legislature is debating setting the state's income tax rate to 16 percent that will be the highest in the U.S. until the rest of the country adopts the trend being set in California.

The voters "bought" into the Democratic platform for higher minimum wages without coinciding economic growth, subsidized health insurance, free higher education, forgiveness of over a trillion dollars in student loan debt and guaranteed weekly income. With higher costs for electricity and fuels coming their way, and more taxes on those "rich" folks which will result in more costs passed through to those voters, it's no "surprise" that further increased relief is on the horizon for the voters that bought into the Democratic platform.

With at least 80 percent of humanity, or almost 6 billion on earth, living on less than $10 a day, the Democratic platform will allow the continuation of 11 million children in the world dying every year from preventable causes of diarrhea, malaria, neonatal infection, pneumonia, preterm delivery, or lack of oxygen at birth as many developing countries have no or minimal access to the thousands of products from oil derivatives enjoyed by the wealthy and healthy countries. Basically, almost 6 billion in the world are already living without fossil fuels, just like Americans did just a few short centuries ago.

The world needs to comprehend that energy is more than intermittent electricity from wind and solar, and battery storage systems for the electrical grid. Ever since the discovery of the versatility of products available from petroleum derivatives, and the beginning of manufacturing and assembly of cars, trucks, airplanes, and military equipment, the world has had almost 200 years to develop clones or generics to replace the crude oil derivatives that are the basis of more than 6,000 products we use, such as: medications, electronics, communications, tires, asphalt, and fertilizers.

All the COVID-19 medical equipment, PPE, and vaccinations would not have been possible without the petroleum derivatives that begin from a barrel of crude oil. Anyone who takes this vaccine is literally having a shot of oil put into their immune system. Funny to comprehend but it is reality.

The social needs of our materialistic societies are most likely going to opt for continuous, uninterruptable, and reliable electricity from coal, natural gas, or nuclear electricity generation, rather than just intermittent electricity from wind and solar. Add the overwhelming need for all those chemicals derivatives that get manufactured out of crude oil, that makes everything else that's part of our daily lifestyles and economies, and we become aware that those thousands of products are something the world cannot live without.

While the "small" country of the USA is trying to reduce emissions, there are 62,500 power plants around the world operating today, all types, generating electricity for the world's inhabitants.[968] Of that total, more than 2,449 are coal-fired power plants. Over half (1,363) of the world's coal power plants (2,449) are in China and India whose populations of mostly poor peoples is roughly 2.7 billion. Together they are in the process of building 284 new ones of the 546 total.[969] [970]

The Democratic platform extols the virtues of climate change and renewables, but the voters who favor that platform have never been told, or considered the fact the U.S., a developed country representing 4 percent of the world's population, literally has a minuscule impact on global emissions and pollution. The U.S.' emissions declined almost 3 percent in 2019. American economic activity and electricity is not the problem for climate change or increasing emissions and pollution, it is the growth of developing China, India, and Africa.

America may take credit for the emission reductions within the boundaries of the nation, but the country's energy and fossil fuel product demands would continue to be met with imported finished products from foreign countries afar, i.e., more "leakage" of emissions to other manufacturing locations to support the demands of the USA as there are more than 700 refineries worldwide of suppliers that can meet global demands and all are located in areas with significantly less stringent environmental controls than America.

The unintended consequences of the Democratic platform of getting rid of fossil fuels in America by 2050 is that it would result in importing the fuels and products from foreign locations that have significantly less stringent to zero environmental controls. That plan will work, but with higher costs to the American consumer, and may put America at a national security risk with increased dependance on foreign countries for the products and fuels for America's economy and our existence.

Green "electricity" from wind, solar, EV batteries, and battery storage systems for electrical grids uses materials from some of the worst ecological and human rights offenders that exist on earth. Thus, more environmental degradation and humanity atrocities in the foreign countries mining for the exotic minerals and metals to support the green movement.

In the future, intermittent electricity from wind and solar may be used to convert fresh or seawater into "green hydrogen" – a form of energy that analysts expect to be in increasing demand as a replacement for coal and natural gas for electricity generation in the years and decades ahead. The use of electrolysis to manufacture hydrogen from the intermittent electricity by renewables could become the cheapest 'transformative fuel' within a decade. Storing that hydrogen for later, to be used to generate truly clean continuous and uninterruptible electricity. With an endless supply of water on this earth, this may be a small step toward clean electricity without any environmental degradation or humanity atrocities now occurring in the worldwide supply chain for exotic minerals and metals to support wind, solar, and EV batteries.

Renewables have a role in our energy usage, but we need to consider what they can do, and what they cannot do. Renewables can only generate electricity, and intermittent electricity at best.

The undisputable science is that renewables CANNOT manufacture any of the oil derivatives that are the basis of more than 6,000 products that we did not have before 1900 that are the foundation of societies and economies around the world.

Indeed, virtually every aspect of our daily lifestyles and economies of all U.S. Democrats, Republicans, western environmentalists, and all the religions around the world, are based on the oil derivatives for those thousands of products we now take for granted.

Imagine how life was without that industry that did not exist before 1900 when we had, NO medications and medical equipment, NO vaccines, NO water filtration systems, NO sanitation systems, NO fertilizers to help feed billions, NO pesticides to control locusts and other pests, NO communications systems, including cell phones, computers, and iPads, NO vehicles, NO airlines, NO cruise ships, NO merchant ships that are now moving billions of dollars of products monthly throughout the world, NO tires for vehicles, and NO asphalt for roads, and NO space program.

Looking back just a few short centuries, we have come a long way since the pioneer days. Disappointingly, people voted in 2020 for a return to a Middle Ages feudal system where only a few rules over the many. Serfdom may see its return from the ashes under the Democratic platform.

In summary, we believe the Democratic platform will not be good for the American economy, and not good for those developing countries that are the basis of the supply chain of exotic minerals and metals for America to go green.

RON & TODD

ENDNOTES

Chapter End Notes:
(Due to the dynamic nature of the Internet, the location of some of the more than 500 items cited in this work—and accessed at the time of writing—may change as menus, homepages, and files are reorganized.)

INTRODUCTION

1 U.S. Congressional Testimony by Kevin D. Dayaratna, PhD, "Methods and Parameters Used to Establish the Social Cost of Carbon Testimony before the Subcommittee on Environment and Oversight," Committee on Science and Technology, U.S. House of Representative, www.Docs.House.Gov, February 24, 2017. https://docs.house.gov/meetings/SY/SY18/20170228/105632/HHRG-115-SY18-Wstate-DayaratnaK-20170228.pdf
2 Loris, Nicolas, "Green New Deal Would Barely Change Earth's Temperature. Here Are the Facts," www.Heritage.org, February 8, 2019. https://www.heritage.org/energy-economics/commentary/green-new-deal-would-barely-change-earths-temperature-here-are-the
3 Forbes, Viv, "The Green Road to Blackouts," www.MichaelSmithNews.com, September 1, 2020. https://www.michaelsmithnews.com/2020/09/viv-forbes-the-green-road-to-blackouts.html
4 Clifford, Catherine, CNBC, Executives from Jeff Bezos to Ford Motor Co.'s Bill Ford tell Joe Biden: Fighting climate change means job creation, Jan 25, 2021, https://www.cnbc.com/2021/01/25/

bezos-benioff-ford-sign-open-letter-to-biden-on-climate-change.html?__source=iosappshare%7Ccom.apple.UIKit.activity.Mail

5 Open Letter, Jan 25, 2021, "PRESIDENT BIDEN: YOU CAN BE THE CLIMATE PRESIDENT", https://www.climatepower2020.org/world-leaders/

6 Mill, Mark P., "A Carbon-Powered Shot," www.City-Journal.org, December 29, 2020. https://www.city-journal.org/a-carbon-powered-shot

7 Stein, Ron, Royal, Todd, Just Green Electricity – Helping Citizens Understand a World without Fossil Fuels, (Archway Publishing, New York, NY) Entire book, June 17, 2020.

8 Rucker, Craig, "Deadly "Green" energy hypocrites," www.cfact.org, August 11, 2020. https://www.cfact.org/2020/08/11/deadly-green-energy-hypocrites/

9 Shellenberger, Michael, "Why Climate Activists Will Go Nuclear – Or Go Extinct," www.Quillette.com, June 25, 2020. https://quillette.com/2020/06/25/why-climate-activists-will-go-nuclear-or-go-extinct/

10 Plumer, Bird, Washington Post, All of the world's power plants, in one handy map, December 8, 2012, https://www.washingtonpost.com/news/wonk/wp/2012/12/08/all-of-the-worlds-power-plants-in-one-handy-map/?noredirect=on

11 Coal-fired Power Stations by Region, Global Coal Plant Tracker, January 2021, https://docs.google.com/spreadsheets/d/1ZPbbwBI1cNoS1NqFEnN8PdXGRO0xuYfza1F0AWUDRuY/edit#gid=739846511

12 Coal-fired Power Stations by Country, Global Coal Plant Tracker, January 2021, https://docs.google.com/spreadsheets/d/1kXtAw6QvhE14_KRn5lnGoVPsHN3fDZHVMlvz_s_ch1w/edit#gid=165011444

13 Desjardins, Jeff, Visual Capitalist, "All of these things can be made with one barrel of oil," www.BusinessInsider.com, September 28, 2016. https://www.businessinsider.com/things-that-can-be-made-with-one-barrel-of-oil-2016-9

14 Driessen, Paul, "The staggering human costs of "renewable" energy," www.cfact.org, August 9, 2020. https://www.cfact.org/2020/08/09/renewable-energys-staggering-human-costs/

15 Driessen. Ibid. August 9, 2020.

16 Driessen. Ibid. August 9, 2020.

17 Driessen, Paul, Heartland Institute Policy Brief, How the Green New Deal's Renewable Energy Mining Would Harm Humans and the Environment, April 2020, https://www.heartland.org/_template-assets/documents/publications/PBdriessenmining2Apr20.pdf

18 Amnesty International, "Amnesty challenges industry leaders to clean up their batteries," www.Amnesty.org, March 21, 2019. https://www.amnesty.org/en/latest/news/2019/03/amnesty-challenges-industry-leaders-to-clean-up-their-batteries/

19 DeVore, Paul, "How Refusing Too Wisely Manage California's Resources Led To Incessant Blackouts," www.TheFederalist.com, September 7, 2020. https://thefederalist.com/2020/09/07/how-refusing-to-wisely-manage-californias-resources-led-to-incessant-blackouts/

20 Koop, Fermin, "Almost two-thirds of the global wildlife population has disappeared since 1970," www.ZmeScience.com, September 10, 2020. https://www.zmescience.com/science/two-thirds-wildlife-gone-053232/

21 Wootton, Hannah, Fowler, Elouise, "Companies, directors, governments face wave of climate change lawfare," www.AFR.com, (Financial Review), January 4, 2021. https://www.afr.com/companies/financial-services/companies-directors-governments-face-wave-of-climate-change-lawfare-20201228-p56qf9

22 Bryce, Robert (Director), Narrated by Robert Bryce, "Juice: How Electricity Explains the World," Available on iTunes and Amazon Prime. Released Summer 2020. www.JuicetheMovie.com, http://juicethemovie.com

23 Stein, Ronald, Royal, Todd, Just Green Electricity: Helping Citizens Understand a World Without Fossil Fuels, (Archway Publishing from Simon & Schuster, New York, NY) June 17, 2020. www.Amazon.com, https://www.amazon.com/Just-Green-Electricity-Citizens-Understand/dp/1480890707#customerReviews (See entire book for paragraph's source)

24 Kreutzer, David, Senior Economist at Institute for Energy Research, Guest Blogger for Natural Gas Now, "Climate Policy or Climate Injustice (Or Just Climate TomFoolery?" www.NaturalGasNow.org, January 7, 2021. https://naturalgasnow.org/climate-policy-or-climate-injustice-or-just-climate-tomfoolery/

25 Driessen, Paul, "Report renewable energy risks, too," www.WattsUpWithThat.com, July 9, 2020. https://wattsupwiththat.com/2020/07/09/report-renewable-energy-risks-too/

26 Africa Oil & Power, "African Lives Matter Too, Energy Policy Decisions Should Consider Their Needs," www.AfricaOilandPower.com, June 22, 2020. https://www.africaoilandpower.com/2020/06/22/african-lives-matter-too-energy-policy-decisions-should-consider-their-needs/?utm
27 Kotkin, Joel, "Triumph of the Woke Oligarchs," www.RealClearEnergy.com, April 27, 2020. https://www.realclearenergy.org/articles/2020/04/27/triumph_of_the_woke_oligarchs_490094.html
28 Stein and Royal. Ibid. Just Green Electricity: Helping Citizens Understand a World Without Fossil Fuels. Entire book. 2020.
29 Stein & Royal. Ibid. Entire book. 2020.
30 Carlton, Jim, "Why Wildfires Are So Bad This Year in California, Oregon and Washington," www.WSJ.com, September 11, 2020. https://www.wsj.com/articles/why-wildfires-are-so-bad-this-year-in-california-oregon-and-washington-11599768604
31 Stein and Royal. Ibid. "Just Green Electricity: Helping Citizens Understand a World Without Fossil Fuels." Entire book. 2020.
32 Bryce, Robert, "The Trillion-Dollar Reason Why Joe Biden Won't Ban Fracking," www.Forbes.com, September 6, 2020. https://www.forbes.com/sites/robertbryce/2020/09/06/the-trillion-dollar-reason-biden-wont-ban-fracking/#1bd0bee2accf
33 Wikipedia.com, "Nuclear Power in China," www.EN.Wikipedia.com, Page last updated September 14, 2020. Page accessed on September 18, 2020. https://en.wikipedia.org/wiki/Nuclear_power_in_China#:~:text=As%20of%20March%202019%2C%20China,for%20an%20additional%2036%20GW.
34 Dr. Constable, John, GWPF Energy Editor, "China's Geostrategic Priorities Become Clear: Oil not Wind..." www.TheGWPF.com (The Global Warming Policy Forum), July 8, 2020. https://www.thegwpf.com/chinas-geostrategic-priorities-become-clear-oil-not-wind/
35 U.S. Energy Information Administration, Independent Statistics & Analysis, Frequently Asked Questions, "What is U.S. electricity generation by energy source?" www.EIA.gov, 2019. https://www.eia.gov/tools/faqs/faq.php?id=427&t=3
36 Jones, Barbara, "Child miners aged four living a hell on Earth so YOU can drive an electric car: Awful human cost in squalid Congo cobalt mine that Michael Grove didn't consider in his 'clean' energy crusade," www.DailyMail.co.uk, August 5, 2017. https://www.

dailymail.co.uk/news/article-4764208/Child-miners-aged-four-living-hell-Earth.html
37 Hill, Alicia, Martinez-Diaz, Leonardo, "Adapt or Perish: Preparing for the Inescapable Effects of Climate Change, Foreign Affairs, www.ForeignAffairs.com, January/February 2020. https://www.foreignaffairs.com/articles/united-states/2019-12-10/adapt-or-perish?utm_medium=newsletters&utm_source=fatoday&utm_campaign=A%20Grand%20Strategy%20of%20Resilience&utm_content=20200911&utm_term=FA%20Today%20-%20112017
38 Lomberg, Bjorn, False Alarm: How Climate Change Panic Costs Us Trillions, Hurts the Poor, and Fails to Fix the Planet, (Basic Books, New York, NY), Entire book. July 14, 2020. https://www.amazon.com/False-Alarm-Climate-Change-Trillions/dp/1541647467/ref=sr_1_1?dchild=1&keywords=False+Alarm%3A+How+Climate+Change+Panic+Costs+Us+Trillions%2C+Hurts+the+Poor%2C+and+Fails+to+Fix+the+Planet&qid=1599840737&sr=8-1
39 Shellenberger, Michael, Apocalypse Never: Why Environmental Alarmism Hurts Us All, (Harper Books, New York, NY), Entire book. June 30, 2020. https://www.amazon.com/Apocalypse-Never-Environmental-Alarmism-Hurts/dp/0063001691/ref=sr_1_1?crid=4E743XN04YM1&dchild=1&keywords=michael+shellenberger&qid=1599840970&sprefix=mchael+she%2Caps%2C176&sr=8-1
40 The Editorial Board, "Big Oil to the Coronavirus Rescue: Look whose products are crucial for fighting off Covid-19," www.WSJ.com, April 23, 2020. https://www.wsj.com/articles/big-oil-to-the-coronavirus-rescue-11587683239
41 New Zealand Ministry for the Environment, First national climate change risk assessment for New Zealand, www.MFE.govt.nz, August 2020. https://www.mfe.govt.nz/climate-change/assessing-climate-change-risk
42 Wojick, David, "Cascading fallacies in climate risk assessment," www.cfact.org, August 10, 2020. https://www.cfact.org/2020/08/10/cascading-fallacies-in-climate-risk-assessment/
43 American Public Transportation Association (APTA), Public Transportation Fact Book, www.APTA.com, website accessed on September 10, 2020. https://www.apta.com/research-technical-resources/transit-statistics/public-transportation-fact-book/

44 Destinvil, Handel, "Obama Administration Introduces New Administrative Rule on Fair Housing," www.AmericanBar.org, (American Bar Association), August 13, 2015. https://www.americanbar.org/groups/litigation/committees/minority-trial-lawyer/practice/2015/obama-administration-introduces-new-administrative-rule-fair-housing/
45 Gates, Bill, "COVID-19 is awful. Climate change could be worse," GatesNotes The Blog of Bill Gates, Climate and the Coronavirus section, www.Gatesnotes.com, August 4, 2020. https://www.gatesnotes.com/Energy/Climate-and-COVID-19?WT.mc_id=20200804100000_COVID19-and-Climate_BG-TW_&WT.tsrc=BGTW
46 Clean Energy Alliance, "Climate Fraud," YouTube.com video presentation. www.CleanEnergyAlliance.com, Video accessed on September 11, 2020. https://clearenergyalliance.com/project/climate-fraud/
47 J.P. Morgan, Eye On The Market, Energy Outlook 2020, Tenth Annual Energy Paper, www.JPMorgan.com, June 2020. https://www.jpmorgan.com/jpmpdf/1320748699400.pdf
48 Kotkin, Joel, "Blackouts to Fires: California's Summer Attractions," www.RealClearEnergy.com, August 25, 2020. https://www.realclearenergy.org/articles/2020/08/25/blackouts_and_fires_californias_summer_attractions_575221.html
49 Dears, Donn, "…It's Time to Abandon Wind Power…" Power for USA, Energy Facts From Oil To Electricity, www.DDears.com, July 21, 2020. https://ddears.com/2020/07/21/its-time-to-abandon-wind-power/
50 Viv Forbes. Ibid. 2020.
51 Bolinger, Mark, Gorman, Will, Millstein, Dev, Jordan, Dirk, "System-level performance and degradation of 21 GW of utility-scale PV plants in the United States," Journal of Renewable Sustainable Energy 12, 043501, www.AIP.scitation.org, July 7, 2020. https://aip.scitation.org/doi/pdf/10.1063/5.0004710
52 I & I Editorial Board, "Will America's Return To Nuclear Power Kill The Dems' Green New Deal? Let's Hope So," www.IssuesInsights.com, September 9, 2020. https://issuesinsights.com/2020/09/09/will-americas-return-to-nuclear-power-kill-the-dems-green-new-deal-lets-hope-so/
53 Pereira, Dioga Santos, Marques, Antonio Cardoso, Fuinhas, Jose Alberto, "Are renewables affecting income distribution and increasing

the risk of household poverty," www.Ideas.Repec.org, 2019. https://ideas.repec.org/a/eee/energy/v170y2019icp791-803.html

54 StopTheseThings.com, "How Green is This? Millions of Toxic Solar Panels & Wind Turbine Blades Destined for Landfill," www.StopTheseThings.com, September 12, 2020. https://stopthesethings.com/2020/09/12/how-green-is-this-millions-of-toxic-solar-panels-wind-turbine-blades-destined-for-landfill/

55 StopTheseThings.com. September 12. 2020. Ibid.

56 StopTheseThings.com. September 12, 2020. Ibid.

57 Kapur, Devesh, "India and China are edging towards a more serious conflict," www.FT.com, September 12, 2020. https://www.ft.com/content/ecd85303-9818-45c7-9e9d-753e88f7b7f5

58 Devesh. 2020. Ibid.

59 Friedberg, Aaron L.,"An Answer to Aggression: How to Push Back Against Beijing," www.ForeignAffairs.com, September/October 2020. https://www.foreignaffairs.com/articles/china/2020-08-11/ccp-answer-aggression?utm_medium=newsletters&utm_source=fatoday&utm_campaign=An%20Answer%20to%20Aggression&utm_content=20200915&utm_term=FA%20Today%20-%20112017

60 Ring, Edward, "Environmentalists Destroyed California's Forests: The catastrophic fires that have immolated millions of acres of forests in the Golden State were preventable, and for decades everyone knew what had to be done," www.AmericanGreatness.com, September 9, 2020. https://amgreatness.com/2020/09/09/environmentalists-destroyed-californias-forests/

61 Shellenberger. Entire book Climate Apocalypse. Ibid. 2020. Mr. Shellenberger an ardent environmentalist advocates nuclear energy to electricity in this books, articles in Forbes Magazine/website, testimony before the U.S. Congress, and his work with the United Nations.

62 World Nuclear Association, Safety of Nuclear Power Reactors, December 2020, https://www.world-nuclear.org/information-library/safety-and-security/safety-of-plants/safety-of-nuclear-power-reactors.aspx#:~:text=The%20evidence%20over%20six%20decades,with%20other%20commonly%20accepted%20risks

63 World Nuclear Organization, Plans For New Reactors Worldwide, January 2021), https://www.world-nuclear.org/information-library/

current-and-future-generation/plans-for-new-reactors-worldwide.aspx
64 World Nuclear Organization, Nuclear-Powered Ships, November 2020, https://www.world-nuclear.org/information-library/non-power-nuclear-applications/transport/nuclear-powered-ships.aspx
65 Stein and Royal. Ibid. Just Green Electricity: Helping Citizens Understand a World Without Fossil Fuels. Chapter One. 2020.
66 Sierra Club | Beyond Coal, "America, Let's Move Beyond Coal," www.Coal.SierraClub.org, 2019. Website accessed September 14, 2020. https://coal.sierraclub.org
67 Shellenberger, Michael, "Stop Letting Your Ridiculous Fears of Nuclear Waste Kill The Planet," www.Forbes.com, June 19, 2018. https://www.forbes.com/sites/michaelshellenberger/2018/06/19/stop-letting-your-ridiculous-fears-of-nuclear-waste-kill-the-planet/#1b7b292b562e
68 Royal, Todd, "Questioning Global Warming," www.AmericanThinker.com, August 13, 2019. https://www.americanthinker.com/articles/2019/04/questioning_global_warming.html
69 StopTheseThings.com, "Simply Staggering: What Weather Dependent Wind & Solar Really Costs," www.StopTheseThings.com, June 8, 2020. https://stopthesethings.com/2020/07/27/simply-staggering-what-weather-dependent-wind-solar-really-costs/
70 CO2 Coalition, Carbon Dioxide Is Essential For Life: Learn the facts about the vital role that CO2 plays in our environment," www.CO2Coalition.org, 2020. Website accessed September 14, 2020. https://co2coalition.org
71 Ring, Edward, "As California's Mismanaged Forests Burn, Governor Newsome Declares 'No More Patience for Climate Deniers,'" www.CaliforniaGlobe.com, September 14, 2020. https://californiaglobe.com/section-2/as-californias-mismanaged-forests-burn-gov-newsom-declares-no-more-patience-for-climate-deniers/
72 Ring. Ibid. 2020.
73 British Petroleum (BP), British Petroleum Statistical Review of World Energy 2020, 69th Edition, Released June 2020. https://www.bp.com/en/global/corporate/energy-economics/statistical-review-of-world-energy.html

CHAPTER ONE: CALIFORNIA

74 Morano, Mark, Climate Depot, No American politician has killed more clean energy than Governor Brown' – 'Benefited his own family financially', January 11, 2018, https://www.climatedepot.com/2018/01/11/report-no-american-politician-has-killed-more-clean-energy-than-gov-brown-benefited-his-own-family-financially/

75 Critical Masses: Opposition to Nuclear Power in California, 1958-1978, https://www.amazon.com/Critical-Masses-Opposition-Nuclear-California/dp/0299158500

76 Gerth, Jeff, New York Times, Governor Brown Supporting Projects That Aid a Mexican Contributor, March 11, 1979, https://www.nytimes.com/1979/03/11/archives/gov-brown-supporting-projects-that-aid-a-mexican-contributor-gov.html

77 Shellenberger, Michael, Environmental Progress, Jerry Brown's Secret War on Clean Energy, January 11, 2018, https://environmentalprogress.org/big-news/2018/1/11/jerry-browns-secret-war-on-clean-energy

78 Resources of the Future, How California Can Prevent CO2 "Leakage", Press Release, May 13, 2016,https://www.rff.org/news/press-releases/how-california-can-prevent-co2-leakage/

79 Union of Concerned Citizens, Each Country's Share of CO2 Emissions, Aug 12, 2020, https://www.ucsusa.org/resources/each-countrys-share-co2-emissions

80 Hannah Ritchie and Max Roser, Our World in Data, CO and Greenhouse Gas Emissions, August 2020, https://ourworldindata.org/co2-and-other-greenhouse-gas-emissions#how-have-global-co2-emissions-changed-over-time

81 California's 2017 Climate Change Scoping Plan, page 70, https://ww2.arb.ca.gov/sites/default/files/classic//cc/scopingplan/scoping_plan_2017.pdf?utm_medium=email&utm_source=govdelivery

82 Reuters, Copper demand for electric cars to rise nine-fold by 2027, JUNE 12, 2017, https://www.reuters.com/article/us-copper-demand-electric-vehicles-idUSKBN1940PC

83 Mills, Mark, Mines, Minerals, and "Green" Energy: A Reality Check, July 9, 2020, https://www.manhattan-institute.org/mines-minerals-and-green-energy-reality-check

84 Barrera, Priscila, Investigating News, Top Cobalt Production by Country, June 16, 2020, https://investingnews.com/daily/

resource-investing/battery-metals-investing/cobalt-investing/top-cobalt-producing-countries-congo-china-canada-russia-australia/

85 Amnesty International, Industry giants fail to tackle child labor allegations in cobalt battery supply chains, November 15, 2017, https://www.amnesty.org/en/latest/news/2017/11/industry-giants-fail-to-tackle-child-labour-allegations-in-cobalt-battery-supply-chains/ and https://www.gettyimages.com/detail/news-photo/child-breaks-rocks-extracted-from-a-cobalt-mining-at-a-news-photo/534434744?adppopup=true

86 Kreiger, Adrenne, ENVIRONMENTAL IMPACTS OF NICKEL MINING AND PROCESSING April 29, 2018, https://everlingjewelry.com/blogs/news/environmental-impacts-of-nickel-mining-and-processing

87 ROB NIKOLEWSKI, Rob, San Diego Tribune, California new vehicle sales cool in 2016, but still top 2 million, FEB. 23, 2017, https://www.sandiegouniontribune.com/business/sd-fi-car-sales-20170223-story.html

88 Marathon News Release, August 3, 2020, https://ir.marathonpetroleum.com/investor/news-releases/news-details/2020/Marathon-Petroleum-Corp.-Reports-Second-Quarter-2020-Results/default.aspx

89 Gas Prices, https://gasprices.aaa.com/state-gas-price-averages/

90 The Airport Authority, List of all airports in California, USA, https://airport-authority.com/browse-US-CA

91 U.S. Energy Information Administration, Jet Fuel Consumption by State 2018, https://www.eia.gov/state/seds/data.php?incfile=/state/seds/sep_fuel/html/fuel_jf.html&sid=CA

92 Federal Highway Administration, State Motor-Vehicle Registrations – 2018, Dec 2019, https://www.fhwa.dot.gov/policyinformation/statistics/2018/mv1.cfm

93 AERA Energy, The numbers add up: California counts on oil and gas, July 16, 2019, https://www.aeraenergy.com/the-numbers-add-up-california-counts-on-oil-and-gas/

94 700 Refineries Supply Oil Products to the World https://hrcak.srce.hr/file/65010

95 Todd, Felix, California imports more electricity than any other US state, says EIA, April 5, 2019, https://www.nsenergybusiness.com/news/california-electricity-imports-us/

96 California Energy Commission, Crude Oil Supply Sources To California Refineries, https://www.energy.ca.gov/data-reports/energy-almanac/californias-petroleum-market/oil-supply-sources-california-refineries
97 700 Refineries Supply Oil Products to the World https://hrcak.srce.hr/file/65010
98 700 Refineries Supply Oil Products to the World https://en.wikipedia.org/wiki/List_of_oil_refineries
99 Plumer, Bird, Washington Post, All of the world's power plants, in one handy map, December 8, 2012, https://www.washingtonpost.com/news/wonk/wp/2012/12/08/all-of-the-worlds-power-plants-in-one-handy-map/?noredirect=on
100 Coal-fired Power Stations by Region, Global Coal Plant Tracker, January 2021, https://docs.google.com/spreadsheets/d/1ZPbbwBI1cNoS1NqFEnN8PdXGRO0xuYfza1F0AWUDRuY/edit#gid=739846511
101 Coal-fired Power Stations by Country, Global Coal Plant Tracker, January 2021, https://docs.google.com/spreadsheets/d/1kXtAw6QvhE14_KRn5lnGoVPsHN3fDZHVMlvz_s_ch1w/edit#gid=165011444
102 Worldometer, China population, https://www.worldometers.info/world-population/china-population/
103 India population 2020, https://worldpopulationreview.com/countries/india-population
104 Just Green Electricity - Helping Citizens Understand a World Without Fossil Fuels, https://www.amazon.com/Just-Green-Electricity-Citizens-Understand/dp/1480890707
105 Specht, Mark, Energy Analyst, Union of Concerned Scientists, February 25, 2019, Natural Gas Power Plant Retirements in California, https://blog.ucsusa.org/mark-specht/gas-retirements-california
106 Roth, Sammy, Los Angeles Times, September 1, 2020, California to let gas plants stay open as time runs low for climate action, https://www.latimes.com/environment/story/2020-09-01/california-gas-plants-stay-open-time-runs-low-for-climate-action
107 AES Redondo Beach, http://www.aescalifornia.com/facilities/redondo-beach#:~:text=AES%20Redondo%20Beach%20is%20an,million%20California%20homes%20and%20businesses
108 Symon, Evan, March 31, 2020, Redondo Beach Waterfront AES Gas Power Plant to Close by 2023, https://californiaglobe.com/section-2/redondo-beach-waterfront-aes-ga

s-power-plant-to-close-by-2023/#:~:text=On%20Monday%2C%20AES%20Corp.,number%20of%20mixed%2Duse%20projects

109 Roth, Sammy, Los Angeles Times, February 12, 2019, How will L.A. replace three gas plants that Mayor Eric Garcetti plans to shut down? https://www.latimes.com/business/la-fi-garcetti-dwp-gas-plants-20190212-story.html

110 O'Brien, Chip, To Dam or Un-Dam California? Now That's a Complicated Question, https://activenorcal.com/to-dam-or-un-dam-california-now-thats-a-complicated-question/

111 California Energy Commission, 2018 Total System Electric Generation, https://www.energy.ca.gov/data-reports/energy-almanac/california-electricity-data/2019-total-system-electric-generation/2018

112 Gough, Matt, Sierra Club, California's Cities Lead the Way to a Gas-Free Future,July 24, 2020, https://www.sierraclub.org/articles/2020/07/californias-cities-lead-way-gas-free-future

113 Nemec, Rich, Natural Gas Intelligence, California Regulators Approve Natural Gas Bans Across State, December 12, 2019, https://www.naturalgasintel.com/california-regulators-approve-natural-gas-bans-across-state/

114 Local Government Decarbonization Ordinances, http://www.buildingdecarb.org/active-code-efforts.html

115 U.S. Energy Information Administration, Natural gas generators make up the largest share of overall U.S. generation capacity, December 18, 2017, https://www.eia.gov/todayinenergy/detail.php?id=34172

116 Weise, Elizabeth, USA Today, "As Earth faces climate catastrophe, US set to open nearly 200 power plants", September 9, 2019, https://www.usatoday.com/story/news/2019/09/09/climate-change-threatens-earth-us-open-nearly-200-power-plants/2155631001/

117 U.S. Energy Information Administration, 2019, What is U.S. electricity generation by energy source? https://www.eia.gov/tools/faqs/faq.php?id=427&t=3

118 Washington state utility report of July 14, 2020 titled "Wind Power and Clean Energy Policy Perspectives," PDF of the 16 page report dated July 14, 2020: https://www.bentonpud.org/getattachment/Safety-Education/Safety/Wind/Wind-Power-and-Clean-Energy-Policy-Perspectives-Report-Benton-PUD-FINAL-July-14-2020-(1).PDF.aspx?lang=en-US

119 New Straits Times, Japan removes two last wind power turbines, December 18, 2020, https://www.nst.com.my/amp/world/region/2020/12/650410/japan-removes-two-last-wind-power-turbines?fbclid=IwAR1tTHSjT7WEv84jzznLwB-FYkTbfyW4U6JhJ0SVfN4pSwVUHULDUncp0i0
120 Renewable Energy Policy, California, USA, https://www.ourenergypolicy.org/wp-content/uploads/2013/08/California-Renewable-Energy-Information.pdf
121 Mathis, Mark, Clean Energy Alliance, 2020, https://www.youtube.com/watch?v=ViOBqC7w-1Q&feature=youtu.be
122 Mayeda, Patrick and Riener, Kenneth In cooperation with: Pacific Gas & Electric Company, Economic Benefits of Diablo Canyon Power Plant An Economic Impact Study, June 2013, https://www.pge.com/includes/docs/pdfs/shared/edusafety/systemworks/dcpp/PGE_Economic_Impact_Report_Final.pdf
123 William Wilkes, Hayley Warren and Brian Parkin, Bloomberg, Germany's Failed Climate Goals A Wake-Up Call for Governments Everywhere, August 15, 2018, https://www.bloomberg.com/graphics/2018-germany-emissions/
124 Ellen Thalman, Benjamin Wehrmann, Clean Energy Wire, What German households pay for power, January 24, 2020, https://www.cleanenergywire.org/factsheets/what-german-households-pay-power
125 JoNova, New Report: Australians pay $1300 in hidden climate bills each year, http://joannenova.com.au/2020/08/new-report-australians-pay-1300-hidden-climate-bill-each-year/
126 William Wilkes, Hayley Warren and Brian Parkin, Bloomberg, Germany's Failed Climate Goals A Wake-Up Call for Governments Everywhere, August 15, 2018, https://www.bloomberg.com/graphics/2018-germany-emissions/
127 William Wilkes, Hayley Warren and Brian Parkin, Bloomberg, Germany's Failed Climate Goals A Wake-Up Call for Governments Everywhere, August 15, 2018, https://www.bloomberg.com/graphics/2018-germany-emissions/
128 Security.org, 32 Cities with the Biggest Homeless Problems, https://www.security.org/resources/homeless-statistics/
129 KATE CIMINI and JACKIE BOTTS, THE CALIFORNIAN, Close quarters: Overcrowded homes fuel spread of coronavirus among workers, June 21, 2020, https://www.mercurynews.com/2020/06/21/

close-quarters-overcrowded-homes-fuel-spread-of-coronavirus-among-workers/

130 Grimes, Katy, California Globe, California Energy Prices Continue to Rise Higher Than Other States, August 6, 2020, https://californiaglobe.com/section-2/california-gas-and-energy-prices-continue-to-rise-higher-than-other-states/

131 Wolfe, Tom, The Electric Kool-Aid Acid Test, 1968, https://www.amazon.com/Electric-Kool-Aid-Acid-Test/dp/031242759X

132 Mark Nelson and Michael Shellenberger, Environmental Progress, Electricity prices in California rose three times more in 2017 than they did in the rest of the United States, February 12, 2018, https://environmentalprogress.org/big-news/2018/2/12/electricity-prices-rose-three-times-more-in-california-than-in-rest-of-us-in-2017

133 Bryce, Robert, Manhattan Institute, How California Promotes Energy Poverty, August 3, 2015, https://www.manhattan-institute.org/html/how-california-promotes-energy-poverty-6168.html

134 Council of Europe Development Bank (CEB), Energy Poverty in Europe https://coebank.org/media/documents/CEB_Study_Energy_Poverty_in_Europe.pdf

135 Voelcker, John, Green Car Reports, July 29, 2014, https://www.greencarreports.com/news/1093560_1-2-billion-vehicles-on-worlds-roads-now-2-billion-by-2035-report

136 MARTIN ZIMMERMAN AND MAURA DOLAN, Los Angeles Times, The end of an era in California automaking, AUG. 28, 2009, https://www.latimes.com/archives/la-xpm-2009-aug-28-fi-toyota-plant28-story.html

137 The Para Rigger, Top Ten Car Producing Countries - 1950 - 2019, July 19, 2020, https://para-rigger.posthaven.com/top-ten-car-producing-countries-1950-2019

138 Worldometer, China population, https://www.worldometers.info/world-population/china-population/

139 India population 2020, https://worldpopulationreview.com/countries/india-population

140 U.S. Energy Information Administration, California was the largest net electricity importer of any state in 2019, December 7, 2020, https://www.eia.gov/todayinenergy/detail.php?id=46156#

141 IANNELLI, JERRY, Miami New Times, Florida Leads Nation in Residents Without Power Over Past Decade, July 30, 2019, https://

www.miaminewtimes.com/news/florida-leads-nation-in-power-outages-11230329
142 U.S. Energy Information Administration, California Price Differences from U.S. average, https://www.eia.gov/state/?sid=CA#tabs-5
143 Johnson, Hans, Public Policy Institute of California, California's Population, April 2020, https://www.ppic.org/publication/californias-population/
144 Hellerstein, Erica, Cal Matters, More than half of Latinos in California struggle to stay afloat, report finds, November 2019, https://calmatters.org/california-divide/2019/10/more-than-half-of-latinos-in-california-struggle-to-stay-afloat-report-finds/
145 Sherter, Alain, CBS News, Nearly 40% of Americans can't cover a surprise $400 expense, May 23, 2019, https://www.cbsnews.com/news/nearly-40-of-americans-cant-cover-a-surprise-400-expense/
146 Duke Stanford, Center for Child and Family Policy, A Third of U.S. Families Face a Different Kind of Poverty, January 6, 2021, https://childandfamilypolicy.duke.edu/news-release/a-third-of-u-s-families-face-a-different-kind-of-poverty/#:~:text=Center%20for%20Child%20%26%20Family%20Policy%20%7C%20Duke%20University&text=In%202019%2C%2057%20percent%20of,in%20terms%20of%20net%20worth.
147 700 Refineries Supply Oil Products to the World https://hrcak.srce.hr/file/65010
148 California Energy Commission, Crude Oil Supply Sources To California Refineries, https://www.energy.ca.gov/data-reports/energy-almanac/californias-petroleum-market/oil-supply-sources-california-refineries
149 Specht, Mark, Energy Analyst, Union of Concerned Scientists, February 25, 2019, Natural Gas Power Plant Retirements in California, https://blog.ucsusa.org/mark-specht/gas-retirements-california
150 Governing California Through Climate Change https://lhc.ca.gov/sites/lhc.ca.gov/files/Reports/221/Report221.pdf
151 National Interagency Fire Center, Total Wildland Fires and Acres (1926-2019), https://www.nifc.gov/fireInfo/fireInfo_stats_totalFires.html
152 Fisher, Rohan, The Conversation, The world's best fire management system is in northern Australia, and it's led by Indigenous land managers, March 10, 2020, https://theconversation.com/the-worlds-best-fire-management-system-is-in-northern-australia-and-its-led-by-indigenous-land-managers-133071

153 Kotlin, Joel, American Affairs, Neo Feudalism in California, May 20, 2020, https://americanaffairsjournal.org/2020/05/neo-feudalism-in-california/

154 Fire on the Mountain: Rethinking Forest Management in the Sierra Nevada, February 2018, https://lhc.ca.gov/report/fire-mountain-rethinking-forest-management-sierra-nevada

155 Fire on the Mountain: Rethinking Forest Management in the Sierra Nevada, February 2018, https://lhc.ca.gov/report/fire-mountain-rethinking-forest-management-sierra-nevada

156 Johnson, Nathanael, Grist,150 million trees died in California's drought, and worse is to come, July 3 2019, https://grist.org/article/150-million-trees-died-in-californias-drought-and-the-worst-is-to-come/

157 Fire on the Mountain: Rethinking Forest Management in the Sierra Nevada, February 2018, https://lhc.ca.gov/report/fire-mountain-rethinking-forest-management-sierra-nevada

158 Kotlin, Joel, Orange County Register, Unsustainable California, November 9, 2019, https://www.ocregister.com/2019/11/09/unsustainable-california-joel-kotkin/

159 California Energy Commission, Weather and Climate Informatics for the Electricity Sector, https://ww2.energy.ca.gov/2020publications/CEC-500-2020-039/CEC-500-2020-039.pdf

160 Vranich, Joseph, San Francisco Business Times, California exodus roared on even before Covid hit, https://www.bizjournals.com/sanfrancisco/news/2020/11/17/california-exodus-roared-on-even-before-covid-hit.html

161 Gas Buddy, Gas Price Charts, https://www.gasbuddy.com/charts

162 American Petroleum Institute, Notes to State Motor Fuel Excise and Other Taxes, January 2021, https://www.api.org/-/media/Files/Statistics/State-Motor-Fuel-Notes-Summary-january-2021.pdf?la=en&hash=9F058B51CBF13367AA91F3D98C4CEDDA15049805

163 Leigh Noda, Stillwater Associates, Sacramento policymakers drive California's high gasoline prices, February 1, 2021, https://stillwaterassociates.com/sacramento-policymakers-drive-californias-high-gasoline-prices/

164 Stein, Ronald, Fox & Hounds, May 17, 2018, Legislatures Kill Transparency Pricing at the Pump, https://www.foxandhoundsdaily.com/2018/05/legislatures-kill-transparency-pricing-pump/

165 Bliss, Steven, Public Policy Institute of California (PPIC), California Poverty Measure Finds 8.1 Million Poor–More Than Official Estimate, https://www.ppic.org/press-release/california-poverty-measure-finds-8-1-million-poor-more-than-official-estimate/
166 ERIN BALDASSARI, Erin, Bay Area News Group, Bay Area workers commuting from edges of 'megaregion, June 29, 2016, https://www.mercurynews.com/2016/06/29/bay-area-workers-commuting-from-edges-of-megaregion-new-report-says/
167 California Energy Commission, 2018 Total System Electric Generation, https://www.energy.ca.gov/data-reports/energy-almanac/california-electricity-data/2019-total-system-electric-generation/2018
168 U.S. Energy Information Administration, Overview, https://www.eia.gov/state/analysis.php?sid=CA
169 California Energy Commission, 2018 Total System Electric Generation, https://www.energy.ca.gov/data-reports/energy-almanac/california-electricity-data/2019-total-system-electric-generation/2018
170 U.S. Energy Information Administration, California imports the most electricity from other states, April 4, 2019, https://www.eia.gov/todayinenergy/detail.php?id=38912
171 Olson, Arne, Long-Run Resource Adequacy under Deep Decarbonization Pathways for California, SB100 Technologies & Scenarios Workshop, November 18, 2019, https://efiling.energy.ca.gov/GetDocument.aspx?tn=230840
172 TONY BARBOZA, Phil Willon, Los Angeles Times, Newsom orders 2035 phaseout of gas-powered vehicles, calls for fracking ban, September 23, 2020, https://www.latimes.com/california/story/2020-09-23/gavin-newsom-fracking-ban-california-zero-emissions-cars
173 Stein, Ronald, CFSACT.org, California to replace Cuba as vintage car capital of the world, September 25, 2020, https://www.cfact.org/2020/09/25/phasing-out-internal-combustion-could-make-california-vintage-car-capital-of-the-world/
174 Frank, Stephen, California Political Review, The Darker Side of Green Technology, August 28, 2020, http://www.capoliticalreview.com/capoliticalnewsandviews/the-darker-side-of-green-technology/
175 CARMAX, 2017 Hybrid & Electric Cars Survey Results, July 18, 2017, https://www.carmax.com/articles/hybrid-electric-2017-survey-results

176 Kandrach, Matthew, Real Clear Energy, EV Subsidies Funnel Taxpayer Money to the Rich, February 12, 2019, https://www.realclearenergy.org/articles/2019/02/12/ev_subsidies_funnel_taxpayer_money_to_the_rich_110392.html

177 Public Policy Institute of California, California's population is diverse, https://www.ppic.org/publication/californias-population/#:~:text=No%20race%20or%20ethnic%20group,the%202018%20American%20Community%20Survey.

178 Mitchell, Russ, Los Angeles Times, Car buyers shun electric vehicles not named Tesla. Are carmakers driving off a cliff?, January 17, 2020, https://www.latimes.com/business/story/2020-01-17/ev-sales-fizzle#:~:text=Despite%20the%20debut%20of%2045,the%20United%20States%20in%202019.https://www.latimes.com/business/story/2020-01-17/ev-sales-fizzle#:~:text=Despite%20the%20debut%20of%2045,the%20United%20States%20in%202019.

179 Mitchell, Russ, Los Angeles Times, Car buyers shun electric vehicles not named Tesla. Are carmakers driving off a cliff?, January 17, 2020, https://www.latimes.com/business/story/2020-01-17/ev-sales-fizzle

180 Nikolewski, Rob, The San Diego Union-Tribune, California vehicle sales exceed 2 million for third straight year, February 22, 2018, https://www.sandiegouniontribune.com/business/energy-green/sd-fi-car-sales-20180222-story.html

181 Federal Highway Administration, State Motor-Vehicle Registrations – 2018, https://www.fhwa.dot.gov/policyinformation/statistics/2018/mv1.cfm

182 Lee, Morgan, The San Diego Union-Tribune, State's gasoline use is on the rise, December 2, 2014, https://www.sandiegouniontribune.com/sdut-california-burns-more-gasoline-2014dec02-story.html

183 Lightman, David, The Sacramento Bee, Californians backed a gas tax to fix roads. What will coronavirus mean for highway projects?, April 28, 2020, https://www.sacbee.com/news/local/transportation/article242235946.html#:~:text=In%202017%2D18%2C%20state%20gasoline%20tax%20revenue%20was%20%246.4%20billion.&text=In%20California%2C%20motor%20fuel%20taxes,multi%2Dmodal%20congestion%20relief%20corridors.

184 Baldassari, Erin, The Mercury News, Gas tax: What Californians actually pay on each gallon of gas, Hint: It's more than you think., September 4, 2018, https://www.mercurynews.com/2018/09/04/gas-tax-what-you-actually-pay-on-each-gallon-of-gas/

185 Carney Kristen, Cubit Planning, Road Miles by State, February 9, 2010, https://blog.cubitplanning.com/2010/02/road-miles-by-state/
186 Hopkins, Hop, Sierra Club, Racism is killing the planet, June 8, 2020, https://www.sierraclub.org/sierra/racism-killing-planet
187 Jennifer L. Hernandez and David Friedman, California, Greenhouse Gas Regulation, and Climate Change, Chapman University Center for Demographics and Policy, 2018, https://www.hklaw.com/en/insights/publications/2018/07/california-greenhouse-gas-regulation-and-climate-c
188 Reuters, Restaurant group sues California city over ban on new natural gas connections, November 21, 2019, https://www.reuters.com/article/us-usa-california-naturalgas/restaurant-group-sues-california-city-over-ban-on-new-natural-gas-connections-idUSKBN1XV2NG
189 Kotlin, Joel, Neo-Feudalism in California, https://americanaffairsjournal.org/2020/05/neo-feudalism-in-california/
190 Climate-Related Deaths Are at Historic Lows, https://ourworldindata.org/natural-disasters
191 Jennifer L. Hernandez and David Friedman, HK Law, In the Name of the Environment: Litigation Abuse Under CEQA, August 2015, https://www.hklaw.com/en/insights/publications/2015/08/in-the-name-of-the-environment-litigation-abuse-un
192 Legislative Analyst's Office, California's High Housing Costs: Causes and Consequences, March 17, 2015, https://lao.ca.gov/Publications/Detail/3214
193 Walter, Dan, Cal Matter, High living costs make people poor, July 19, 2020, https://calmatters.org/commentary/dan-walters/2020/07/california-living-costs-make-people-poor/
194 Bryce, Robert, The High Cost of California Electricity Is Increasing Poverty, July 8, 2020, https://freopp.org/the-high-cost-of-california-electricity-is-increasing-poverty-d7bc4021b705
195 Jennifer L. Hernandez and David Friedman, California, Greenhouse Gas Regulation, and Climate Change, Chapman University Center for Demographics and Policy, 2018, https://www.hklaw.com/en/insights/publications/2018/07/california-greenhouse-gas-regulation-and-climate-c
196 MIT, The Future of the Electrical Grid, https://energy.mit.edu/wp-content/uploads/2011/12/MITEI-The-Future-of-the-Electric-Grid.pdf

197 Council of Europe Development Bank (CEB), Energy Poverty in Europe, https://coebank.org/media/documents/CEB_Study_Energy_Poverty_in_Europe.pdf
198 United states Census Bureau, The Supplemental Poverty Measure: 2018, OCTOBER 07, 2019, https://www.census.gov/library/publications/2019/demo/p60-268.html#:~:text=The%20SPM%20rate%20for%202018,million%20individuals%20out%20of%20poverty.
199 World Population Review, Welfare Recipients by State 2020 Gggg, https://worldpopulationreview.com/state-rankings/welfare-recipients-by-state
200 Langlois, Shawn, Market Watch, No other state comes close to California in terms of total welfare spending, November 28, 2017, https://www.marketwatch.com/story/no-other-state-comes-close-to-california-when-it-comes-to-welfare-spending-2017-11-28#:~:text=California%2C%20with%20its%20suffocating%20cost,two%20on%20the%20list%20combined
201 ELENA KRIEGER, PHD, BORIS LUKANOV, PHD, and SETH B.C. SHONKOFF, PHD, PSE, Net Zero Carbon California by 2045: What Will It Take, OCTOBER 2, 2018 https://www.psehealthyenergy.org/news/blog/net-zero-carbon-california-by-2045-what-will-it-take/
202 Gough, Matt, Sierra Club, California's Cities Lead the Way to a Gas-Free Future, July 24, 2020, https://www.sierraclub.org/articles/2020/07/californias-cities-lead-way-gas-free-future
203 Sobey, Rick, Boston Herald, Brookline bans natural gas, heating oil pipes for new buildings: 'Gas is the past', November 21, 2019, https://www.bostonherald.com/2019/11/21/brookline-bans-natural-gas-heating-oil-pipes-for-new-buildings-gas-is-the-past/
204 Mingle, Jonathan, Yale Environment, To Cut Carbon Emissions, a Movement Grows to 'Electrify Everything', APRIL 14, 2020 https://e360.yale.edu/features/to-cut-carbon-emissions-a-movement-grows-to-electrify-everything
205 Bryce, Robert, The High Cost of California Electricity Is Increasing Poverty, July 8, 2020, https://freopp.org/the-high-cost-of-california-electricity-is-increasing-poverty-d7bc4021b705
206 Stein, Ronald, The Heartland Institute, CALIFORNIA PUBLIC ADVOCATES OFFICE ENTERS A "COMMON INTEREST AGREEMENT" WITH THE SIERRA CLUB TO HELP THE CPUC BAN THE USE OF NATURAL GAS, December 2, 2020, https://

www.heartland.org/news-opinion/news/california-public-advocates-office-enters-a-common-interest-agreement-with-the-sierra-club-to-help-the-cpuc-ban-the-use-of-natural-gas
207 Todd, Felix, California imports more electricity than any other US state, says EIA, April 5, 2019, https://www.nsenergybusiness.com/news/california-electricity-imports-us/
208 Top 10 Largest Companies Based in Silicon Valley, October 2017, https://learn.stash.com/largest-companies-based-in-silicon-valley
209 Stangel, Luke, Silicon Valley Business Journal, Los Altos Hills is officially the wealthiest town in America, May 31, 2018, https://www.bizjournals.com/sanjose/news/2018/05/31/los-altos-hills-wealthiest-town-us-calif-homes.html
210 Schwarzenegger, Arnold, Protecting the Environment and Promoting Clean Energy, 2008, http://www.schwarzenegger.com/issues/milestone/protecting-the-environment-and-promoting-clean-energy
211 Baker, David, San Francisco Chronicle, California may reach 50% renewable power goal by 2020 — 10 years early, https://www.sfchronicle.com/business/article/California-may-reach-50-renewable-power-goal-by-12354313.php
212 California Legislative Information, Senate Bill No. 100, 2018, https://leginfo.legislature.ca.gov/faces/billTextClient.xhtml?bill_id=201720180SB100
213 Todd, Felix, California imports more electricity than any other US state, says EIA, April 5, 2019, https://www.nsenergybusiness.com/news/california-electricity-imports-us/

CHAPTER TWO: GERMANY

214 Four Shellenberger, Michael, Environmental Progress, Jerry Brown's Secret War on Clean Energy, January 11, 2018, https://environmentalprogress.org/big-news/2018/1/11/jerry-browns-secret-war-on-clean-energy
215 Robin McKie in New York, Interview of Dr. James Hansen, "President 'has four years to save Earth' US must take the lead to avert eco-disaster," www.TheGuardian.com, January 17, 2009. https://www.theguardian.com/environment/2009/jan/18/jim-hansen-obama
216 Folley, Aris, "Trump nominated a third time for Nobel Peace Prize," www.TheHill.com, September 29, 2020. https://thehill.com/

homenews/administration/518842-trump-nominated-a-third-time-for-nobel-peace-prize

217 Williams D., Thomas, PH.D., "Climate Apocalypse: Trump's Reelection Could Mean End Of Human Race," www.ClimateChangeDispatch.com, September 25, 2020. https://climatechangedispatch.com/climate-apocalypse-trumps-reelection-could-mean-end-of-human-race/

218 Moran, Alan, "The low emissions technology statement: a (hydrogen) bomb," www.Spectator.com.AU, September 23, 2020. https://www.spectator.com.au/2020/09/the-low-emissions-technology-statement-a-hydrogen-bomb/

219 Jordans, Frank, Associated Press, "Germany is first major economy to phase out coal and nuclear," www.ABCNews.Go.com, July 3, 2020. https://abcnews.go.com/International/wireStory/germany-finalizing-plan-phase-coal-energy-71591216According

220 Jordans. Ibid. 2020.

221 Jordans. Ibid. 2020.

222 Proctor, Darrell, "Germany Brings Last New Coal Plant Online," www.PowerMag.com, June 2, 2020. https://www.powermag.com/germany-brings-last-new-coal-plant-online/

223 Shellenberger, Michael, "The Reason Renewables Can't Power Modern Civilization Is Because They Were Never Meant To," www.Forbes.com, May 6, 2019. https://www.forbes.com/sites/michaelshellenberger/2019/05/06/the-reason-renewables-cant-power-modern-civilization-is-because-they-were-never-meant-to/?sh=119f1885ea2b

224 Shellenberger, Michael, Apocalypse Never: Why Environmental Alarmism Hurts Us All, (Harper Collins Publishers, New York, NY), www.Amazon.com, Entire book for the source, June 30, 2020. https://www.amazon.com/Apocalypse-Never-Environmental-Alarmism-Hurts/dp/0063001691/ref=sr_1_1?crid=22UXCIMDWCHI&dchild=1&keywords=apocalypse+never&qid=1607532688&sprefix=apoca%2Caps%2C198&sr=8-1

225 UNICEF, Goal: Reduce child mortality, https://sites.unicef.org/mdg/childmortality.html

226 Monckton, Christopher, "Lancing the Lancet's global-warming pustule," www.WattsUpWIthThat.com, December 5, 2020. https://wattsupwiththat.com/2020/12/05/lancing-the-lancets-global-warming-pustule/

227 L. Gosselin, Pierre, "German Electricity Imports Hit New Record, Rise 43.3 Percent in First Half Of 2020!" www.NoTricksZone.com, September 16, 2020. https://notrickszone.com/2020/09/16/german-electricity-imports-hit-new-record-rise-43-3-percent-in-first-half-of-2020/

228 McKinsey & Co., lead partner on the report, Kirsten Best-Werbunat, Energiewende-Index, www.McKinsey.de, August 2019. https://www.mckinsey.de/branchen/chemie-energie-rohstoffe/energiewende-index

229 von Dohmen, Frank, Jung, Alexander, Schultz, Stefan, Traufetter, Gerald, "German Failure on the Road to a Renewable Future," Climate Stasis section, www.Spiegel.de, May 13, 2019. https://www.spiegel.de/international/germany/german-failure-on-the-road-to-a-renewable-future-a-1266586.html

230 Hassan, Jennifer, "How long until it's too late to save Earth from climate disaster? This clock is counting down," www.WashingtonPost.com, September 21, 2020. https://www.washingtonpost.com/climate-environment/2020/09/21/climate-change-metronome-clock-nyc/

231 Perry, Mark J., "50 years of failed doomsday, eco-poclyptic predictions; the so-called 'experts' are 0-50," www.AEI.org, September 23, 2019. https://www.aei.org/carpe-diem/50-years-of-failed-doomsday-eco-pocalyptic-predictions-the-so-called-experts-are-0-50/ (Additionally, for further verification see Ron and I's first two books: Energy Made Easy & Just Green Electricity for sources verifying this sentence and the fraudulent behavior of global warming and renewable advocates and the questionability that man is warming the earth irredeemably.

232 Mathis, Mark, Video titled, "Baseload Electricity," www.CleanEnergyAlliance.com, September 21, 2020. https://clearenergyalliance.com/project/baseload-electricity/

233 Bloomberg News, "How Al Gore amassed a $200-million fortune after presidential defeat," www.FinancialPost-com.cdn.ampproject.org, May 6, 2013. https://financialpost-com.cdn.ampproject.org/c/s/financialpost.com/news/how-al-gore-amassed-a-200-million-fortune-after-presidential-defeat/wcm/da139d1a-fa74-4fac-9171-427798151a0c/amp/

234 Mills, Mark, Senior Fellow at the Manhattan Institute video for PragerU.com, "What's Wrong with Wind and Solar?" www.PragerU.com, September 14, 2020. https://www.youtube.com/watch?v=RqppRC37OgI&feature=emb_rel_pause

235 C40 Cities Media, "Mayors of 12 Major Cities to Divest From Fossil Fuel Companies, Invest in Green and Just Recovery from COVID-19 Crisis," www.C40.org, September 22, 2020. https://www.c40.org/press_releases/cities-commit-divest-invest

236 McKitrick, Ross, "Ditch the fashionable green recovery plans," www.FinancialPost.com, August 19, 2020. https://financialpost.com/opinion/ross-mckitrick-ditch-the-fashionable-green-recovery-plans

237 The Editorial Board, Wall Street Journal, "Big Oil to the Coronavirus Rescue: Look whose products are crucial for fighting off Covid-19," www.WSJ.com, April 23, 2020. https://www.wsj.com/articles/big-oil-to-the-coronavirus-rescue-11587683239

238 Pryor, Steve, "A Partial list of the over 6,000 products made from one barrel of oil (after creating 19 gallons of gasoline," www.Linkedin.com, Justly 26, 2016. https://www.linkedin.com/pulse/partial-list-over-6000-products-made-from-one-barrel-oil-steve-pryor/

239 Ranken Energy, "Products made from petroleum: With over 6,000 products and counting, petroleum continues to be a crucial requirement for all consumers," www.Ranken-Energy.com, page accessed September 23, 2020. https://www.ranken-energy.com/index.php/products-made-from-petroleum/

240 Parts of a Wind turbine https://www.horizoncurriculum.com/supportmaterial/parts-of-a-wind-turbine/

241 Parts of a Solar System https://www.cleanenergyauthority.com/solar-energy-resources/components-of-a-residential-solar-electric-system

242 Stone, Maddie, "Solar panels are starting to die. What will we do with the megaton of toxic trash?" www.Grist.org, August 13, 2020. https://grist.org/energy/solar-panels-are-starting-to-die-what-will-we-do-with-the-megatons-of-toxic-trash/

243 Folk, Emily, "Waste in the Renewable Energy Industry and How We Can Sustainably Power Our World," www.RenewableEnergyMagazine.com, March 5, 2020. https://www.renewableenergymagazine.com/emily-folk/waste-in-the-renewable-energy-industry-and-20200305

244 Benton P.U.D., Wind Power and Clean Energy Perspectives, www.BentonPUD.org, July 14, 2020. https://www.bentonpud.org/getattachment/Safety-Education/Safety/Wind/Wind-Power-and-Clean-Energy-Policy-Perspectives-Report-Benton-PUD-FINAL-July-14-2020-(1).PDF.aspx?lang=en-US

245 Virgin, Bill, "'No more wind.' WA state utility questions efficacy of wind farms for power generation," www.TheNewsTribune.com, September 19, 2020. https://www.thenewstribune.com/news/business/biz-columns-blogs/article245808810.html
246 Widdershoven, Cyril, "The Unintended Of Fossil Fuel Divestment," www.OilPrice.com, September 27, 2020. https://oilprice.com/Energy/Crude-Oil/The-Unintended-Consequences-Of-Fossil-Fuel-Divestment.html
247 Vazquez, Joseph, "Soros/Gates-Funded Org Says World May Need 'Climate Lockdown,'" www.ClimateChangeDispatch.com, September 25, 2020. https://climatechangedispatch.com/soros-gates-funded-org-says-world-may-need-climate-lockdown/
248 Guest Post by Eschenbach, Willis, "40 NASA Scientists Tell The Truth," www.WattsUpWithThat.com, September 10, 2020. https://wattsupwiththat.com/2020/09/10/49-nasa-scientists-call-bs/
249 Moore, Patrick Albert, Confessions of a Greenpeace Dropout: The Making of a Sensible Environmentalist, (Beatty Street Publishing, Vancouver, British Columbia, Canada), www.Amazon.com, page 242, November 22, 2010. https://www.amazon.com/Confessions-Greenpeace-Dropout-Sensible-Environmentalist/dp/0986480827/ref=sr_1_2?crid=7ICS4D6ZYCLK&dchild=1&keywords=confessions+of+a+greenpeace+dropout&qid=1607439210&sprefix=confessions+of+a+green%2Caps%2C177&sr=8-2
250 Davis Hanson, Victor, PhD, "When Conventional Wisdom Gets Downright Dangerous," www.NationalReview.com, September 22, 2020. https://www.nationalreview.com/2020/09/when-conventional-wisdom-gets-downright-dangerous/ch
251 Tverberg, Gail, "Why a Great Reset Based on Green Energy Isn't Possible," www.OurFiniteWorld.com, July 17, 2020. https://ourfiniteworld.com/2020/07/17/why-a-great-reset-based-on-green-energy-isnt-possible/
252 Pryor, Steve, "A Partial list of the over 6,000 products made from one barrel of oil (after creating 19 gallons of gasoline), www.Linkedin.com, July 26, 2016. https://www.linkedin.com/pulse/partial-list-over-6000-products-made-from-one-barrel-oil-steve-pryor/
253 Constable, John Ph.D., The Brink of Darkness: Britain's Fragile Power Grid, The Global Warming Policy Foundation, Briefing 47, www.TheGWPF.com, June 2020. https://www.thegwpf.org/content/uploads/2020/06/Constable-Brink-of-Darkness.pdf

254 The Emergency Email & Wireless Network, General News, Reporting by David Lawder, "Climate change since 2000 will cut U.S. growth over next 30 years," www.EmergencyMail.org, https://www.emergencyemail.org/newsemergency/anmviewer.asp?a=23228&z=1 No date given for article in September 2020. Website accessed for source on September 24, 2020.
255 Roberts, Andrew, Churchill Walking With Destiny, (Viking Press Publishing Company, New York, NY), 5th printing edition, Page 419, November 6, 2018. https://www.amazon.com/Churchill-Walking-Destiny-Andrew-Roberts/dp/1101980990/ref=sr_1_2?crid=2T3J2Z78R8F7H&dchild=1&keywords=churchill+walking+with+destiny&qid=1600971852&sprefix=churchill+%2Caps%2C186&sr=8-2
256 Roberts. Ibid. Page 420. 2018.
257 Davis Hanson, Victor, "Civilization Requires Collective Common Sense," www.Townhall.com, September 24, 2020. https://townhall.com/columnists/victordavishanson/2020/09/24/civilization-requires-collective-common-sense-n2576737
258 VOS News, Europe, "Germany Threatens Sanctions on Russia over Navalny Poisoning," www.VoaNews.com, September 6, 2020. https://www.voanews.com/europe/germany-threatens-sanctions-russia-over-navalny-poisoning
259 Steidler, Paul, "Point: Zealous Rush to Renewable Energy is Hurting America's Poor," www.InsideSource.com, July 27, 2020. https://www.insidesources.com/point-zealous-rush-to-renewable-energy-is-hurting-americas-poor/
260 Wojick, David, "California secretly struggles with renewables," www.Cfact.org, January 16, 2021. https://www.cfact.org/2021/01/16/california-secretly-struggles-with-renewables/
261 Mackinnon, Amy, "Could Navalny's Poisoning Spell the End for Nord Stream 2? Calls grow for Angela Merkel to halt the controversial Russian gas pipeline" www.ForeignPolicy.com, September 4, 2020. https://foreignpolicy.com/2020/09/04/could-navalnys-poisoning-spell-the-end-for-nord-stream-2/
262 Butler, Michael, "Michael Butler: The greatest story Pennsylvanians were never told. The U.S is leading the world in cutting air-polluting emissions," www.Post-Gazette.com, September 14, 2020. https://www.post-gazette.com/opinion/2020/09/15/Michael-Butler-The-greatest-story-Pennsylvanians-were-never-told/stories/202009150025

263 Lewis, C.S., The Abolition of Man, (Harper One, a division of Harper Collins, New York, NY), 1944 and 1977, page 77. https://www.amazon.com/Abolition-Man-C-S-Lewis/dp/0060652942/ref=sr_1_1?crid=1OGB31K0HK7HL&dchild=1&keywords=the+abolition+of+man+by+c.s.+lewis&qid=1600973925&sprefix=the+abolition%2Caps%2C204&sr=8-1
264 Gilham, Preston, Swagger: Keeping Your Wits When Others Are Not, (Bonefish Publications, 2020 Wilshire Boulevard, Fort Worth, TX., 76110), page 47, August 16, 2020. https://www.amazon.com/Swagger-Keeping-Your-Wits-Others/dp/0984510354/ref=sr_1_1?crid=11RAIMZHWUM7K&dchild=1&keywords=swagger+preston+gillham&qid=1600973969&sprefix=swagger+pre%2Caps%2C196&sr=8-1
265 Nolte, John, "Blackouts force Newsome to admit green energy falls short," www.Breitbart.com, August 17, 2020. https://www.breitbart.com/politics/2020/08/17/nolte-blackouts-force-newsom-to-admit-green-energy-falls-short/
266 Travers, Mike, "The Hidden Cost of Net-Zero: Rewiring the UK," www.TheGWPF.org, (The Global Warming Policy Foundation), July 2020. https://www.thegwpf.org/content/uploads/2020/07/Travers-Net-Zero-Distribution-Grid-Replacement.pdf
267 European Environment Agency (EEA), The European environment – state and outlook 2020: knowledge for transition to a sustainable Europe, www.EEA.Europa.EU, June 8, 2020. https://www.eea.europa.eu/soer/2020/intro
268 Legates, David, "It's Not About the Climate – It Never Was," www.Townhall.com, March 1, 2019. https://townhall.com/columnists/davidlegates/2019/03/01/its-not-about-the-climateit-never-was-n2542428
269 Katwala, Amit, Wired on Energy, "The spiraling environmental cost of our lithium battery addiction," www.Wired.Co.UK, August 5, 2018. https://www.wired.co.uk/article/lithium-batteries-environment-impact
270 MINING.com, "Trump Signs Emergency Order To Bolster Rare Earth Mining," www.OilPrice.com, October 1, 2020. https://oilprice.com/Energy/Energy-General/Trump-Signs-Emergency-Order-To-Bolster-Rare-Earth-Mining.html
271 Chupka, Marc, Vice President of Research & Programs at the U.S. Energy Storage Association (ESA), End-of-Life Management of

Lithium-ion Energy Storage Systems, www.EnergyStorage.org, April 22, 2020. https://energystorage.org/wp/wp-content/uploads/2020/04/ESA-End-of-Life-White-Paper-CRI.pdf
272 Jamasmie, Cecilia, "Demand for battery metals, renewables fueling rights violations – report," www.Mining.com, September 5, 2019. https://www.mining.com/demand-for-battery-metals-renewables-fuelling-human-rights-abuse-report/
273 Ring, Edward, "Saving the Golden State will Save America," www.AmGreatness.com, October 1, 2020. https://amgreatness.com/2020/10/01/saving-the-golden-state-will-save-america/
274 Ring. Ibid. 2020.
275 Catanoso, Justin, "Are forests the new coal? Global alarm sounds as biomass burning surges," www.News.Mongabay.com, August 31, 2020. https://news.mongabay.com/2020/08/are-forests-the-new-coal-global-alarm-sounds-as-biomass-burning-surges/
276 Catanoso. Ibid. 2020.
277 Kulhmann, Wolfgang, Putt, Pegg, Are Forests the New Coal? A Global Threat Map of Biomass Energy Development, November 2018. www.EnvironmentalPaper.org, (Environmental Paper Network). https://environmentalpaper.org/wp-content/uploads/2018/11/Threat-Map-Briefing-Are-Forests-the-New-Coal-01.pdf
278 Catanoso. Ibid. 2020.
279 How Green Is My Industrial Wind Turbine? https://climatism.blog/2016/03/27/how-green-is-my-industrial-wind-turbine/
280 Shellenberger, Michael, "New Michael Moore-Backed Documentary On YouTube Reveals Massive Ecological Impacts Of Renewables," www.Forbes.com, April 21, 2020. https://www.forbes.com/sites/michaelshellenberger/2020/04/21/new-michael-moore-backed-documentary-on-youtube-reveals-massive-ecological-impacts-of-renewables/#6d6ccaa36c96
281 StopTheseThings.com, "Blood & Gore: Mike Moore's 'Planet of The Humans' Unmasks The Power & Money Behind Renewables Scam," www.StopTheseThings.com, April 25, 2020. https://stopthesethings.com/2020/04/25/blood-gore-mike-moores-planet-of-the-humans-unmasks-the-power-money-behind-renewables-scam/
282 StopTheseThings.com, "How Much CO2 Gets Emitted to Build a Wind Turbine?" www.StopTheseThings.com, August 16, 2014. https://stopthesethings.com/2014/08/16/how-much-co2-gets-emitted-to-build-a-wind-turbine/

283 StopTheseThings.com, "Slash & Burn: Environmentalists Start Counting Destructive Costs of Wind & Solar," www.StopTheseThings.com, September 25, 2020. https://stopthesethings.com/2020/09/25/slash-burn-environmentalists-start-counting-destructive-costs-of-wind-solar/

284 Gosselin, Pierre, "Green Dream Arrives In Germany! But Repowering Obstacles Pose "Imminent Catastrophe" For Wind Power," www.NoTricksZone.com, September 9, 2020. https://notrickszone.com/2020/09/09/green-dream-arrives-in-germany but-repowering-obstacles-pose-imminent-catastrophe-for-wind-power/

285 StopTheseThings.com, "Grinding Halt: Germany's Wind Industry Faces Armageddon As Turbine Running Costs Escalate," www.StopTheseThings.com, September 15, 2020. https://stopthesethings.com/2020/09/15/grinding-halt-germanys-wind-industry-faces-armageddon-as-turbine-running-costs-escalate/

286 Gosselin. Ibid. 2020.

287 Sonter, L.J., Dade, M.C., Watson, J.E.M. et al. Renewable energy production will exacerbate mining threats to biodiversity. Nat Commun 11, 4174 (2020). https://doi.org/10.1038/s41467-020-17928-5 https://www.nature.com/articles/s41467-020-17928-5#citeas (The formatting of this end note was different than the others per Nature Communications sourcing formatting request.)

288 Sonter, et al., Ibid. 2020.

289 Sonter, et al., Ibid. 2020.

290 Perry, Mark J., "Six facts about the non-problem of global warming," www.AEI.org, January 20, 2020. https://www.aei.org/carpe-diem/six-facts-about-the-non-problem-of-global-warming/

291 Harianne, Atte, Aalto University, Korhonen M., Ianne, Abandoning the concept of renewable energy, Energy Policy (127, pp.330-340), www.ResearchGate.net, December 2018. https://www.researchgate.net/publication/329964319_Abandoning_the_concept_of_renewable_energy

292 Pullella, Philip, "Pope says free market, 'trickle-down' policies fail society," www.Reuters.com, October 4, 2020. https://www.reuters.com/article/us-pope-encyclical-idUSKBN26P0E1?taid=5f79a7b6f056530001bf9b51&utm_campaign=trueAnthem:+Trending+Content&utm_medium=trueAnthem&utm_source=twitter

293 AllThatsInteresting.com, "20 Powerful Quotes By Pope Francis On Climate Change And The Environment," www.

AllThatsInteresting.com, June 18, 2015. https://allthatsinteresting.com/pope-francis-climate-change-quotes

294 Mills, Mark P., Senior Fellow at the Manhattan Institute, YouTube.com video for PragerU, "What's Wrong with Wind and Solar," www.YouTube.com via www.PragerU.com, September 14, 2020. https://www.youtube.com/watch?v=RqppRC37OgI&feature=emb_rel_pause (PLEASE NOTE: Entire lengthy paragraph is based on this source. Thank you.)

295 Fairless, Tom, "China, Once Germany's Partner in Growth Turns Into a Rival," www.WSJ.com, September 17, 2020. https://www.wsj.com/articles/china-once-germanys-partner-in-growth-turns-into-a-rival-11600338663

296 Mills. Ibid. PragerU video 2020.

297 Mills. Ibid. PragerU video 2020.

298 Mills. Ibid. PragerU video 2020.

299 Husock, Howard, "New York Politicians Decide to Keep a Brooklyn Neighborhood Poor: A proposed waterfront redevelopment would have created thousands of jobs and cost the city nothing," www.WSJ.com, October 2, 2020. https://www.wsj.com/articles/new-york-politicians-decide-to-keep-a-brooklyn-neighborhood-poor-11601678296

300 Reuters Staff, "Vietnam city says approves LNG project to be developed by Exxon Mobil," www.Reuters.com, October 2, 2020. https://www.reuters.com/article/us-vietnam-exxon-mobil-lng/vietnam-city-says-approves-lng-project-to-be-developed-by-exxon-mobil-idUSKBN26N1KZ

301 Romei, Valentina, Reed, John, "The Asian century is set to begin," www.FT.com, March 25, 2019. https://www.ft.com/content/520cb6f6-2958-11e9-a5ab-ff8ef2b976c7

302 Reuters Staff. Ibid. 2020.

303 Wiggins, Brandon, "The World's Wealthiest Consume 20 Times More Energy Than the World's Poorest," www.GlobalCitizen.org, March 18, 2021. https://www.globalcitizen.org/en/content/wealthy-people-climate-change-energy-consumption/#:~:text=The%20wealthiest%2010%25%20of%20the,widely%20between%20and%20within%20countries

304 Reuters Staff, "Eastern Mediterranean states formally establish Egypt-based gas forum," www.Reuters.com, September 22, 2020. https://www.reuters.com/article/us-mideast-energy/east-mediterranean-states-formally-establish-egypt-based-gas-forum-idUSKCN26D14D

305 Reuters Staff. Ibid to “Easter Mediterranean” article. 2020.

306 Devkota, Sisir, “How Turk Stream is forcing Europe on its heels,” www.ModernDiplomacy.eu, February 12, 2020. https://moderndiplomacy.eu/2020/02/12/how-turk-stream-is-forcing-europe-on-its-heels/

307 Reuters via Insurance Journal, “Germany’s Merkel Urges Stronger Global Effort on Climate Change,” www.InsuranceJournal.com, August 28, 2020. https://www.insurancejournal.com/news/international/2020/08/28/580740.htm

308 U.S. Energy Information Administration (EIA), Independent Statistics & Analysis, “U.S. natural gas production, consumption, and exports set new records in 2019,” www.EIA.gov, October 5, 2020. https://www.eia.gov/todayinenergy/detail.php?id=45377

309 Russell Mead, Walter, “How American Fracking Changes the World: Low energy prices enhance U.S. power at the expense of Moscow and Tehran,” www.WSJ.com, November 26, 2018. https://www.wsj.com/articles/how-american-fracking-changes-the-world-1543276935

310 Shiryaevskaya, Anna, Khrennikova, Dina, “Why the World Worries About Russia’s Natural Gas Pipeline,” www.Bloomberg.com, June 13, 2019. Updated on December 23, 2019. https://www.bloomberg.com/news/articles/2019-06-13/why-world-worries-about-russia-s-natural-gas-pipeline-quicktake

311 Yergin, Daniel, “How the shale revolution has redrawn the global political map,” www.DallasNews.com, September 20, 2020. https://www.dallasnews.com/opinion/commentary/2020/09/20/how-the-shale-revolution-has-redrawn-the-global-political-map/

312 Aslund, Anders, Russia’s Crony Capitalism: The Path from Market Economy to Kleptocracy, (Yale University Press, New Haven, CT.), Entire book. May 21, 2019. https://www.amazon.com/Russias-Crony-Capitalism-Economy-Kleptocracy/dp/030024309X/ref=tmm_hrd_swatch_0?_encoding=UTF8&qid=1602087333&sr=8-1

313 Yergin. Ibid. 2020.

314 Yergin. Ibid. 2020.

315 Yergin. Ibid. 2020.

316 Yergin. Ibid. 2020.

317 Borger, Julian, “US to pull 12,000 troops out of Germany as Trump blasts ‘delinquent’ Berlin,” www.TheGuardian.com, July 29, 2020. https://www.theguardian.com/us-news/2020/jul/29/us-germany-troop-withdrawal-donald-trump

318 Yergin. Ibid. 2020.
319 Yergin. Ibid. 2020.
320 Fairless, Tom, Hua, Sha, Pancevski, Bojan, "U.S.-China Tensions Leave Germany Squirming in the Middle," www.WSJ.com, June 24, 2020. https://www.wsj.com/articles/u-s-china-tensions-leave-germany-squirming-in-the-middle-11592989132
321 Stein, Ron, Royal, Todd, Just Green Electricity – Helping Citizens Understand a World without Fossil Fuels, (Archway Publishing, New York, NY) Entire book, June 17, 2020.

CHAPTER THREE: AUSTRALIA

322 Dunn, Katherine, "Shell announces big layoffs as the price of its net zero restructuring plan becomes clearer," www.Forbes.com, September 30, 2020. https://fortune.com/2020/09/30/net-zero-shell-layoffs/
323 Dunn. Ibid. 2020.
324 Mathis, Mark, YouTube.com video for Clean Energy Alliance, titled, "4-Quadrant Model," www.CleanEnergyAlliance.com, October 12, 2020. https://clearenergyalliance.com/project/4-quadrant-model/
325 StopTheseThings.com linked to a YouTube.com video titled, "The Impossibility of Windmills," from Jan Smelik, September 8, 2020. Post from October 11, 2020. www.StopTheseThings.com, https://stopthesethings.com/2020/10/11/impossible-dream-why-promise-of-all-wind-sun-powered-future-is-complete-nonsense/
326 U.S. Representative Will Hurd, Texas Congressional District 23, "Hurd on the Hill: Keeping an enemy out of our backyard," www.Hurd.House.Gov, September 10, 2020. https://hurd.house.gov/media-center/hurd-hill/hurd-hill-keeping-enemy-out-our-backyard
327 Holmes, James, "Preparing for War": What Is China's Xi Jingping Trying to Tell Us?" www.NationalInterest.org, October 18, 2020. https://nationalinterest.org/feature/preparing-war-what-chinas-xi-jinping-trying-tell-us-170939
328 Isaac, Jason, "Industry layoffs threaten the green energy fairy tale," www.WashingtonExaminer.com, October 12, 2020. https://www.washingtonexaminer.com/opinion/industry-layoffs-threaten-the-green-energy-fairy-tale
329 Shellenberger, Michael, "If Solar And Wind Are SO Cheap, Why Are They Making Electricity So Expensive?" www.Forbes.com, April 23,

2018. https://www.forbes.com/sites/michaelshellenberger/2018/04/23/if-solar-and-wind-are-so-cheap-why-are-they-making-electricity-more-expensive/#79036cd71dc6
330 Isaac. Ibid. 2020.
331 CNN Town Hall moderated by Anderson Cooper, CNN Video via Twitter, "Biden says 'fracking has to continue because we need a transition' when challenged about his position at CNN town hall," www.CNN.com, & www.Twitter.com, September 17, 2020. https://twitter.com/i/events/1306771888157290503
332 U.S. Energy Information Administration (EIA), Monthly Energy Review, Table 1.2 Primary Energy Production by Source (Quadrillion Btu), www.EIA.gov, September 2020. https://www.eia.gov/totalenergy/data/monthly/pdf/sec1_5.pdf
333 International Energy Agency, World Energy Outlook 2020, Flagship Report, www.IEA.org, October 2020. https://www.iea.org/reports/world-energy-outlook-2020
334 Kimani, Alex, "5 Major Takeaways From The IEA's World Energy Outlook 2020," www.OilPrice.com, October 14, 2020. https://oilprice.com/Energy/Energy-General/5-Major-Takeaways-From-The-IEAs-World-Energy-Outlook-2020.html
335 Meliksetian, Vanand, "India Continues To Bet Big On Coal Despite Renewable Energy Ambitions," www.OilPrice.com, October 14, 2020. https://oilprice.com/Energy/Coal/India-Continues-To-Bet-Big-On-Coal-Despite-Renewable-Energy-Ambitions.html
336 Dears, Donn, "Don't Ignore Coal," www.DDears.com, July 14, 2020. https://ddears.com/2020/07/14/dont-ignore-coal/
337 Ranken Energy, "Products made from petroleum: With Over 6000 products and counting, petroleum continues to be a crucial requirement for all consumers," www.Ranken-Energy.com, 2017. https://www.ranken-energy.com/index.php/products-made-from-petroleum/
338 Life Without Oil, Chart, https://tallbloke.files.wordpress.com/2018/07/oila.jpg?w=640
339 Life without Oil and Products, YouTube video, https://www.facebook.com/watch/?v=1058773337581878
340 StopTheseThings.com, "Keep It Simple Stupid: Why No Country Will Ever Be Powered By Wind & Solar Alone," linked to an article by David Bidstrup, "The quest for complexity," www.StopTheseThings.com, September 3, 2020. https://stopthesethings.

com/2020/09/03/keep-it-simple-stupid-why-no-country-will-ever-be-powered-by-wind-solar-power-alone/

341 Isaac. Ibid. 2020.

342 U.S. Energy Information Administration (EIA), Independent Statistics & Analysis, Today in Energy, "U.S. utility-scale battery storage power capacity to grow substantially by 2023," www.EIA.gov, July 10, 2019. https://www.eia.gov/todayinenergy/detail.php?id=40072

343 The New York Independent System Operator, Inc., (ISO), 2019 Load & Capacity Data, A report by The New York Independent System Operator, Inc. Gold Book, www.ISO.gov, April 2019. https://www.nyiso.com/documents/20142/2226333/2019-Gold-Book-Final-Public.pdf/a3e8d99f-7164-2b24-e81d-b2c245f67904?t=1556215322968

344 Bennett, Brett, "To Believe or Not to Believe? Breaking Down the Hype Around Tesla's Battery Day," www.RealClearEnergy.org, September 30, 2020. https://www.realclearenergy.org/articles/2020/09/30/to_believe_or_not_to_beli eve_breaking_down_the_hype_around_teslas_battery_day_579092.html

345 Dyke, Geoff, Energy Policy Institute of Australia, Public Policy Paper, Paper 2/2020, Nuclear Power Through The Lens of An Australian Trade Union, www.EnergyPolicyInstitute.com.au, July 2020. http://energypolicyinstitute.com.au/images/2-20__Geoff_Dyke_PP.pdf

346 Rochester, Chris, Dr. Considine, Timothy, MacIver Institute, Study: Renewable Energy Mandates Come Up Short On Economic Promises, www.MacIverInstitute.com, July 7, 2016. https://www.maciverinstitute.com/2016/07/study-renewable-energy-mandate s-come-up-short-on-economic-promises/

347 Driessen, Paul, "Green New Deal ideologies, fantasies and realities," www.Cfact.org, January 15, 2021. https://www.cfact.org/2021/01/15/green-new-deal-ideologies-fantasies-and-realities/

348 Mills, Mark P., "The Green New Deal Can't Break The Laws Of Physics," www.DailyCaller.com, October 27, 2020. https://dailycaller.com/2020/10/27/green-new-deal-laws-of-physics/

349 Watts, Anthony, "USA cost to go without fossil fuels: $18-29 trillion," www.WattsUpWithThat.com, September 22, 2020. https://wattsupwiththat.com/2020/09/22/usa-cost-to-go-without-fossi l-fuels-18-29-trillion/

350 Needham, Kirsty, A Reuters Special Report: "Australia faces down China in high-stakes strategy: The government of Scott Morrison has

reached a stark new view of China. It's not just a top trading partner, but also a threat to the nation's sovereignty. The dramatic shift shows how countries are struggling to cope with China's growing might," www.Reuters.com, September 4, 2020. https://www.reuters.com/investigates/special-report/australia-china-relations/

351 The Editors of Encyclopedia Britannica, "Faustian Bargain," www.Britannica.com, Site accessed October 16, 2020. https://www.britannica.com/topic/historical-injustice

352 Pielke, Roger, "How Billionaire Tom Steyer and Michael Bloomberg Corrupted Climate Science," www.Forbes.com, January 2, 2020. https://www.forbes.com/sites/rogerpielke/2020/01/02/how-billionaires-tom-steyer-and-michael-bloomberg-corrupted-climate-science/#3ff4db0702c6

353 Pielke, Roger, The Climate Fix: What Scientists and Politicians Won't Tell You About Global Warming, (Basic Books, New York, NY), www.Amazon.com, December 6, 2011. Entire book for sourced sentence. https://www.amazon.com/Climate-Fix-Roger-Pielke-Jr/dp/0465025196/ref=tmm_pap_swatch_0?_encoding=UTF8&qid=&sr=

354 Delbert, Caroline, "It's Official: Solar Is the Cheapest Electricity in History," www.PopularMechanics.com, October 22, 2020. https://www.popularmechanics.com/science/a34372005/solar-cheapest-energy-ever/

355 Nova, Joanne, "Extension cord to rescue renewable South Australia will now cost $2.4 billion," www.JoanneNova.com.au, October 10, 2020. http://joannenova.com.au/2020/10/extension-cord-to-rescue-renewable-south-australia-will-now-cost-2-4-billion/

356 Moran, Alan, "New wind and solar generation being built in spite of low prices," www.CatallaxyFiles.com, October 5, 2020. https://catallaxyfiles.com/2020/10/05/new-wind-and-solar-generation-being-built-in-spite-of-low-prices/

357 Haubert, Jon, Institute for Energy Research (IER), "Far Left Foundations Funnell $500m+ to Greens, Cities, and Media," www.InstituteforEnergyResearch.org, October 7, 2020. https://www.instituteforenergyresearch.org/big-green/far-left-foundations-funnel-500m-to-greens-cities-and-media/

358 Darwall, Rupert, Green Tyranny: Exposing the Totalitarian Roots of the Climate Industrial Complex, (Encounter Books, New York, NY), www.Amazon.com, March 26, 2019. https://

www.amazon.com/Green-Tyranny-Exposing-Totalitarian-Industrial/dp/1641770449/ref=sr_1_1?dchild=1&keywords=Green+Tyranny+Exposing+the+Totalitarian+Roots+of+the+Climate+Industrial+Complex&qid=1603492861&sr=8-1

359 Polumbo, Brad, "New Stanford Study Suggests Biden's Agenda Will Have 4 Devastating Economic Consequences," www.FEE.com, (Foundation for Economic Education), October 18, 2020. https://fee.org/articles/new-stanford-study-suggests-bidens-agenda-will-have-4-devastating-economic-consequences/

360 Pielke, Roger, "The Incredible Story Of How Climate Change Became Apocalyptic," www.Forbes.com, October 19, 2019. https://www.forbes.com/sites/rogerpielke/2019/12/06/the-incredible-story-of-how-climate-change-became-apocalyptic/#2d7dbae2789d

361 Pielke, Roger, "In 2020 Climate Science Needs To Hit The Reset Button, Part One," www.Forbes.com, December 22, 2019. https://www.forbes.com/sites/rogerpielke/2019/12/22/in-2020-climate-science-needs-to-hit-the-reset-button-part-one/#4a2faa60482c

362 Heim, Burt, "Climate Change's Bottom Line," www.NYTimes.com, January 31, 2015. https://www.nytimes.com/2015/02/01/business/energy-environment/climate-changes-bottom-line.html

363 Heim. Ibid. 2020.

364 Pielke. Ibid. 2020.

365 Gordon, Kate, Executive Director of the Risky Business Project, A Climate Risk Assessment of the United States: The Economic Risks of Climate Change in the United States, www.RiskyBusiness.org, June 2014. http://riskybusiness.org/site/assets/uploads/2015/09/RiskyBusiness_Report_WEB_09_08_14.pdf

366 Gordon. Ibid. 2014.

367 Moss, Richard H., et al., The next generation of scenarios for climate change research and assessment, www.Nature.com, February 11, 2010. https://www.nature.com/articles/nature08823

368 Gordon, Kate, et al., Risky Business Presentation to the American Climate Prospectus, Economic Risks in the United States to the 2014 AGU Fall Meeting. https://rhg.com/wp-content/uploads/2014/12/ACP-Research-Team-talks-at-AGU2014.pdf

369 Tverberg, Gail, "The Ten Reasons Why Intermittency is a Problem for Renewable Energy," www.OilPrice.com, January 30, 2014. https://oilprice.com/Alternative-Energy/Renewable-Energy/The-Ten-Reasons-Why-Intermittency-is-a-Problem-for-Renewable-Energy.html (Ms.

Tverberg is used, because she is a scientist, whereas Mr. Driessen is more well-known. Ms. Tverberg has many articles on the nature of renewables for further exploration if needed.)

370 Curry, Judith, "Science and politics," Climate Etc., www.JudithCurry.com, October 26, 2020. https://judithcurry.com/2020/10/26/science-and-politics/

371 Carleton, Tamma A., Hsiang, Solomon M., Social and economic impacts of climate, Science 09 Sep 2016; Vol. 353, Issue 6304, aad9837, DOI: 10.1126/science.aad9837. www.Science.ScienceMag.org, https://science.sciencemag.org/content/353/6304/aad9837 (Formatting done differently per source request.)

372 Pielke. Ibid. 2020

373 Pielke. Ibid. 2020.

374 Perry, Mark J., "50 years of failed doomsday, eco-pocalyptic predictions; the so-called 'experts' are 0-50," www.AEI.org, September 23, 2019. https://www.aei.org/carpe-diem/50-years-of-failed-doomsday-eco-pocalyptic-predictions-the-so-called-experts-are-0-50/

375 Faught, Andrew, "Profiles, Kate Gordon '94 Brings An Astute Policy Mind To Climate-Change Economics," www.Magazine.Blogs.Wesleyan.Edu, September 15, 2015. https://magazine.blogs.wesleyan.edu/2015/09/15/kate-gordon-94-brings-an-astute-policy-mind-to-climate-change-economics/

376 Study by Tata Centre for Development (TCD) at the University of Chicago for the Economic Times, 1.5 million more people may die in India by 2100 due to extreme heat by climate change: Study, www.EconomicTimes.IndiaTimes.com, October 31, 2019. https://economictimes.indiatimes.com/news/politics-and-nation/1-5-million-more-people-may-die-in-india-by-2100-due-to-extreme-heat-by-climate-change-study/articleshow/71842685.cms?mc_cid=dec4abb97e&mc_eid=89ee108d17

377 Jeffrey-Wilensky, Jaclyn, Freeman, David, "Rising sea levels could swamp major cities and displace almost 200 million people, scientists say," www.NBCNews.com, May 22, 2019. https://www.nbcnews.com/mach/science/rising-sea-levels-could-swamp-major-cities-displace-almost-200-ncna1008846

378 McMahon, Jeff, "Rise In Climate-Related Deaths Will Surpass All Infectious Diseases, Economist Testifies," www.Forbes.com, December 27, 2019. https://www.forbes.com/sites/jeffmcmahon/2019/12/27/

climate-related-deaths-in-2100-will-surpass-current-mortality-from-all-infectious-diseases-economist-testifies/#1804bfa42225
379 Wallace, Jeremy, "Rep. Dan Crenshaw calls wind and solar energy 'silly solutions' to climate change," www.HoustonChronicle.com, October 1, 2020. https://www.houstonchronicle.com/politics/texas/article/Rep-Dan-Crenshaw-calls-wind-and-solar-energy-15613815.php
380 Phillips, Morgan, "Trump calls Biden 'transition from' oil comments 'perhaps most shocking admission ever made.'" www.FoxNews.com, October 23, 2020. https://www.foxnews.com/politics/trump-biden-transition-from-oil-shocking-admission
381 Stein, Ronald, Breibart, October 27, 2020, Biden's quote "We're not getting rid of fossil fuels for a long time... probably 2050". Biden has yet to comprehend the limitations of just electricity from renewables to support worldwide economies and lifestyles https://www.breitbart.com/2020-election/2020/10/27/expert-biden-clueless-about-role-oil-gas-domestic-global-health-security-prosperity/
382 Middleton, David, "Solar Power Costs 2-3 Times As Much As Wind, Fossil Fuels and Nuclear," www.WattsUpWithThat.com, October 13, 2020. https://wattsupwiththat.com/2020/10/13/solar-power-costs-2-3-times-as-much-as-wind-fossil-fuels-and-nuclear/
383 Doshi, Tilak, "The West Intends Energy Suicide: Will It Succeed?" www.Forbes.com, October 10, 2020. https://www.forbes.com/sites/tilakdoshi/2020/10/10/the-west-intends-energy-suicide-will-it-succeed/#5491a28e490e
384 Darwall. Ibid. 2017. Entire book.
385 McDermott Hughes, David, "To Save the Climate, Give Up the Demand for Constant Electricity," www.BostonReview.com, October 5, 2020. http://bostonreview.net/science-nature/david-mcdermott-hughes-save-climate-give-demand-constant-electricity (Entire paragraph comes from this source, which I obviously have wide disdain for this type of garbage.)
386 The Wall Street Journal Editorial Board, "Big Oil to the Coronavirus Rescue: Look whose products are crucial for fighting off COVID-19," www.WSJ.com, August 23, 2020. https://www.wsj.com/articles/big-oil-to-the-coronavirus-rescue-11587683239
387 Doshi. Ibid. 2020.
388 Gosselin, Pierre, "German Prof: Climate Science Politicized, Exaggerated, Filled With "Fantasy", "Fairy Tales"..."Paris Accord Already Dead!" www.NoTricksZone.com, October 7, 2020. https://

notrickszone.com/2020/10/07/german-prof-climate-science-politicized-exaggerated-filled-with-fantasy-fairy-tales-paris-accord-already-dead/

389 Doshi. Ibid. 2020.

390 Geopolitical Intelligence Services (GIS), "Professor Dr. Fritz Vahrenholt," www.GISReportnline.com, 2020. https://www.gisreportsonline.com/professor-dr-fritz-vahrenholt,143,expert.html (Website accessed October 27, 2020.)

391 White, Chris, "'Gaps' In Renewable Energy Led To Blackouts For Millions Of Californians, Gov Newsome Says," www.DailyCaller.com, August 17, 2020. https://dailycaller.com/2020/08/17/california-blackouts-renewable-energy-california-gavin-newsom/

392 Pearce, Tim, "California's Newsome Wants To Ban Gas-Powered Cars But 'Can't Even Keep The Lights On Today,' EPA Says," www.DailyCaller.com, September 30, 2020. https://www.dailywire.com/news/california-newsom-ban-epa

393 Watts, Anthony, "Michael E. Mann – wrong again," www.WattsUpWithThat.com, July 5, 2020. https://wattsupwiththat.com/2020/07/05/michael-e-mann-wrong-again/

394 Doshi. Ibid. 2020.

395 Doshi. Ibid. 2020.

396 Adams, Mike, "Climate change hoax collapses as Michael Mann's bogus "hockey stick" graph defamation lawsuit dismissed by the Supreme Court of British Columbia," www.Climate.News, August 26, 2019. https://climate.news/2019-08-26-climate-change-hoax-collapses-as-michael-mann-bogus-hockey-stick-graph.html

397 Nova, Joanne, "How to lose the unlosable election: be anti-coal. The climate vote evaporated," www.JoanneNova.com.au, May 20, 2019. http://joannenova.com.au/2019/05/how-to-lose-the-unloseable-election-be-anti-coal-how-the-climate-vote-evaporated/

398 Life Without Oil https://tallbloke.files.wordpress.com/2018/07/oila.jpg?w=640

399 Ring, Edward, "The Battle for California Is the Battle for America," www.AmGreatness.com, October 17, 2020. https://amgreatness.com/2020/10/17/the-battle-for-california-is-the-battle-for-america/

400 Jayaraj, Vijay, "The Myth of Glorious Renewables," www.WattsUpWithThat.com, October 26, 2020. https://wattsupwiththat.com/2020/10/26/the-myth-of-glorious-renewables/

401 Stein, Ronald, CFACT, September 19, 2020, https://www.cfact.org/2020/09/19/californias-energy-scorecard-fails-on-the-world-stage/
402 Ring. Ibid. 2020.
403 Ring. Ibid. 2020.
404 Ring Ibid. 2020.
405 Brilliant, Mark, "The prescient politics and policies of Jerry Brown," www.WashingtonPost.com, July 10, 2020. https://www.washingtonpost.com
406 Schoffstall, Joe, "EPA Requests DOJ Investigate Foreign Funding of Environmental Groups," www.FreeBeacon.com, October 27, 2020. https://freebeacon.com/national-security/epa-requests-doj-investigate-foreign-funding-of-environmental-groups/
407 Ring. Ibid. 2020.
408 Sommer, Lauren, "California Governor Signs Order Banning Sales Of New Gasoline Cars By 2035," www.NPR.org, September 23, 2020. https://www.npr.org/2020/09/23/916209659/california-governor-signs-order-banning-sales-of-new-gasoline-cars-by-2035
409 Royal, Todd, "Questioning Global Warming," www.AmericanThinker.com, August 13, 2019. https://www.americanthinker.com/articles/2019/04/questioning_global_warming.html
410 Patrick, Stewart M., "California Is a Preview of Climate Change's Devastation for the Entire World," www.CFR.org, (Council on Foreign Relations), August 24, 2020. https://www.cfr.org/blog/california-preview-climate-changes-devastation-entire-world
411 Kotkin, Joel, "An "Ecotopian" Future: Can California's Green Extremism Go National?" www.RealClearEnergy.org, September 29, 2020. https://www.realclearenergy.org/articles/2020/09/29/an_ecotopian_future_can_californias_green_extremism_go_national_578968.html
412 Kotkin. Ibid. 2020.
413 Agag, Alejandro, Formula E, "Covid-19 is a test run for when we need to act on climate change – that time is now," www.BusinessGreen.com, September 29, 2020. https://www.businessgreen.com/opinion/4014252/covid-19-test-run-act-climate-change
414 Stein, Ronald, "Newsome promotes California to be the Vintage Car Capital.com, September 24, 2020. https://www.foxandhoundsdaily.com/2020/09/newsom-promotes-california-to-be-the-vintage-car-capital/

415 Friedman, David, Hernandez, Jennifer, Editor, Kotkin, Joel, Center for Demographics & Policy, Research Brief, Chapman University, California Greenhouse Gas Regulation, And Climate Change, www.NewGeography.com, Report accessed on October 29, 2020. http://www.newgeography.com/files/California%20GHG%20Regulation%20Final.pdf

416 Morrison, Richard, "Climate change cronyism. Big businesses tailor policy to benefit themselves at your cost," www.WashingtonExaminer.com, September 19, 2020. https://www.washingtonexaminer.com/opinion/climate-change-cronyism-big-businesses-tailor-policy-to-benefit-themselves-at-your-cost/

417 Kotkin. Ibid. 2020.

418 Hodges, Jeremy, "Wind, Solar Are Cheapest Power Source In Most Places, BNEF Says," www.Bloomberg.com, October 19, 2020. https://www.bloomberg.com/news/articles/2020-10-19/wind-solar-are-cheapest-power-source-in-most-places-bnef-says?mkt_tok=eyJpIjoiT0RRNVpUVTBPRGRoTXpFNCIsInQiOiJhSFBMUEZGTjNKMUE4blR1QUVBbjdjaGdTbHhBZUVweENjSDBwRUtPZ1wvKysrWjZ6UlhXXC9WdGRcL0htcHJha1BRYmU4T0hxRTd3dlBwM01XNU9TUXE2VlFCNGtBRTg1MEE2dUs5RHhVVHcrXC80R3F4d1psRmpnSDRcL2NjXC94aGJvNSJ9

419 Lazard, Insights, Lazard's Levelized Cost of Energy Analysis, Version 14.0 & Lazard's Levelized Cost of Storage Analysis, Version 6.0, www.Lazard.com, October 19, 2020. https://www.lazard.com/perspective/lcoe2020?utm_source=newsletter&utm_medium=email&utm_campaign=newsletter_axiosgenerate&stream=top

420 Press Release, Buckley, Tim, Trivedi, Saurabh, "IEEFA: From zero to fifty, global financial corporations get cracking on major oil/gas lending exits," www.IEEFA.org, (Institute for Energy Economics and Financial Analysis), October 19, 2020. https://ieefa.org/ieefa-from-zero-to-fifty-global-financial-corporations-get-cracking-on-major-oil-gas-lending-exits/

421 Matthews, Merrill, "Debunking Democrats' claims about fossil fuel tax breaks," www.TheHill.com, June 16, 2019. https://thehill.com/opinion/energy-environment/448794-debunking-democrats-claims-about-fossil-fuel-tax-breaks

422 Kazmin, Amy, "India takes its tussle with China to the high seas," www.FT.com, (Financial Times), October 27, 2020. https://www.ft.com/content/bbe548c4-b437-4965-a651-592d0e969890

423 Lin, Anastasia, "The Cultural Revolution Comes to North America," www.WSJ.com, April 7, 2019. https://www.wsj.com/articles/the-cultural-revolution-comes-to-north-america-11554661623
424 Watts, Anthony, "The BBC formally bans climate skeptics," www.WattsUpWithThat.com, September 27, 2018. https://wattsupwiththat.com/2018/09/27/the-bbc-formally-bans-climate-skeptics/
425 Jackson, Kerry, "The Global Warming Thought Police Wants Skeptics In 'Jail,'" www.Investors.com, (Investor's Business Daily), October 24, 2017. https://www.investors.com/politics/commentary/the-global-warming-thought-police-want-skeptics-in-jail/
426 Hasemyer, David, "Fossil Fuels on Trial: Where the Major Climate Change Lawsuits Stand Today: Some of the biggest oil and gas companies are embroiled in legal disputes with cities, states, and children over the industry's role in global warming," www.InsideClimateNews.com, January 17, 2020. https://insideclimatenews.org/news/04042018/climate-change-fossil-fuel-company-lawsuits-timeline-exxon-children-california-cities-attorney-general (Special note in my private work life I have worked on this issue. In each case the plaintiffs are left-leaning U.S. Democrats.)
427 Smith, Jennifer, McIntyre, Daniel, "Socialist AOC appears in Vanity Fair in $14,000 of designer suits and Louboutins to call Trump a motherf***** for not paying tax – as she compares herself to Hillary and Pelosi (and gets to keep a $3,000 outfit)," www.DailyMail.co.UK, October 28, 2020. https://www.dailymail.co.uk/news/article-8889119/AOC-appears-cover-Vanity-Fair-says-shes-boogeyman-Dems.html

CHAPTER FOUR: ENVIRONMENTAL DEGRADATION

428 Stein, Ronald, CFACT, Aha! "Renewable energy" is only renewable ELECTRICITY, July 23, 2020, https://www.cfact.org/2020/07/23/aha-renewable-energy-is-only-kind-of-renewable-electricity/
429 Nuclear power in China, https://en.wikipedia.org/wiki/Nuclear_power_in_China#:~:text=As%20of%20March%202019%2C%20China,for%20an%20additional%2036%20GW.
430 U.N. News, UN highlights urgent need to tackle impact of likely electric car battery production boom, June 28, 2020, https://news.un.org/en/story/2020/06/1067272

431 McLellan, Benjamin, One Earth, Dark" Dark" Materials for a Brighter Energy Future, December 20, 2019, https://www.cell.com/one-earth/fulltext/S2590-3322(19)30219-2?_returnURL=https%3A%2F%2Flinkinghub.elsevier.com%2Fretrieve%2Fpii%2FS2590332219302192%3Fshowall%3Dtrue
432 United Nations Conference on Trade and Development, UNCTAD, Commodities at a Glance: Special issue on strategic battery raw materials, https://unctad.org/en/pages/PublicationWebflyer.aspx?publicationid=2797
433 Williams, Thomas, U.N. Warns of Devastating Environmental Side Effects of Electric Car Boom, June 29, 2020, https://www.breitbart.com/environment/2020/06/29/u-n-warns-of-devastating-environmental-side-effects-of-electric-car-boom-2/?fbclid=IwAR1YPpANmgx3raELMdD10lTh_eOKZ-xJdqCDPTeZRWYNE5JNEY6zaTvtryg
434 Pryor, Steve, A Partial list of the over 6,000 products made from one barrel of oil (after creating 19 gallons of gasoline), July 26, 2016, https://www.linkedin.com/pulse/partial-list-over-6000-products-made-from-one-barrel-oil-steve-pryor/
435 Solaris, Solar Components, https://www.solaris-shop.com/solar-components/
436 Smil, Vaclav, IEEE Spectrum, To Get Wind Power You Need Oil, February 29, 2016, https://spectrum.ieee.org/energy/renewables/to-get-wind-power-you-need-oil
437 Kochhar, Ajay, Li-Cycle featured in a Financial Times article focused on electric vehicle battery recycling, September 5, 2017, https://li-cycle.com/2017/09/05/li-cycle-featured-in-a-financial-times-article-focused-on-electric-vehicle-battery-recycling/
438 Esther de Haan and Vincent Kiezebrink, Cobalt blues - Environmental pollution and human rights violations in Katanga's copper and cobalt mines, April 1, 2016, https://www.somo.nl/cobalt-blues/
439 Lithiummine.com, Lithium Mining and Environmental Impact, http://www.lithiummine.com/lithium-mining-and-environmental-impact
440 Luhn, Alec, The Guardian, Where the river runs red: can Norilsk, Russia's most polluted city, come clean?, September 15, 2016, https://www.theguardian.com/cities/2016/sep/15/norilsk-red-river-russias-most-polluted-city-clean
441 China: Production of "clean" lithium-ion batteries reportedly causes heavy air & water pollution with graphite in northeast provinces, October 4, 2016, https://www.business-humanrights.

org/en/latest-news/china-production-of-clean-lithium-ion-batteries-reportedly-causes-heavy-air-water-pollution-with-graphite-in-northeast-provinces/
442 Mills, Mark, Mines, Minerals, and "Green" Energy, July 9, 2020, https://www.manhattan-institute.org/mines-minerals-and-green-energy-reality-check
443 Eckart, Jonathan, World Economic Forum, November 28, 2017 https://www.weforum.org/agenda/2017/11/battery-batteries-electric-cars-carbon-sustainable-power-energy/
444 Orr, Isaac, Energy, As $7,500 Federal Subsidies Expire, Electric Vehicles Are About to Get More Expensive, September 30, 2019, tps://www.americanexperiment.org/2019/09/as-7500-federal-subsidies-expire-electric-vehicles-are-about-to-get-more-expensive/
445 EVANNEX, Just How Long Will An EV Battery Last?, September 3, 2019, https://insideevs.com/news/368591/electric-car-battery-lifespan/
446 Orr, Isaac, Energy, As $7,500 Federal Subsidies Expire, Electric Vehicles Are About to Get More Expensive, September 30, 2019, tps://www.americanexperiment.org/2019/09/as-7500-federal-subsidies-expire-electric-vehicles-are-about-to-get-more-expensive/
447 Lithiummine.com, Lithium Mining and Environmental Impact, http://www.lithiummine.com/lithium-mining-and-environmental-impact
448 Lithiummine.com, Lithium Mining in China, https://sites.google.com/site/lithiumminecom/lithium-mining-in-china
449 Lithiummine.com, Lithium Mining in Russia, https://sites.google.com/site/lithiumminecom/lithium-mining-in-russia
450 Lithiummine.com, Salt Evaporation Ponds, https://sites.google.com/site/lithiumminecom/salt-evaporation-ponds
451 Lithiummine.com, Lithium Mining in Chile, https://sites.google.com/site/lithiumminecom/lithium-mining-in-chile
452 Lithiummine.com, Lithium Mining in Argentina, https://sites.google.com/site/lithiumminecom/lithium-mining-in-argentina
453 Lithiummine.com, Lithium Mining in Bolivia, https://sites.google.com/site/lithiumminecom/lithium-mining-in-bolivia
454 Lithiummine.com, Lithium Mining China, https://sites.google.com/site/lithiumminecom/lithium-mining-in-china
455 Stacey, Johnathan, Levin Sources, WHITE PAPER: LITHIUM MINING IN THE HIGH PUNA OF THE ANDES, June 2019, https://www.levinsources.com/knowledge-centre/publications/whit

e-paper-lithium-mining-high-puna-andeshttps://www.levinsources.com/knowledge-centre/publications/white-paper-lithium-mining-high-puna-andes
456 Spence, Katie, The Motley Fool, Tesla Motors' Dirty Little Secret Is a Major Problem - Think Tesla's Model S is the green car of the future? Think again., October 29, 2018, https://www.fool.com/investing/general/2014/01/19/tesla-motors-dirty-little-secret-is-a-major-proble.aspx
457 Cision Newswire, Study Identifies Benefits and Potential Environmental/Health Impacts of Lithium-ion Batteries for Electric Vehicles - Life Cycle Assessment Highlights Ways to Reduce Global Warming Emissions, Addresses Nanotechnology Innovations to Improve Battery Performance, May 28, 2013, https://www.prnewswire.com/news-releases/study-identifies-benefits-and-potential-environmentalhealth-impacts-of-lithium-ion-batteries-for-electric-vehicles-209193531.html
458 Frank, Stephen, California Political review, The Darker Side of Green Technology, August 18, 2019, http://www.capoliticalreview.com/capoliticalnewsandviews/the-darker-side-of-green-technology/http://www.capoliticalreview.com/capoliticalnewsandviews/the-darker-side-of-green-technology/
459 West, Karl, The Guardian, Carmakers' electric dreams depend on supplies of rare minerals, July 29, 2017, https://www.theguardian.com/environment/2017/jul/29/electric-cars-battery-manufacturing-cobalt-mining
460 Amnesty International, Industry giants fail to tackle child labour allegations in cobalt battery supply chains, November 15, 2017, https://www.amnesty.org/en/latest/news/2017/11/industry-giants-fail-to-tackle-child-labour-allegations-in-cobalt-battery-supply-chains/https://www.amnesty.org/en/latest/news/2017/11/industry-giants-fail-to-tackle-child-labour-allegations-in-cobalt-battery-supply-chains/
461 Bell, Terence, The World's 20 Largest Copper Mines, The top mines produce about 40 percent of the world's copper, November 25, 2019, https://www.thoughtco.com/the-world-s-20-largest-copper-mines-2014-2339745
462 Sousa, Gregory, The World Atlas, The Top Lithium Producing Countries In The World, April 25 2017, https://www.worldatlas.com/articles/the-top-lithium-producing-countries-in-the-world.html

463 Troy R. Hawkins, Bhawna Singh, Guillaume Majeau-Bettez, and Anders Hammer Strømman. Comparative Environmental Life Cycle Assessment of Conventional and Electric Vehicles, https://onlinelibrary.wiley.com/doi/pdf/10.1111/j.1530-9290.2012.00532.x
464 Lomberg, Bjorn, Wall Street Journal, Bjorn Lomborg: Green Cars Have a Dirty Little Secret, Producing and charging electric cars means heavy carbon-dioxide emissions. Bjorn Lomborg, March 11, 2014, https://www.wsj.com/articles/SB10001424127887324128504578346913994914472
465 Shah, Anup, Global Issues, Poverty Facts and Stats, January 7, 2013, https://www.globalissues.org/article/26/poverty-facts-and-stats
466 Peiser, Benny, The Global Warming Policy Forum, https://www.thegwpf.com/christopher-booker-the-tyranny-of-groupthink/
467 Amnesty International, THIS IS WHAT WE DIE FOR: HUMAN RIGHTS ABUSES IN THE DEMOCRATIC REPUBLIC OF THE CONGO POWER THE GLOBAL TRADE IN COBALT, January 15, 2016, https://www.amnestyusa.org/reports/this-is-what-we-die-for-human-rights-abuses-in-the-democratic-republic-of-the-congo-power-the-global-trade-in-cobalt/
468 Katwala, Amit, The spiralling environmental cost of our lithium battery addiction, August 5, 2018, https://www.wired.co.uk/article/lithium-batteries-environment-impact
469 Vasil, Adria, The EV revolution will take batteries, but are they ethical?, January 20, 2020, https://www.corporateknights.com/channels/transportation/ev-revolution-needs-batteries-ethical-15795118/
470 LePan, Nicholas, Visual Capitalisy, How Much Oil is in an Electric Vehicle?, May 20, 2019, https://www.visualcapitalist.com/how-much-oil-electric-vehicle/
471 Tullo, Alexander, Chemical and Engineering News, Plastics makers plot the future of the car, Plastics in our vehicles will become increasingly specialized as cars lose their drivers and go electric, November 13, 2017, https://cen.acs.org/articles/95/i45/Plastics-makers-plot-future-car.html
472 Industry Week, Lithium Batteries' Dirty Secret: Manufacturing Them Leaves Massive Carbon Footprint, Once in operation, electric cars certainly reduce your carbon footprint, but making the lithium-ion batteries could emit 74% more CO2 than for conventional cars. October 16, 2018, https://www.industryweek.com/technology-and-iiot/article/22026518/lithium-batteries-dirty-secret-manufacturing-

them-leaves-massive-carbon-footprint#:~:text=Technology%20and%20IIoT-,Lithium%20Batteries'%20Dirty%20Secret%3A%20Manufacturing%20Them%20Leaves%20Massive%20Carbon%20Footprint,CO2%20than%20for%20conventional%20cars.

473 Stop These Things, Digging It: Mines, Minerals, and "Green" Energy: A Reality Check, August 16, 2020, https://stopthesethings.com/2020/08/16/digging-it-mines-minerals-and-green-energy-a-reality-check/

474 Jacoby, Mitch, Chemical & Engineering News, It's time to get serious about recycling lithium-ion batteries. A projected surge in electric-vehicle sales means that researchers must think about conserving natural resources and addressing battery end-of-life issues., July 14, 2109, https://cen.acs.org/materials/energy-storage/time-serious-recycling-lithium/97/i28

475 Sarah King, Naomi J. Boxall, Anand I. Bhatt, Lithium battery recycling in Australia, A CSIRO Report, April 2018, https://www.csiro.au/~/media/EF/Files/Lithium-battery-recycling-in-Australia

476 Jacoby, Mitch, Chemical & Engineering News (C&EN), It's time to get serious about recycling lithium-ion batteries, July 14, 2019, https://cen.acs.org/materials/energy-storage/time-serious-recycling-lithium/97/i28

477 Loris, Nicolas, The Heritage Foundation, Electric Cars: Owned by Few, Subsidized by All, July 24, 2019, https://www.heritage.org/energy-economics/commentary/electric-cars-owned-few-subsidized-all

478 Severin Borenstein and Lucas W. Davis, The University of Chicago Press Journals, The Distributional Effects of US Clean Energy Tax Credits, https://www.journals.uchicago.edu/doi/full/10.1086/685597

479 Loris, Nicholas, Energy Economic, "Electric Cars: Owned by Few, Subsidized by All", Jun 24, 2019, https://www.heritage.org/energy-economics/commentary/electric-cars-owned-few-subsidized-all

480 Pu Liu and Claire Y Barlow, Wind Turbine Blade Waste in 2050, https://www.repository.cam.ac.uk/bitstream/handle/1810/263878/Liu_and_Barlow-2017-Waste_Management-AM.pdf?sequence=1&isAllowed=y

481 McLellan, Benjamin, One Earth, "Dark" Materials for a Brighter Energy Future, October 2016, https://www.cell.com/one-earth/fulltext/S2590-3322(19)30219-

2?_returnURL=https%3A%2F%2Flinkinghub.elsevier.com%2Fretrieve%2Fpii%2FS2590332219302192%3Fshowall%3Dtrue
482 Amnesty International, Industry giants fail to tackle child labour allegations in cobalt battery supply chains, November 15, 2017, https://www.amnesty.org/en/latest/news/2017/11/industry-giants-fail-to-tackle-child-labour-allegations-in-cobalt-battery-supply-chains/ https://www.amnesty.org/en/latest/news/2017/11/industry-giants-fail-to-tackle-child-labour-allegations-in-cobalt-battery-supply-chains/
483 DELINGPOLE, JAMES, Breitbart, Delingpole: Michael Moore Has Become a 'Hero' to 'Climate Deniers', Complains Guardian, 9 May 2020, https://www.breitbart.com/europe/2020/05/09/delingpole-michael-moore-has-become-a-hero-to-climate-deniers-complains-guardian/
484 Monbiot, George, The Guardian, How did Michael Moore become a hero to climate deniers and the far right? May 7, 2020, https://www.theguardian.com/commentisfree/2020/may/07/michael-moore-far-right-climate-crisis-deniers-film-environment-falsehoods
485 Shellenberger, Michael, Forbes, New Michael Moore-Backed Documentary On YouTube Reveals Massive Ecological Impacts Of Renewables, April 21, 2020, https://www.forbes.com/sites/michaelshellenberger/2020/04/21/new-michael-moore-backed-documentary-on-youtube-reveals-massive-ecological-impacts-of-renewables/#71c9b6916c96
486 Hinderaker, John, Powerline, THE EPIC HYPOCRISY OF TOM STEYER, April 20, 2014, https://www.powerlineblog.com/archives/2014/04/the-epic-hypocrisy-of-tom-steyer.php?fbclid=IwAR29pPARynySQkVBxNcSt5RYRKWeDkba47kFAJhLhlcul-3Fn5goJmCsdrI
487 Minerals Education Coalition, Fact Sheets, http://mineralseducationcoalition.org/mining-minerals-information/minerals-in-your-life/
488 Mathis, Mark, Rare Earth Emergency, May 6, 2020, https://www.youtube.com/watch?v=fGxqPf4_hmA&feature=em-uploademail
489 Jacobson, Rebecca, Inside Energy, Where Do Wind Turbines Go To Die?, September 9, 2016, http://insideenergy.org/2016/09/09/where-do-wind-turbines-go-to-die/#:~:text=The%20lifespan%20of%20the%20average,age%20of%2018%20years%20old.
490 Rolling, Mitchell, Center of the American Experiment, Limited Lifespans of Wind Turbines Result in Higher Costs of Energy, June

26, 2018, https://www.americanexperiment.org/2018/06/limited-lifespans-wind-turbines-result-higher-costs-energy/

491 U.S. Energy Information Administration, Repowering wind turbines adds generating capacity at existing sites, November 6, 2017, https://www.eia.gov/todayinenergy/detail.php?id=33632

492 XcelEnergy, Ponnequin Wind Farm, https://www.xcelenergy.com/energy_portfolio/electricity/power_plants/ponnequin

493 Stop These Things, Hopeless Joke: Australia's Wind Industry Keeps On Failing to Deliver the Goods, August 17, 2020, https://stopthesethings.com/2020/08/17/hopeless-joke-australias-wind-industry-keeps-on-failing-to-deliver-the-goods/

494 Paterson. Leigh, Inside Energy, n Coal Country, No Cash In Hand For Billions In Cleanup, November 14, 2015, http://insideenergy.org/2015/11/14/in-coal-county-no-cash-in-hand-for-billions-in-cleanup/

495 Wikipedia, Diablo Canyon Power Plant, https://en.wikipedia.org/wiki/Diablo_Canyon_Power_Plant

496 Fox News, Final wind-turbine rule permits thousands of eagle deaths, December 14, 2016, https://www.foxnews.com/politics/final-wind-turbine-rule-permits-thousands-of-eagle-deaths

497 Fox News, Final wind-turbine rule permits thousands of eagle deaths, December 14, 2016, https://www.foxnews.com/politics/final-wind-turbine-rule-permits-thousands-of-eagle-deaths

498 Wind Energy Technologies Office, Environmental Impacts and Siting of Wind Projects, https://www.energy.gov/eere/wind/environmental-impacts-and-siting-wind-projects#:~:text=As%20with%20all%20energy%20supply,wildlife%20like%20birds%20and%20bats.

499 Stop These Things, Filthy Facts: Wind Industry's Environmental Footprint Starts With Bird & Bat Carnage & Finishes In Landfills, August 29, 2020, https://stopthesethings.com/2020/08/29/filthy-facts-wind-industrys-environmental-footprint-starts-with-bird-bat-carnage-finishes-in-landfills/

500 British Ecological Society, Wind turbines cause functional habitat loss for migratory soaring birds, February 14, 2019, https://besjournals.onlinelibrary.wiley.com/doi/full/10.1111/1365-2656.12961

501 Collopy, Michael, Big One Complete, Golden Eagles in a Changing World, September 1, 2017, https://bioone.org/journals/journal-of-raptor-research/volume-51/issue-3/0892-1016-51.3.193/

Golden-Eagles-in-a-Changing-World/10.3356/0892-1016-51.3.193.full
502 Institute for Energy Research, Obama Allows Wind Turbines to Legally Kill Eagles, January 9, 2017, https://www.instituteforenergyresearch.org/renewable/wind/obama-allows-wind-turbines-legally-kill-eagles/
503 National Public Radio, Unfurling The Waste Problem Caused By Wind Energy, September 10, 2019, https://www.npr.org/2019/09/10/759376113/unfurling-the-waste-problem-caused-by-wind-energy
504 Issues & Insights, That Dirty Green Energy, August 24, 2020, https://issuesinsights.com/2020/08/24/that-dirty-green-energy/
505 Stein, Ronald, Eurasia Review, Blown Away By Wind Farm Capacity Versus Actual Output, July 25, 2019, https://www.eurasiareview.com/25072019-blown-away-by-wind-farm-capacity-versus-actual-output-oped/
506 Washington state utility report, "Wind Power and Clean Energy Policy Perspectives," PDF of the 16 page report dated July 14, 2020: https://www.bentonpud.org/getattachment/Safety-Education/Safety/Wind/Wind-Power-and-Clean-Energy-Policy-Perspectives-Report-Benton-PUD-FINAL-July-14-2020-(1).PDF.aspx?lang=en-US
507 Stop These Things, 'Green' Energy's Poisonous Legacy: Millions of Toxic Turbine Blades Destined for African Landfills, May 10, 2018, https://stopthesethings.com/2018/05/10/green-energys-poisonous-legacy-millions-of-toxic-turbine-blades-destined-for-african-landfills/
508 Stop These Things, 'Green' Energy's Toxic Legacy: Millions of Wind Turbine Blades Destined for Landfills, September 27, 2019, https://stopthesethings.com/2019/09/27/green-energys-toxic-legacy-millions-of-wind-turbine-blades-destined-for-landfills/
509 Wikipedia, Bisphenol, https://en.wikipedia.org/wiki/Bisphenol_A
510 Goggin, Michel, AWEA, Wind energy saves 2.5 billion gallons of water annually in drought-parched California, April 2, 2015, https://www.aweablog.org/wind-energy-saves-2-5-billion-gallons-of-water-annually-in-drought-parched-california/#:~:text=By%20displacing%20generation%20from%20these,285%20billion%20bottles%20of%20water.
511 The DARK side of Big Wind and Solar Skillets, https://saveourskylineohio.com/2020/06/12/the-dark-side-of-bigwind-and-solarskillets/
512 The DARK side of Big Wind and Solar Skillets, https://saveourskylineohio.com/2020/06/12/the-dark-side-of-bigwind-and-solarskillets/

513 Rolling, Mitchell, American Experiment, Limited Lifespans of Wind Turbines Result in Higher Costs of Energy, June 26, 2018, https://www.americanexperiment.org/2018/06/limited-lifespans-wind-turbines-result-higher-costs-energy/
514 Griffith, Keith, Dailymail and Associated Press, Not so green energy: Hundreds of non-recyclable fiberglass wind turbine blades are pictured piling up in landfill, May 6, 2020, https://www.dailymail.co.uk/news/article-8294057/Hundreds-non-recyclable-fiberglass-wind-turbine-blades-pictured-piling-landfills.html
515 NPR, Unfurling The Waste Problem Caused By Wind Energy, September 10, 2019, https://www.npr.org/2019/09/10/759376113/unfurling-the-waste-problem-caused-by-wind-energy
516 Martin, Chris, Bloomberg Green, Wind Turbine Blades Can't Be Recycled, So They're Piling Up in Landfills, February 5, 2020, https://www.bloomberg.com/news/features/2020-02-05/wind-turbine-blades-can-t-be-recycled-so-they-re-piling-up-in-landfills
517 Griffith, Keith, Daily Mail, Not so green energy: Hundreds of non-recyclable fiberglass wind turbine blades are pictured piling up in landfill, May 6, 2020, https://www.dailymail.co.uk/news/article-8294057/Hundreds-non-recyclable-fiberglass-wind-turbine-blades-pictured-piling-landfills.html
518 Driessen, Paul, Heartland Institute, POLICY BRIEF: PROTECTING THE ENVIRONMENT FROM THE GREEN NEW DEAL, DECEMBER 20, 2019, https://www.heartland.org/publications-resources/publications/policy-brief-protecting-the-environment-from-the-green-new-deal
519 Paul Driessen, Jim Lakely, Heartland Institute,, PRESS RELEASE: HOW THE GREEN NEW DEAL'S RENEWABLE ENERGY MINING WOULD HARM HUMANS AND THE ENVIRONMENT, APRIL 16, 2020, https://www.heartland.org/news-opinion/news/press-release-how-the-green-new-deals-renewable-energy-mining-would-harm-humans-and-the-environment
520 America's Power, It's Time to End Subsidies for Renewable Energy, April 17, 2020, http://www.americaspower.org/its-time-to-end-subsidies-for-renewable-energy/
521 Michael Greenstone and Ishan Nath, Do Renewable Portfolio Standards Deliver?, MAY 2019, https://epic.uchicago.edu/research/publications/do-renewable-portfolio-standards-deliver

522 Detsch, Jack, Gramer, Robbie, "Deep in the Heart of Texas, a Chinese Wind Farm Raises Eyebrows," www.ForeignPolicy.com, June 25, 2020. https://foreignpolicy.com/2020/06/25/texas-chinese-wind-farm-national-security-espionage-electrical-grid/

523 Driessen, Paul, Heartland Institute, POLICY BRIEF: PROTECTING THE ENVIRONMENT FROM THE GREEN NEW DEAL, DECEMBER 20, 2019, https://www.heartland.org/publications-resources/publications/policy-brief-protecting-the-environment-from-the-green-new-deal

524 Stein, Ronald, Amazon, https://www.amazon.com/Just-Green-Electricity-Citizens-Understand/dp/1480890707

525 Driessen, Paul, Heartland Institute, POLICY BRIEF: PROTECTING THE ENVIRONMENT FROM THE GREEN NEW DEAL, DECEMBER 20, 2019, https://www.heartland.org/publications-resources/publications/policy-brief-protecting-the-environment-from-the-green-new-deal

526 Paul Driessen, Jim Lakely, Heartland Institute,, PRESS RELEASE: HOW THE GREEN NEW DEAL'S RENEWABLE ENERGY MINING WOULD HARM HUMANS AND THE ENVIRONMENT, APRIL 16, 2020, https://www.heartland.org/news-opinion/news/press-release-how-the-green-new-deals-renewable-energy-mining-would-harm-humans-and-the-environment

527 Energy Sage, How long do solar panels last?, https://news.energysage.com/how-long-do-solar-panels-last/#:~:text=As%20a%20general%20solar%20industry,to%20be%20a%20significant%20amount.

528 Mow, Benjamin, Lifetime of PV Panels. April 23, 2018, https://www.nrel.gov/state-local-tribal/blog/posts/stat-faqs-part2-lifetime-of-pv-panels.html

529 Gambone, Sara, Solar Panel Degradation and The Lifespan of Solar Panels, April 24, 2019, https://www.paradisesolarenergy.com/blog/solar-panel-degradation-and-the-lifespan-of-solar-panels?hs_amp=true

530 Solar Energy Development Environmental Considerations, https://solareis.anl.gov/guide/environment/index.cfm

531 Union of Concerned Scientists, Environmental Impacts of Solar Power, March 5, 2013, https://www.ucsusa.org/resources/environmental-impacts-solar-power

532 Solar Energy Development Environmental Considerations, https://solareis.anl.gov/guide/environment/index.cfm

533 TESLA MEGA-BATTERY IN AUSTRALIA ACTIVATED, https://smartcity.lv/tesla-mega-battery-in-australia-activated/https://smartcity.lv/tesla-mega-battery-in-australia-activated/
534 Spector, Julian, The Guardian, California's big battery experiment: a turning point for energy storage?, September 15, 2017, https://www.theguardian.com/sustainable-business/2017/sep/15/californias-big-battery-experiment-a-turning-point-for-energy-storage
535 Penn, Ivan, Los Angeles Times, California invested heavily in solar power. Now there's so much that other states are sometimes paid to take it, June 22, 2017, https://www.latimes.com/projects/la-fi-electricity-solar/
536 Mills, Mark, Inconvenient Energy Realities, July 1, 2019, https://economics21.org/inconvenient-realities-new-energy-economy
537 Green Match, Impact Of Solar Energy On The Environment, October 1, 2019, https://www.greenmatch.co.uk/blog/2015/01/impact-of-solar-energy-on-the-environment#:~:text=Impacts%20to%20Soil%2C%20Water%20and,drainage%20channels%20and%20increased%20erosion.
538 Jeffrey E. Lovich, Joshua R. Ennen , ,American Institute of Biological Science, Wildlife Conservation and Solar Energy Development in the Desert Southwest, United States, Volume 61, Issue 12, Pages 982–992, December 2011, https://academic.oup.com/bioscience/article/61/12/982/392612
539 Black & Veatch, Impact of Solar Energy on Wildlife Is an Emerging Environmental Issue, January 1, 2017, https://www.bv.com/perspectives/impact-solar-energy-wildlife-emerging-environmental-issue
540 Laine, David, Sciencing, Effects of Solar Power Farms on the Environment, April 24, 2017, https://sciencing.com/effects-solar-power-farms-environment-13547.html
541 I & I Editorial Board, Issues & Insights, That Dirty Green Energy, August 24, 2020, https://issuesinsights.com/2020/08/24/that-dirty-green-energy/
542 Shellenberger, Michael, If Solar Panels Are So Clean, Why Do They Produce So Much Toxic Waste?, May 23, 2018, https://www.forbes.com/sites/michaelshellenberger/2018/05/23/if-solar-panels-are-so-clean-why-do-they-produce-so-much-toxic-waste/#20e177ce121c
543 An Inconvenient Truth: Solar Panels Wear Out and They're a Potent Source of Hazardous Waste, April 3, 2020, https://www.

hazardouswasteexperts.com/solar-panels-wear-out-hazardous-waste/#:~:text=%E2%80%9CSolar%20panels%20create%20300%20times,than%20do%20nuclear%20power%20plants.

544 I & I Editorial Board, Issues & Insights, , Elizabeth Warren's Energy Plan Is Unplugged From Reality, September 6, 2019, https://issuesinsights.com/2019/09/06/elizabeth-warrens-energy-plan-is-unplugged-from-reality/

545 Stone, Maddie, Solar panels are starting to die. What will we do with the megatons of toxic trash?, August 13, 2020, https://grist.org/energy/solar-panels-are-starting-to-die what-will-we-do-with-the-megatons-of-toxic-trash/

546 I & I Editorial Board, , Issue & Insights, California's Power Failure Is A Frightening Preview Of Democrats' 'Green New Deal', August 20, 2020, https://issuesinsights.com/2020/08/20/californias-power-failure-is-a-frightening-preview-of-democrats-green-new-deal/

547 I & I Editorial Board, Issue & Insights, That Dirty Green Energy, August 24, 2020, https://issuesinsights.com/2020/08/24/that-dirty-green-energy/

548 Solar Energy Development Environmental Considerations, https://solareis.anl.gov/guide/environment/index.cfm

549 Driessen, Paul, Heartland Institute, POLICY BRIEF: PROTECTING THE ENVIRONMENT FROM THE GREEN NEW DEAL, DECEMBER 20, 2019, https://www.heartland.org/publications-resources/publications/policy-brief-protecting-the-environment-from-the-green-new-deal

550 Paul Driessen, Jim Lakely, Heartland Institute,, PRESS RELEASE: HOW THE GREEN NEW DEAL'S RENEWABLE ENERGY MINING WOULD HARM HUMANS AND THE ENVIRONMENT, APRIL 16, 2020, https://www.heartland.org/news-opinion/news/press-release-how-the-green-new-deals-renewable-energy-mining-would-harm-humans-and-the-environment

551 Flanakin, Duggan, CFACT, The high costs and low benefits of electric vehicles, August 6th, 2020, https://www.cfact.org/2020/08/06/the-high-costs-and-low-benefits-of-electric-vehicles/

552 Pickerel, Kelly, Solar Power World, It's time to plan for solar panel recycling in the United States, April 2, 2018, https://www.solarpowerworldonline.com/2018/04/its-time-to-plan-for-solar-panel-recycling-in-the-united-states/

553 Amy Joi O'Donoghue, The dark side of 'green energy' and its threat to the nation's environment, Jan 30, 2021, https://www.deseret.com/utah/2021/1/30/22249311/why-green-energy-isnt-so-green-and-poses-harm-to-the-environment-hazardous-waste-utah-china-solar?fbclid=IwAR3q1CMrpkZRVCT_NJ0BvKHUkAaVjMTmG8hqUb5PR_oNTx6kE1ZRdJowm-xk
554 U.S. Environmental Protection Agency, RENEWABLE ENERGY WASTE STREAMS - PREPARING FOR THE FUTURE Briefing Paper, January 2021, https://www.epa.gov/sites/production/files/2021-01/documents/renewable_energy_waste_briefing_paper_january_2021.pdf.
555 Vekony, Atila, Green Match, The Opportunities of Solar Panel Recycling, What Happens to PV Panels When Their Life Cycle Ends, March 10, 2020, https://www.greenmatch.co.uk/blog/2017/10/the-opportunities-of-solar-panel-recycling
556 Stone, Maddie, Solar panels are starting to die. What will we do with the megatons of toxic trash?, August 13, 2020, https://grist.org/energy/solar-panels-are-starting-to-die-what-will-we-do-with-the-megatons-of-toxic-trash/
557 Flanakin, Duggan, "Biden should let us mine the rare earths his energy plans depend on," www.Cfact.org, January 24, 2021. https://www.cfact.org/2021/01/24/biden-should-let-us-mine-the-rare-earths-his-energy-plans-depend-on/?fbclid=IwAR2wBLSsF2JF_YYB3JSeUubj8lvGzX5u5ttcEvdVKrv3Ei-dxP9xUOBqwgc
558 Just GREEN Electricity – Helping Citizens Understand a World without Fossil Fuel https://www.amazon.com/Just-Green-Electricity-Citizens-Understand/dp/1480890707
559 Reuters, Copper demand for electric cars to rise nine-fold by 2027, JUNE 12, 2017, https://www.reuters.com/article/us-copper-demand-electric-vehicles-idUSKBN1940PC
560 Miles. Mark, Mines, Minerals, and "Green" Energy: A Reality Check, July 9, 2020, https://www.manhattan-institute.org/mines-minerals-and-green-energy-reality-check
561 Kelly, Michael, Until we get a proper roadmap, Net Zero is a goal without a plan, June 8, 2020, https://capx.co/until-we-get-a-proper-roadmap-net-zero-is-a-goal-without-a-plan/

562 Henk Car Sales Statistics, January 16, 2020, https://www.best-selling-cars.com/international/2019-full-year-international-worldwide-car-sales/
563 Green Car Reports, https://www.greencarreports.com/news/1093560_1-2-billion-vehicles-on-worlds-roads-now-2-billion-by-2035-report
564 Driessen, Paul, The Green New Deal Dress Rehearsal, June 6, 2020, https://townhall.com/columnists/pauldriessen/2020/06/06/the-green-new-deal-dress-rehearsal-n2570153
565 Lasley, Shane, Mining News, Pebble hosts huge critical rhenium lode, August 21, 2020, https://www.miningnewsnorth.com/story/2020/08/21/news/pebble-hosts-huge-critical-rhenium-lode/6406.html
566 Roberts, Joshua, Rueters, Pruitt's Bad Pebble Precedent, Jan. 30, 2018, https://www.wsj.com/articles/pruitts-bad-pebble-precedent-1517357679
567 Cohen, Bonner, CFACT, Alaska's Pebble Mine no threat to salmon, July 27th, 2020, https://www.cfact.org/2020/07/27/army-corps-of-engineers-concludes-alaska/?mc_cid=ce2acbb206&mc_eid=e99ae00b3b
568 Driessen, Paul and Mamula, Ned, Democrats' Green New Deal would make US reliance on China much worse, July 24, 2020, https://wattsupwiththat.com/2020/07/24/democrats-green-new-deal-would-make-us-reliance-on-china-much-worse/
569 Parry, Simon, and Douglas, Ed, In China, the true cost of Britain's clean, green wind power experiment: Pollution on a disastrous scale, January 26, 2011, https://www.dailymail.co.uk/home/moslive/article-1350811/In-China-true-cost-Britains-clean-green-wind-power-experiment-Pollution-disastrous-scale.html
570 Dickerson, Kelly, The world's lust for new technology is creating a 'hell on Earth' in Inner Mongolia, May 12, 2015, https://www.businessinsider.com/the-worlds-tech-waste-lake-in-mongolia-2015-5
571 AMIT KATWALA, Amit, WIRED on Energy. The spiraling environmental cost of our lithium battery addiction, August 5, 2018, https://www.wired.co.uk/article/lithium-batteries-environment-impact
572 Jones, Barbara, Daily Mail, Child miners aged four living a hell on Earth so YOU can drive an electric car: Awful human cost in squalid Congo cobalt mine that Michael Gove didn't consider in his 'clean' energy crusade, August 5, 2017, https://www.dailymail.co.uk/news/article-4764208/Child-miners-aged-four-living-hell-Earth.html

573 Olander, Eric, China's Cobalt Mining Giant in the DR Congo is in Trouble (Again), December 17, 2019, https://chinaafricaproject.com/2019/12/17/chinas-cobalt-mining-giant-in-the-dr-congo-is-in-trouble-again/
574 Dreissen, Paul, Heartland Institute Policy Brief, How the Green New Deal's Renewable Energy Mining Would Harm Humans and the Environment, April 2020, https://www.heartland.org/_template-assets/documents/publications/PBdriessenmining2Apr20.pdf
575 Gunasekaracon, Mandy, Daily Caller, May 22, 2019, https://dailycaller.com/2019/05/22/gunasekara-california-vehicle/?utm_campaign=Liberty%20
576 WSJ, The Green New Deal in Action - Democrats kill a child-labor measure for green-energy minerals., July 1, 2020, https://www.wsj.com/articles/the-green-new-deal-in-action-11593645373?mod=opinion_lead_pos3
577 EIA, Global EV Outlook 2018, May 2018, https://www.iea.org/reports/global-ev-outlook-2018
578 Global car market remains stable during 2018, as continuous demand for SUVs offsets decline in sales of Compact cars and MPVs, February 21, 2019, https://www.jato.com/global-car-market-remains-stable-during-2018-as-continuous-demand-for-suvs-offsets-decline-in-sales-of-compact-cars-and-mpvs/
579 Broom, Douglas, World Economic Forum, The dirty secret of electric vehicles, March 19, 2019, https://www.weforum.org/agenda/2019/03/the-dirty-secret-of-electric-vehicles/
580 20 In-Depth Global and US Auto Sales Statistics For 2020, Oct 20, 2020, https://policyadvice.net/insurance/insights/auto-sales-statistics/
581 Roland Irle, Roland, EV-volumes.com, Global BEV and PHEV Volumes for 2020 H1, http://www.ev-volumes.com/country/total-world-plug-in-vehicle-volumes/
582 Pyper, Julia, GTM, US Electric Vehicle Sales Increased by 81% in 2018 - Tesla had a banner year. But it's not all good news for the U.S. EV market, JANUARY 07, 2019, https://www.greentechmedia.com/articles/read/us-electric-vehicle-sales-increase-by-81-in-2018
583 Berggreen, Jesper, Clean Technica, Almost One Third Of All New Car Sales In Norway In 2018 Were For Pure Electric Vehicles, January 3, 2019, https://cleantechnica.com/2019/01/03/almost-one-third-of-all-new-car-sales-in-norway-in-2018-were-for-pure-electric-vehicles/
584 EV Market Share – California,https://evadoption.com/ev-market-share/ev-market-share-california/

585 Voelker, John, Green Car Reports, 1.2 billion vehicles on the world's roads with projections of 2 billion by 2035, July 29, 2014, https://www.greencarreports.com/news/1093560_1-2-billion-vehicles-on-worlds-roads-now-2-billion-by-2035-report

586 Gray, Alex, World Economic Forum, Countries are announcing plans to phase out petrol and diesel cars. Is yours on the list?, September 27, 2017, https://www.weforum.org/agenda/2017/09/countries-are-announcing-plans-to-phase-out-petrol-and-diesel-cars-is-yours-on-the-list/

587 Stein, Ronald, CFACT, Ramifications of California Governor Newsom's ban on gas-powered vehicles,| October 3, 2020, https://www.cfact.org/2020/10/03/ramifications-of-california-governor-newsoms-ban-on-gas-powered-vehicles/

588 Berman, Bradley, Insideevs, Lessons From Norway: Dispatch From Electric Car Revolution, March 1, 2019, https://insideevs.com/news/343106/lessons-from-norway-dispatch-from-electric-car-revolution/

589 Desjardins, Jeff, Visual Capitalist, Visualizing the Rise of the Electric Vehicle, September 27, 2018, https://www.visualcapitalist.com/rise-electric-vehicle/

590 Morgan Stanley, Batteries May Power Future of Auto Industry, May 19, 2017, https://www.morganstanley.com/ideas/electric-cars-sales-growth

591 Desjardins, Jeff, World Economic Forum, Electric vehicles should overtake traditional sales in just 20 years, October 1, 2018, https://www.weforum.org/agenda/2018/10/visualizing-the-rise-of-the-electric-vehicle

592 Amnesty International, Amnesty challenges industry leaders to clean up their batteries, March 19, 2019, https://www.amnesty.org/en/latest/news/2019/03/amnesty-challenges-industry-leaders-to-clean-up-their-batteries/

593 Bloomberg, Lithium Batteries' Dirty Secret: Manufacturing Them Leaves Massive Carbon Footprint, October 16, 2018, https://www.industryweek.com/technology-and-iiot/article/22026518/lithium-batteries-dirty-secret-manufacturing-them-leaves-massive-carbon-footprint

594 Amnesty International, DEMOCRATIC REPUBLIC OF CONGO: "THIS IS WHAT WE DIE FOR": HUMAN RIGHTS ABUSES IN THE DEMOCRATIC REPUBLIC OF THE CONGO POWER THE

GLOBAL TRADE IN COBALT, Janaury 19, 2016, https://www.amnesty.org/en/documents/afr62/3183/2016/en/
595 Amnesty International, Industry giants fail to tackle child labour allegations in cobalt battery supply chains, November 15, 2017, https://www.amnesty.org/en/latest/news/2017/11/industry-giants-fail-to-tackle-child-labour-allegations-in-cobalt-battery-supply-chains/
596 Amnesty International, DEMOCRATIC REPUBLIC OF CONGO: "THIS IS WHAT WE DIE FOR": HUMAN RIGHTS ABUSES IN THE DEMOCRATIC REPUBLIC OF THE CONGO POWER THE GLOBAL TRADE IN COBALT, Janaury 19, 2016, https://www.amnesty.org/en/documents/afr62/3183/2016/en/
597 Amnesty International, Amnesty challenges industry leaders to clean up their batteries, Without radical changes, the batteries which power green vehicles will continue to be tainted by human rights abuses, March 21, 2019, https://www.amnesty.org/en/latest/news/2019/03/amnesty-challenges-industry-leaders-to-clean-up-their-batteries/
598 Amnesty International, DEMOCRATIC REPUBLIC OF CONGO: "THIS IS WHAT WE DIE FOR": HUMAN RIGHTS ABUSES IN THE DEMOCRATIC REPUBLIC OF THE CONGO POWER THE GLOBAL TRADE IN COBALT, Janaury 19, 2016, https://www.amnesty.org/en/documents/afr62/3183/2016/en/
599 Shah, Anup, Global Issues, Poverty Facts and Stats, January 7, 2013, https://www.globalissues.org/article/26/poverty-facts-and-stats
600 Goad, Robin, New cobalt supply central to growing electric vehicle market, https://s1.q4cdn.com/337451660/files/doc_downloads/in-the-media/190101-Canadian-Mining-Journal-Cobalt-Commentary.pdf
601 USGS, Cobalt Statistics and Information, https://www.usgs.gov/centers/nmic/cobalt-statistics-and-information
602 Amnesty International, Industry giants fail to tackle child labour allegations in cobalt battery supply chains, November 15, 2017, https://www.amnesty.org/en/latest/news/2017/11/industry-giants-fail-to-tackle-child-labour-allegations-in-cobalt-battery-supply-chains/
603 Goad, Robin, New cobalt supply central to growing electric vehicle market, https://s1.q4cdn.com/337451660/files/doc_downloads/in-the-media/190101-Canadian-Mining-Journal-Cobalt-Commentary.pdf
604 LME PROPOSES REQUIREMENTS FOR THE RESPONSIBLE SOURCING OF METAL IN LISTED BRANDS, October 15, 2018, https://www.lme.com/News/Press-room/Press-releases/

Press-releases/2018/10/LME-proposes-requirements-for-the-respo nsible-sourcing-of-metal-in-listed-brands
605 World Economic Forum, https://www.weforum.org/global-battery-alliance/action
606 Amnesty International Press Release, Electric vehicle companies called on to produce 'ethical battery' within five years, March 21, 2019, https://www.amnesty.org.uk/press-releases/electric-vehicle-companie s-called-produce-ethical-battery-within-five-years
607 Broom, Douglas, World Economic Forum, The dirty secret of electric vehicles, March 27, 2019, https://www.weforum.org/agenda/2019/03/the-dirty-secret-of-electric-vehicles/
608 Ritchie, Hannah, Our World in Data, Cars, planes, trains: where do CO2 emissions from transport come from?, October 6, 2020, https://ourworldindata.org/co2-emissions-from-transport
609 Bloomberg, Battery Pack Prices Fall As Market Ramps Up With Market Average At $156/kWh In 2019, December 3, 2019, https://about.bnef.com/blog/battery-pack-prices-fall-as-market-ramps-up-with-market-average-at-156-kwh-in-2019/
610 World Bank Group, The Growing Role of Minerals and Metals for a Low Carbon Future, https://openknowledge.worldbank.org/handle/10986/28312
611 CandriAm, COBALT, THE NEW BLOOD DIAMOND, October 23, 2018, https://www.candriam.com/en/private/market-insights/topics/sri/cobalt-the-new-blood-diamond/
612 Mining watch, International Conference, November 14 & 15, 2019, https://miningwatch.ca/turning-down-the-heat?__cf_chl_jschl_tk__=76f0d5735bb1e966becf36259e02f2f4715440a6-1604425109-0-AYdzMswBwLO2n6LSiQqOfMIhZShqE2UU2lC5jwkIeOZHNB1Olslhcfea5mlQxUW_4O-knILKJHenL7zXAA-2CC2fcMk63GiNagzPuHIaoMEkVv-j0gqVsz3YoIUOijHkRwH78LS_KfKUaJbje6N4EZA_gHVJc_xrn7aQZMnTOSt3vlgCQyIytg-lFPQYRtU5QJ3hO9jWAe2VnpOaEVBeBW7aj7CVGoG2IS3sB4iBTQLwDAY1aQLCxi1U8sGhPVV0mgM3i8DYsN0D1EWPjS2yRqlAMacgZ-PKFkG1K-7jAWCC
613 Church, Claire and Crawford, Alec, International Institute for Sustainable Development , Green Conflict Minerals - The fuels of conflict in the transition to a low-carbon economy, August 2018 , https://www.iisd.org/story/green-conflict-minerals/

614 Voelk, Tom, New York Times, Rise of S.U.V.s: Leaving Cars in Their Dust, With No Signs of Slowing, May 21, 2020,https://www.nytimes.com/2020/05/21/business/suv-sales-best-sellers.html

615 IBM, Revolutionize your supply chain with trusted data from end to end, https://www.ibm.com/blockchain/industries?p1=Search&p4=43700050370610197&p5=b&cm_mmc=Search_Google-_-1S_1S-_-WW_NA-_-%2Bblockchain%20%2Bindustries_b&cm_mmca7=71700000060890036&cm_mmca8=aud-382859943522:kwd-447097811179&cm_mmca9=Cj0KCQjwwOz6BRCgARIsAKEG4FWy0GedFyqLEzYoFqcazh9zJskXoiyyVy3l3_9eU5PUzaMAO4iUyWYaAk4sEALw_wcB&cm_mmca10=406207199924&cm_mmca11=b&gclid=Cj0KCQjwwOz6BRCgARIsAKEG4FWy0GedFyqLEzYoFqcazh9zJskXoiyyVy3l3_9eU5PUzaMAO4iUyWYaAk4sEALw_wcB&gclsrc=aw.ds

616 Hawkins, Andrew, The Verge, VOLVO UNVEILS ITS FIRST FULLY ELECTRIC CAR — AND A BOLD PLEDGE TO GO CARBON-NEUTRAL, October 16, 2019, https://www.theverge.com/2019/10/16/20915841/volvo-xc40-recharge-electric-suv-specs-miles-range-reveal

617 Volvo, 1 million Volvo Cars has committed to putting one million electrified cars on the road by 2025. https://group.volvocars.com/company/innovation/electrification

618 Glencore Press Release, Glencore to join The Responsible Sourcing Blockchain Network, December 12, 2019, https://www.glencore.com/media-and-insights/news/glencore-joins-responsible-sourcing-blockchain-network

619 Vasil, Adria, Corporate Knights, The EV revolution will take batteries, but are they ethical?, January 20, 2020, https://www.corporateknights.com/channels/transportation/ev-revolution-needs-batteries-ethical-15795118/

620 U.S. National Library of Medicine, Adverse health effects of industrial wind turbines, May 2013, https://www.ncbi.nlm.nih.gov/pmc/articles/PMC3653647/#__sec3title

621 Adverse health effects of industrial wind turbines , http://documents.dps.ny.gov/public/Common/ViewDoc.aspx?DocRefId=%7B07354F74-908F-4EA3-A2D7-F846C7CEED8C%7D

622 Glencore Facts about Industrial Wind Turbine Noise, https://lincoln.ne.gov/city/plan/dev/wind/working-group-info/chapman-6.pdf

623 World Health Organization, Guidelines for community noise, https://apps.who.int/iris/handle/10665/66217
624 Lange, Sherri, Master Resource, World Health Organization: Wind Turbine Noise as a Health Hazard (opening recognition likely to lead to more acknowledgement), October 17, 2018, https://www.masterresource.org/wind-turbine-noise-issues/wto-wind-turbine-noise-as-a-health-hazard/
625 Stop These Things, America's Rural Communitics Go On The Offensive Against Big Wind's Turbine Onslaught, October 18, 2020, https://stopthesethings.com/2020/10/18/americas-rural-communities-go-on-the-offensive-against-big-winds-turbine-onslaught/
626 Haggerty, Frank, TOWN HID WARNING TURBINES 6 DECIBELS HIGHER THAN ORIGINAL KEMA GE NOISE MODEL, December 2, 2015, https://patch.com/massachusetts/falmouth/town-hid-warning-turbines-6-decibels-higher-original-kema-ge-noise-model-0
627 Better Plan, Wisconsin, ICE THROW AND BLADE THROW, http://betterplan.squarespace.com/todays-special/2008/12/11/121208-part-3-of-our-look-at-the-history-and-content-of-the.html#:~:text=ICE%20THROW%20AND%20BLADE%20THROW,blade%20throws%20can%20reach%202500.
628 Overton, Thomas, Power, U.S. Fish & Wildlife Service Finalizes Rule on Wind Turbine Eagle Deaths, December 14, 2016, https://www.powermag.com/u-s-fish-wildlife-service-finalizes-rule-on-wind-turbine-eagle-deaths/
629 Daly, Matthew, AP News, Final wind-turbine rule permits thousands of eagle deaths, December 14, 2016, https://apnews.com/article/b8dd6050c702467e8be4b1272a3adc87#:~:text=WASHINGTON%20(AP)%20%E2%80%94%20The%20Obama,protected%20bald%20and%20golden%20eagles.
630 Institute for Energy Research, Obama Allows Wind Turbines to Legally Kill Eagles, January 9, 2017, https://www.instituteforenergyresearch.org/renewable/wind/obama-allows-wind-turbines-legally-kill-eagles/
631 Conca, James, Forbes, Forget Eagle Deaths, Wind Turbines Kill Humans, https://www.forbes.com/sites/jamesconca/2013/09/29/forget-eagle-deaths-wind-turbines-kill-humans/?sh=d620af546739
632 Caithness Windfarm Information Forum 2020. Wind Farm Accidents and Fatalities, Summary of Wind Turbine Accident data, September 30, 2020, http://www.caithnesswindfarms.co.uk/AccidentStatistics.htm

633 Unicef, Reduce child mortality, https://sites.unicef.org/mdg/childmortality.html
634 Unicef, Reduce child mortality, https://sites.unicef.org/mdg/childmortality.html
635 Worldometers, Abortions worldwide, https://www.worldometers.info/abortions/
636 Energy Central, Deaths from Nuclear Energy Compared with Other Causes, Feb 26, 2013, https://energycentral.com/c/ec/deaths-nuclear-energy-compared-other-causes
637 Energy Central, Deaths from Nuclear Energy Compared with Other Causes, Feb 26, 2013, https://energycentral.com/c/ec/deaths-nuclear-energy-compared-other-causes
638 World Health Organization, Household air pollution and health, May 8, 2018, https://www.who.int/news-room/fact-sheets/detail/household-air-pollution-and-health
639 International Labour Organization, The enormous burden of poor working conditions, https://www.ilo.org/moscow/areas-of-work/occupational-safety-and-health/WCMS_249278/lang--en/index.htm
640 COVID-19 Dashboard by the Center for Systems Science and Engineering (CSSE) at Johns Hopkins University (JHU), https://gisanddata.maps.arcgis.com/apps/opsdashboard/index.html#/bda7594740fd40299423467b48e9ecf6
641 Energy Central, Deaths from Nuclear Energy Compared with Other Causes, Feb 26, 2013, https://energycentral.com/c/ec/deaths-nuclear-energy-compared-other-causes
642 World Health Organization, More than 270 000 pedestrians killed on roads each year, https://www.who.int/mediacentre/news/notes/2013/make_walking_safe_20130502/en/
643 International Overdose Awareness Day, https://www.overdoseday.com/facts-stats/
644 Conca, James, How Deadly Is Your Kilowatt? We Rank The Killer Energy Sources, June 10, 2012, https://www.forbes.com/sites/jamesconca/2012/06/10/energys-deathprint-a-price-always-paid/?sh=7164dcab709b
645 Los Angeles Times, The Blade, More accidents feared as wind, solar-power installations spread, August 13, 2011, https://www.toledoblade.com/Energy/2011/08/14/More-accidents-feared-as-wind-solar-power-installations-spread.html

646 NCBI, Adverse health effects of industrial wind turbines, https://www.ncbi.nlm.nih.gov/pmc/articles/PMC3653647/
647 Fraser Institute, Ontario's Green Energy Act a Bad Bargain for Ontarians, https://www.fraserinstitute.org/article/ontarios-green-energy-act-bad-bargain-ontarians
648 French Academy of Medicine Declare Wind Turbines Health Nuisance, https://patch.com/massachusetts/falmouth/french-academy-medicine-declare-wind-turbines-health-nuisance
649 NCBI, Adverse health effects of industrial wind turbines, ttps://www.ncbi.nlm.nih.gov/pmc/articles/PMC3653647/
650 Ontario's Green Energy Act a Bad Bargain for Ontarians, https://www.fraserinstitute.org/article/ontarios-green-energy-act-bad-bargain-ontarians
651 Environmental and Economic Consequences of Ontario's Green Energy Act, https://www.fraserinstitute.org/studies/environmental-and-economic-consequences-ontarios-green-energy-act
652 Haggerty, Frank, French Academy of Medicine Declare Wind Turbines Health Nuisance, May 24, 2017, https://patch.com/massachusetts/falmouth/french-academy-medicine-declare-wind-turbines-health-nuisance
653 Bryce, Robert, Forbes Magazine, Warren Buffett's Iowa Wind Power Expansion Derailed By The Bridges Of Madison County, https://www.forbes.com/sites/robertbryce/2021/01/13/iowa-wind-expansion-derailed-by-the-bridges-of-madison-county/?sh=6c9f2a29a0ce
654 Madison County IA Wind Ordinance, Madison County Board of Supervisors , December 22, 2020, http://www.windaction.org/posts/52008-madison-county-ia-wind-ordinance#.YAH7U-iQHro
655 BIDEN-SANDERS UNITY TASK FORCE RECOMMENDATIONS, https://joebiden.com/wp-content/uploads/2020/08/UNITY-TASK-FORCE-RECOMMENDATIONS.pdf
656 Wehrmann, Benjamin, Clean Energy Wire, Limits to growth: Resistance against wind power in Germany, March 27, 2019, https://www.cleanenergywire.org/factsheets/fighting-windmills-when-growth-hits-resistance
657 Stop These Things, Renewables 'Transition' Hits Roadblock: Germans Vote 25 to 1 to Reject Giant Wind Power Project, March 19, 2020, https://stopthesethings.com/2020/03/19/renewables-transition-hits-roadblock-germans-vote-25-to-1-to-reject-giant-wind-power-project/

658 Industrial Wind Turbines and Health, Wind Turbines Can Harm Humans if too close to Residents1, http://wiseenergy.org/Energy/Health/Summary_references_wind_turbines_and_health_April_2015.pdf
659 Roth, Sammy, Los Angeles Times, California's San Bernardino County slams the brakes on big solar projects, February 28, 2019, https://www.latimes.com/business/la-fi-san-bernardino-solar-renewable-energy-20190228-story.html
660 Stop These Things, America's Rural Communities Go On The Offensive Against Big Wind's Turbine Onslaught, October 18, 2020, https://stopthesethings.com/2020/10/18/americas-rural-communities-go-on-the-offensive-against-big-winds-turbine-onslaught/
661 Mothers Against Wind Turbines, Insurance Company Balks at Claims for Damages Due to Wind Turbine Noise..., September 27, 2016, https://mothersagainstwindturbines.com/tag/acoustics-experts/
662 Prohaska, Thomas, National Wind Watch, Survey: Opposition to wind power project grows in Somerset, Yates , October 27, 2020, https://www.wind-watch.org/news/2020/10/28/survey-opposition-to-wind-power-project-grows-in-somerset-yates/
663 Hubbuch, Chris, Wisconsin State Journal, Controversial Green County wind farm scrapped , October 26, 2020, https://www.wind-watch.org/news/2020/10/27/controversial-green-county-wind-farm-scrapped/
664 National Wind Watch , Backlash against Shetland wind farm proposal , October 28, 2020, https://www.wind-watch.org/news/2020/10/29/backlash-against-shetland-wind-farm-proposal/
665 Mexico News Daily, New 132-turbine Oaxaca wind farm is largest in Latin America, The farm has been controversial for years because of opposition from indigenous groups, May 29, 2019, https://mexiconewsdaily.com/news/wind-farm-is-largest-in-latin-america/
666 McGovern, Michael, Wind Power Monthly, Kenya, 61MW Kinangop project cancelled, February 25, 2016, https://www.windpowermonthly.com/article/1385206/61mw-kinangop-project-cancelled
667 Roy D. Jeffery, MD, Adverse health effects of industrial wind turbines, May 2013, https://www.ncbi.nlm.nih.gov/pmc/articles/PMC3653647/
668 Ross McKitrick and Kenneth Green, Fraser Institute, Ontario's Green Energy Act a Bad Bargain for Ontarians, https://www.fraserinstitute.org/article/ontarios-green-energy-act-bad-bargain-ontarians

669 Haggerty, Frank, French Academy of Medicine Declare Wind Turbines Health Nuisance, ADVERSE HEALTH EFFECTS, DIRECT EFFECTS, HEALTH, NOISE, WHO, FRENCH ACADEMY OF MEDICINE DECLARE WIND TURBINES HEALTH NUISANCE, May 24, 2017, https://patch.com/massachusetts/falmouth/french-academy-medicine-declare-wind-turbines-health-nuisance

670 The Guardian, Renewable energy to expand by 50% in next five years – report, https://www.theguardian.com/environment/2019/oct/21/renewable-energy-to-expand-by-50-in-next-five-years-report

671 International Renewable Energy Agency, IRENA, Roadmap to 2050, 2018, https://www.irena.org/publications/2019/Apr/Global-energy-transformation-A-roadmap-to-2050-2019Edition

672 Karlsson, Carl-Johan and Zimmer, Katrina, Foreign Policy, Green Energy's Dirty Side Effects, The global transition to renewables could lead to human rights abuses and risks exacerbating inequalities between the West and the developing world, June 18, 2020, https://foreignpolicy.com/2020/06/18/green-energy-dirty-side-effects-renewable-transition-climate-change-cobalt-mining-human-rights-inequality/

673 Gearino, Dan, Inside Climate News, Inside Clean Energy: 6 Things Michael Moore's 'Planet of the Humans' Gets Wrong, The documentary's "facts" are deceptive and misleading, not to mention way out of date, April 30, 2020, https://insideclimatenews.org/news/29042020/inside-clean-energy-michael-moore-planet-of-the-humans-review

674 What is blockchain technology?, https://www.ibm.com/blockchain/what-is-blockchain?p1=Search&p4=43700054784999107&p5=e&cm_mmc=Search_Google-_-1S_1S-_-WW_NA-_-blockchain%20technology_e&cm_mmca7=71700000061222563&cm_mmca8=aud-382859943522:kwd-296141061820&cm_mmca9=Cj0KCQjw8rT8BRCbARIsALWiOvRS0NfQvgjVT0_14uZgtzw88f0RRC-jzLOvc0Uw0sWZM8-jilx0TDsaAr8eEALw_wcB&cm_mmca10=406205800052&cm_mmca11=e&gclid=Cj0KCQjw8rT8BRCbARIsALWiOvRS0NfQvgjVT0_14uZgtzw88f0RRC-jzLOvc0Uw0sWZM8-jilx0TDsaAr8eEALw_wcB&gclsrc=aw.ds

675 Karlsson, Carl-Johan and Zimmer, Katrina, Foreign Policy, Green Energy's Dirty Side Effects, The global transition to renewables could lead to human rights abuses and risks exacerbating inequalities between the West and the developing world, June 18, 2020, https://foreignpolicy.com/2020/06/18/green-energy-dirty-side-effects-re

newable-transition-climate-change-cobalt-mining-human-rights-inequality/
676 Global Environmental Change, The decarbonisation divide: Contextualizing landscapes of low-carbon exploitation and toxicity in Africa, January 2020, https://www.sciencedirect.com/science/article/pii/S0959378019305886
677 Minter, Adam, Smithsonian Magazine, The Burning Truth Behind an E-Waste Dump in Africa, Ending the toxic smoke rising from an iconic dump in Ghana will take more than curbing Western waste, January 13, 2016, https://www.smithsonianmag.com/science-nature/burning-truth-behind-e-waste-dump-africa-180957597/
678 Ebenezer Forkuo, AmankwaaaKwame, A.Adovor Tsikudob, Jay A.Bowmanb, ScienceDirect, 'Away' is a place: The impact of electronic waste recycling on blood lead levels in Ghana, https://www.sciencedirect.com/science/article/abs/pii/S0048969717313979
679 U.N. Environmental Programme, Mineral Resource Governance in the 21st Century Gearing Extractive Industries Towards Sustainable Development, https://www.resourcepanel.org/reports/mineral-resource-governance-21st-century
680 International Council on Mining & Minerals (ICCM), Health and Safety, https://www.icmm.com/en-gb/health-and-safety
681 Glencore, What we do Metals & Minerals, https://www.glencore.com/what-we-do/metals-and-minerals
682 Glencore Sustainability Report,2019, https://www.glencore.com/dam/jcr:c7f6c5fc-b275-4e12-9448-de5302805781/2019-Glencore-Sustainability-Highlights--.pdf
683 Gold Fields, South Africa region, https://www.goldfields.com/south-africa-region.php
684 Lavietes, Mathew, Reuters, Tesla, Apple among firms accused of aiding child labor in Congo, December 16, 2019, https://www.reuters.com/article/us-usa-mining-children-trfn/tesla-apple-among-firms-accused-of-aiding-child-labor-in-africa-idUSKBN1YK24F
685 Ali, Saleem, Professor of Energy and the Environment, https://www.udel.edu/faculty-staff/experts/saleem-ali/
686 Benjamin K. Sovacool, Saleem H. Ali, Science, Sustainable minerals and metals for a low-carbon future, January 3, 2020, https://science.sciencemag.org/content/367/6473/30

687 ScienceDirect, An overview of solar photovoltaic panels' end-of-life material recycling, https://www.sciencedirect.com/science/article/pii/S2211467X19301245

688 UN report: Time to seize opportunity, tackle challenge of e-waste, January 22, 2019, https://www.unenvironment.org/news-and-stories/press-release/un-report-time-seize-opportunity-tackle-challenge-e-waste

689 Chris Jochnick, Phil Bloomer, BUSINESS & HUMAN RIGHTS RESOURCE CENTRE, https://www.unenvironment.org/news-and-stories/press-release/un-report-time-seize-opportunity-tackle-challenge-e-waste

690 Violations of Indigenous Peoples' and local communities' rights and steps towards reform in 27 countries. For Sustainable Development, Violations of Indigenous Peoples' and local communities' rights and steps towards reform in 27 countries, https://www.indigenouspeoples-sdg.org/index.php/english/all-global-news/779-violations-of-indigenous-peoples-and-local-communities-rights-and-steps-towards-reform-in-27-countries

691 Environmental Justice Atlas, Corporate Wind Farms in Ixtepec vs community's initiative, Oaxaca, Mexico, March 29, 2017, https://ejatlas.org/conflict/communal-members-of-ixtepec-contending-to-develop-a-wind-farm-cooperative

692 Stein, Ronald, CFACT, THE DARK SIDE OF RENEWABLE ELECTRICITY, May 19, 2020, https://www.heartland.org/news-opinion/news/the-dark-side-of-renewable-electricity

693 Service, Robert, Science Magazine, Can the world make the chemicals it needs without oil?, September 19, 2019, https://www.sciencemag.org/news/2019/09/can-world-make-chemicals-it-needs-without-oil

694 Life Without Oil Chart, https://tallbloke.files.wordpress.com/2018/07/oila.jpg?w=640

695 The Environmental Literacy Council, Petroleum History, https://enviroliteracy.org/energy/fossil-fuels/petroleum-history/

696 Minerals Education Coalition, Minerals In Your Life, http://mineralseducationcoalition.org/mining-minerals-information/minerals-in-your-life/

697 Rare Earth Emergency #1, YouTube video, May 6, 2020, https://www.youtube.com/watch?v=fGxqPf4_hmA&feature=em-uploademail

698 Components of Wind Machines, https://mragheb.com/NPRE%20 475%20Wind%20Power%20Systems/Components%20of%20 Wind%20Machines.pdf
699 Leena Grandell and Mikael Höök, Assessing Rare Metal Availability Challenges for Solar Energy Technologies, 2015, https://www.mdpi. com/2071-1050/7/9/11818/pdf
700 Hongqiao, Liu, The dark side of renewable energy, August 25, 2016, https://earthjournalism.net/stories/the-dark-side-of-renewable-energy
701 Paris Agreement, https://en.wikipedia.org/wiki/Paris_Agreement
702 WSJ, The Best-Laid Energy Plans, January 16, 2020, https://www. wsj.com/articles/the-best-laid-energy-plans-11579219416?mod= opinion_lead_pos1
703 Rotter, Charles, U.S. Government continues to dump funds into an electrical sinkhole, January 24, 2020, https://wattsupwiththat. com/2020/01/24/u-s-government-continues-to-dump-funds-in to-an-electrical-sinkhole/
704 Crescent Dunes Solar Energy Project, https://en.wikipedia.org/wiki/ Crescent_Dunes_Solar_Energy_Project
705 Concentrated solar power, https://en.wikipedia.org/wiki/ Concentrated_solar_power
706 Molten Salt, https://en.wikipedia.org/wiki/Molten_salt
707 Pielke, Roger, Forbes, The World Is Not Going To Halve Carbon Emissions By 2030, So Now What?, October 27, 2019, https:// www.forbes.com/sites/rogerpielke/2019/10/27/the-world-is-not-going-to-reduce-carbon-dioxide-emissions-by-50-by-2030-now-what/?sh=33d6494f3794
708 William Wilkes, Hayley Warren and Brian Parkin, August 15, 2018, Germany's Failed Climate Goals A Wake-Up Call for Governments Everywhere https://www.bloomberg.com/ graphics/2018-germany-emissions/
709 Ellen Thalman and Benjamin Wehrmann, Clean Energy Wire, What German households pay for power, January 4, 2020, https://www. cleanenergywire.org/factsheets/what-german-households-pay-power
710 Hongqiao, Liu, The dark side of renewable energy, August 25, 2016, https://earthjournalism.net/stories/the-dark-side-of-renewable-energy
711 Bushong, Steven, Windpower Engineering Development, Rare earths, minerals used in windpower technology, could fall into short supply, June 4, 2013, https://www.windpowerengineering.

com/rare-earths-minerals-used-in-windpower-technology-could-fall-into-short-supply/

712 Institute of Rare Earths and Metals , China is suffering from severe pollution from rare earth mining, https://en.institut-seltene-erden.de/China-is-suffering-from-heavy-pollution-from-mining-rare-earths/

713 Mancheri, Nabeel, East Asia Forum, China's white paper on rare earths, August 16, 2012, https://www.eastasiaforum.org/2012/08/16/chinas-white-paper-on-rare-earths/

714 Hongqiao, Liu, Wrong Kind of Green, THE DARK SIDE OF RENEWABLE ENERGY: THE BOTTLENECK OF A LOW-CARBON FUTURE, August 25, 2016, http://www.wrongkindofgreen.org/tag/medium-and-heavy-rare-earth-elements-mhrees/

715 China Water Risk, Rare Earths: Shades Of Grey, July 20, 2016, https://www.chinawaterrisk.org/notices/cwr-rare-earths-shades-of-grey/

716 Schultz, Colin, Smithsonian Magazine, High-Tech's Crucial Rare Earth Elements Are Already Running Low, June 22, 2012, https://www.smithsonianmag.com/smart-news/high-techs-crucial-rare-earth-elements-are-already-running-low-133603653/?no-ist

717 Yan, Zhou, China Daily Europe, Plan to oversee Ganzhou rare earths industry, July 3, 2012, http://europe.chinadaily.com.cn/business/2012-07/03/content_15545079.htm

718 Hongqiao, Liu, The dark side of renewable energy, August 25, 2016, https://earthjournalism.net/stories/the-dark-side-of-renewable-energy

719 U.S. Geological Survey, Rare Earths Statistics and Information, https://www.usgs.gov/centers/nmic/rare-earths-statistics-and-information

720 Liu, Hongqiao, CWR, Rare Earth Black Market: An Open Dirty Secret, July 20, 2016, https://www.chinawaterrisk.org/resources/analysis-reviews/rare-earth-black-market-an-open-dirty-secret/

721 Jie, Chen, China Dialogue, Death of the desert, July 14, 2015, https://chinadialogue.net/en/pollution/8015-death-of-the-desert/

722 Maughan, Tim, BBC, The dystopian lake filled by the world's tech lust, April 2, 2015, https://www.bbc.com/future/article/20150402-the-worst-place-on-earth

723 Els, Frik, Mining.com, Is the rare earth price slump finally over?, October 5, 2015, http://www.mining.com/is-the-rare-earth-price-slump-finally-over/

724 Renewable energy generation is expected to dominate, http://finance.sina.com.cn/chanjing/cyxw/20141219/022421113824.shtml

725 "Made in China 2025" Industrial Policies: Issues for Congress, https://fas.org/sgp/crs/row/IF10964.pdf
726 The Guardian, Chinese premier declares war on pollution in economic overhaul, 2014, https://www.theguardian.com/world/2014/mar/05/china-pollution-economic-reform-growth-target
727 Mills, Mark, Manhattan Institute, The "New Energy Economy": An Exercise in Magical Thinking, March 26, 2019, https://www.manhattan-institute.org/green-energy-revolution-near-impossible
728 William Wilkes, Hayley Warren and Brian Parkin, August 15, 2018, Germany's Failed Climate Goals A Wake-Up Call for Governments Everywhere https://www.bloomberg.com/graphics/2018-germany-emissions/
729 Driessen, Paul, Heartland Institute, POLICY BRIEF: PROTECTING THE ENVIRONMENT FROM THE GREEN NEW DEAL, December 20, 2019, https://www.heartland.org/publications-resources/publications/policy-brief-protecting-the-environment-from-the-green-new-deal
730 Lakely, Jim and Driessen Paul, Heartland Institute, PRESS RELEASE: HOW THE GREEN NEW DEAL'S RENEWABLE ENERGY MINING WOULD HARM HUMANS AND THE ENVIRONMENT, April 16, 2020, https://www.heartland.org/news-opinion/news/press-release-how-the-green-new-deals-renewable-energy-mining-would-harm-humans-and-the-environment

CHAPTER SIX: CHINA AND INDIA

731 Bickerton, James, "WW3 fear explode as US, India, Japan and Australia unite against China in military drills," www.Express.Co.UK.com, October 24, 2020. https://www.express.co.uk/news/world/1350257/WW3-news-China-US-America-India-Japan-Australia-military-drills-navy-south-china-sea-ont
732 Sheffield, Hazel, "'Carbon-neutrality is a fairy tale'" how the race for renewables is burning Europe's forest – Wood pellets are sold as a clean alternative to coal. But is the subsidized bioenergy boom accelerating the climate crisis?" www.TheGuardian.com, January 14, 2021. https://www.theguardian.com/world/2021/jan/14/carbon-neutrality-is-a-fairy-tale-how-the-race-for-renewables-is-burning-europes-forests

733 Hill, Geoff, Heart of Darkness: Why Electricity For Africa Is A Security Issue, Essay 13, www.GWPF.org, (The Global Warming Policy Foundation), 2020. https://www.thegwpf.org/content/uploads/2020/06/Heart-Darkness-Africa-Energy-Poverty.pdf
734 Mathis, Mark for the Clear Energy Alliance, China CO2 Pledge (Video for YouTube.com), www.YouTube.com, January 25, 2021. https://www.youtube.com/watch?v=mtKplzTh61s&feature=youtu.be
735 McCarthy, Gerard, Haigh, Ivan, "The Atlantic is entering a cool phase that will change the world's weather," www.TheConversation.com, May 29, 2015. https://theconversation.com/the-atlantic-is-entering-a-cool-phase-that-will-change-the-worlds-weather-42497
736 Richard, Kenneth, "Over 400 Scientific Papers Published In 2020 Support A Skeptical Position On Climate Alarm," www.NoTricksZone.com, January 29, 2021. https://notrickszone.com/2021/01/29/over-400-scientific-papers-published-in-2020-support-a-skeptical-position-on-climate-alarm/
737 Lifson, Thomas, "120 years of climate scares," www.AmericanThinker.com, August 4, 2014. https://www.americanthinker.com/blog/2014/08/120_years_of_climate_scares.html
738 WiseEnergy.org, "Sample Books Related To Anthropogenic Global Warming (listed alphabetically by primary author)," www.WiseEnergy.org, Page accessed February 2, 2021. http://wiseenergy.org/Energy/AGW/Sample_AGW_Books.pdf
739 Global Temperature Trends From 2500 B.C. To 2040 A.D. By Climatologist Cliff Harris and Meteorologist Randy Mann Article and Chart Updated: April 8, 2020. http://www.longrangeweather.com/global_temperatures.htm
740 Homewood, Paul, "China's Dystopian Lake-Courtesy Of The World's Lust For Rare Earths," www.NotALotOfPeopleKnowThat.wordpress.com, December 24, 2020. https://notalotofpeopleknowthat.wordpress.com/2020/12/24/chinas-dystopian-lake-courtesy-of-the-worlds-lust-for-rare-earths/
741 Mills, Mark P., "The Green New Deal Can't Break The Laws Of Physics," www.DailyCaller.com, October 27, 2020. https://dailycaller.com/2020/10/27/green-new-deal-laws-of-physics/
742 Cheng, Evelyn, "China's electric car strategy is starting to go global – and the U.S. is lagging behind," www.CNBC.com, October 22, 2020. https://www.cnbc.com/2020/10/23/chinas-electric-car-strategys-implications-for-us-energy-security.html

743 Mills, Mark P., Mines, Minerals, and "Green" Energy: A Reality Check, Report, Energy & Environment Regulations, Manhattan Institute, www.Manhattan-Institute.com, July 9, 2020. https://www.manhattan-institute.org/mines-minerals-and-green-energy-reality-check
744 Mills. Ibid. Daily Caller. 2020.
745 Burton, Mark, Cherry, Libby, Stringer, David, "Elon Musk Is Going to Have a Hard Time Finding Clean Nickel," www.Bloomberg.com, August 21, 2020. https://www.bloomberg.com/news/articles/2020-08-22/elon-musk-is-going-to-have-a-hard-time-finding-clean-nickel
746 Mills. Ibid. Daily Caller. 2020.
747 Mills. Ibid. Manhattan Institute Report. 2020.
748 Moss, Trefor, "Electric-Car Sales Shrink in China but Remain a Government Priority," www.WSJ.com, May 26, 2020. https://www.wsj.com/articles/electric-car-sales-shrink-in-china-but-remain-a-government-priority-11590514874
749 United States Department of Energy, Quadrennial Technology Review, AN ASSESSMENT OF ENERGY TECHNOLOGIES AND RESEARCH OPPORTUNITIES, www.Energy.gov, September 2015. https://www.energy.gov/sites/prod/files/2017/03/f34/quadrennial-technology-review-2015_1.pdf
750 China Power, "Unpacking the complexity of China's rise," www.ChinaPower.CSIS.org, Data Sources from China's Rare Earth Exports, 2017-2019. https://chinapower.csis.org/china-rare-earths/#:~:text=Beijing's%20policies%20allowed%20China%20to,%2C%20alloys%2C%20and%20permanent%20magnets.
751 Peterson, John, "The Cobalt Cliff Could Eradicate Non-Chinese EV Manufacturing Before 2030," www.SeekingAlpha-com.cdn.ampproject.org, July 3, 2019. https://seekingalpha-com.cdn.ampproject.org/c/s/seekingalpha.com/amp/article/4273346-cobalt-cliff-eradicate-non-chinese-ev-manufacturing-2030
752 U.S. Energy Information Administration (EIA), Independent Statistics & Analysis, Today in Energy, "U.S. energy-related carbon dioxide emission fell in 2019, mainly in electric generation," www.EIA.gov, November 10, 2020. https://www.eia.gov/todayinenergy/detail.php?id=45836
753 EIA. Ibid. 2020.

754 Paulitz, Heinrik, Sonne, Kalte, ""8 January 2021: Europe just skirted blackout disaster, On 8 January 2021, the European electricity grid only just missed a large-scale collapse. Around 13.04 p.m. there was a sharp drop in frequency that could have paralysed Europe," www.GWPF.com (Global Warming Policy Foundation), January 8, 2021. https://www.thegwpf.com/8-january-2021-europe-just-skirted-blackout-disaster/

755 Tinker, Scott, "We must go honest to 'go green,'" www.TheHill.com, January 20, 2021. https://thehill.com/opinion/energy-environment/534914-we-must-go-honest-to-go-green?rl=1

756 Dr. R. van der Vaart, Donald, "Battery storage is not what you think," www.JohnLocke.org, January 22, 2021. https://thehill.com/opinion/energy-environment/534914-we-must-go-honest-to-go-green?rl=1

757 Kawate, Iori, Nikkei staff writer, "China readies new law to ban companies on national security grounds," www.Asia.Nikkei.com, October 9, 2020. https://asia.nikkei.com/Politics/International-relations/US-China-tensions/China-readies-new-law-to-ban-companies-on-national-security-grounds?utm_campaign=R...

758 Associated Press (AP), Global carbon emissions projected to hit record levels, due to big jumps by China, India," www.MarketWatch.com, December 3,

759 Stumvoll, Ashley, "Are There Potential Downsides Of Going To 100 Percent Renewable Energy," www.PSMag.com, June 20, 2019. https://psmag.com/environment/what-are-the-downsides-to-renewable-energy

760 Broom, Douglas, "The dirty secret of electric vehicles," www.WEForum.org, March 27, 2019. https://www.weforum.org/agenda/2019/03/the-dirty-secret-of-electric-vehicles/

761 Hodal, Kate, "'Most renewable energy companies' linked with claims of abuses in mines," www.TheGuardian.com, September 5, 2019. https://www.theguardian.com/global-development/2019/sep/05/most-renewable-energy-companies-claims-mines

762 Reuters Staff, "Automakers pledge ethical minerals sourcing for electric cars," www.Reuters.com, November 29, 2017. https://www.reuters.com/article/us-autos-minerals/automakers-pledge-ethical-minerals-sourcing-for-electric-cars-idUSKBN1DT1SK

763

764 Cheng, Vincent K.M., Hammond, Geoffrey, Life-cycle energy densities and land take requirements of various power generators:

A UK perspective, www.ScienceDirect.com, Journal of Energy Institute, Volume 90, Issue 2, April 2017, Pages 201-213. https://www.sciencedirect.com/science/article/pii/S1743967115300921

765 Royal, Todd, "China And India Will Watch The West Destroy Itself – Op-ed," www.EurasiaReview.com, June 5, 2019. https://www.eurasiareview.com/05062019-china-and-india-will-watch-the-west-destroy-itself-oped/#comments

766 International Energy Agency (IEA), World Energy Outlook 2020, www.IEA.org, Part of the World Energy Outlook, Flagship Report – October 2020. https://www.iea.org/reports/world-energy-outlook-2020

767 Mills. Ibid. Daily Caller. 2020.

768 Our Bureau, New Delhi, "In just 8 years, India will overtake China as world's most populous country: UN report," www.TheHinduBusinessLine.com, June 17, 2019. https://www.thehindubusinessline.com/economy/india-will-overtake-china-as-worlds-most-populous-country-in-2027-un-report/article28021947.ece Note the date says 2019, which would make the above sourced sentence incorrect since 8 years was written instead of 7. UN demographic reports, and demographic reports in general need to be given lag time and are sometimes overly optimistic in their assertions has been my takeaway from examining them for this third book.

769 Varadhan, Sudarshan, "Coal to be India's energy mainstay for 30 years – NITI Aayog report," www.In.Reuters.com, May 15, 2017. https://in.reuters.com/article/india-coal-energy/coal-to-be-indias-energy-mainstay-for-next-30-years-niti-aayog-report-idINKCN18B1XE

770 International Energy Agency (IEA), Renewables 2020, Analysis and Forecast to 2025, www.IEA.org, Fuel Report – November 2020. https://www.iea.org/reports/renewables-2020 Please note this source is for lines 2-4 in the paragraph.

771 Cater, Nick, "Gas is fracking hell to protesters," www.TheAustralian.com.au, October 20, 2020. https://www.theaustralian.com.au/commentary/gas-is-fracking-hell-to-protesters/news-story/2fb1fb147cf8d41f666b5d7149751886

772 Guruswamy, Mohan, "Why is India still the world's dirtiest and most polluted country," www.NationalHeraldIndia.com, December 10, 2018. https://www.nationalheraldindia.com/opinion/why-is-india-still-the-worlds-dirtiest-and-most-polluted-country

773 OhAiseadha, C.; Quinn, G.' Connolly, R." Connolly, M.; Soon, W. Energy and Climate Policy-An Evaluation of Global Climate Change

Expenditures 2011-2018. Energies 2020, 13, 4839. https://www.ceres-science.com/content/Renewables.html Please note this is how the source wanted this written out, which is different from previous entries in this chapter.

774 OhAiseadha, Quinn, Connolly. Ibid. 2020.

775 Roberts, Andrews, Churchill: Walking With Destiny, (Penguin Group (USA) LLC, New York, NY), Page 721, November 6, 2018, www.Amazon.com. https://www.amazon.com/Churchill-Walking-Destiny-Andrew-Roberts-ebook/dp/B079R3VH13/ref=sr_1_3?dchild=1&keywords=churchill&qid=1605894413&sr=8-3

776 Roberts. Ibid. Page 721. 2018.

777 Roberts. Ibid. Page 721. 2018.

778 Roberts. Ibid. Page 721. 2018.

779 Roberts. Ibid. Page 721. 2018.

780 Roberts. Ibid. Page 721. 2018.

781 Roberts. Ibid. Page 721. 2018.

782 Guruswamy. Ibid. 2018.

783 Varadhan, Sudarshan, "India's coal-fired power output picks up as industrial use rises," www.Reuters.com, September 16, 2020. https://www.reuters.com/article/india-electricity/indias-coal-fired-power-output-picks-up-as-industrial-use-rises-idUSKBN2671DU

784 U.S. Energy Information Administration (EIA), Independent Statistics & Analysis, India, www.EIA.gov, September 30, 2020. https://www.eia.gov/international/analysis/country/IND

785 EIA. Ibid. 2020.

786 Dr. Schernikau, Lars, "The truth behind renewable energy," www.WattsUpWithThat.com, republished from the October 2020 issue of International Cement Review. https://wattsupwiththat.com/2020/10/18/the-truth-behind-renewable-energy/

787 International Energy Outlook (IEA), World Energy Outlook 2019, www.IEA.org, Flagship Report – November 2019, Page 734. https://www.iea.org/reports/world-energy-outlook-2019

788 EIA. Ibid. 2020.

789 EIA. Ibid. 2020.

790 O'Brien, Chip, "To Dam Or Un-Dam California? Now That's a Complicated Question," www.ActiveNorcal.com, November 22, 2020. https://activenorcal.com/to-dam-or-un-dam-california-now-thats-a-complicated-question/ (Entire paragraph comes from this source.)

791 O'Brien. Ibid. 2020.

792 Ridler, Keith, "US gives first-ever OK for small commercial nuclear reactor," www.APNews.com, September 2, 2020. https://apnews.com/article/910766c07afd96fbe2bd875e16087464

793 Columbia Law School, Sabin Center for Climate Change Law, Climate Regulations in a Biden Administration, www.Climate.Law.Columbia.Edu, August 2020. https://climate.law.columbia.edu/sites/default/files/content/Climate%20Reregulation%20in%20a%20Biden%20Administration.pdf

794 Rudgard, Olivia, "Climate change makes solar less reliable in sunniest parts of the world," www.Telegraph.co.UK, October 8, 2020. https://www.telegraph.co.uk/news/2020/10/08/climate-change-make-sola r-less-reliable-sunniest-parts-world/

795 Dears, Don, "The Limits of Solar Power," www.DDears.com, September 25, 2020. https://ddears.com/2020/09/25/the-limits-of-solar-power/

796 StopTheseThings.com, "How Much CO2 Gets Emitted to Build a Wind Turbine?" www.StopTheseThings.com, August 16, 2014. https://stopthesethings.com/2014/08/16/how-much-co2-gets-emitte d-to-build-a-wind-turbine/

797 Ambrose, Jillian, "Investors plan major move into renewable energy infrastructure," www.TheGuardian.com, November 23, 2020. https://www.theguardian.com/environment/2020/nov/23/investors-plan-major-move-into-renewable-energy-infrastructure?mkt_tok=eyJpIjoiWmpWa1ltUXlZakl5WTJJdyIsInQiOiI3Ykx3NnRvcGw5WXF1ZXVJVVh6M1wvXC81eDlKeEt0dkoraTZiYktqQm9JQzhSTkxHKzhlbG9Ha3JpcXpKMDBkOVRJQUIwcG5jVW5Ud0ZDSzBnWXByeklqVDVPcHZGRTFnZEhpWCt5blZYZE40RUgyTDVUdzhKVWRPb1wva2ZSWFpGbyJ9

798 Reuters Staff, "Modi says India set to double oil refining capacity in five years, earlier than expected," www.Reuters.com, November 21, 2020. https://www.reuters.com/article/us-india-energy-modi/modi-says-india-set-to-double-oil-refining-capacity-in-5 -years-earlier-than-expected-idUSKBN2810FN

799 Slav, Irina, "China Looks To Surpass The U.S. As The World's Biggest Oil Refiner," www.OilPrice.com, November 23, 2020. https://oilprice.com/Energy/Energy-General/China-Looks-To-Surpass-Th e-US-As-The-Worlds-Biggest-Oil-Refiner.html

800 Tamborrino, Kelsey, "Biden's trans-Atlantic truce?" www.Politico.com, November 23, 2020. https://www.politico.com/newsletters/morning-energy/2020/11/23/bidens-trans-atlantic-truce-791829
801 Reuters Staff, "Germany earmarks $10 billion for hydrogen," www.Reuters.com, June 4, 2020. https://www.reuters.com/article/us-health-coronavirus-germany-stimulus/germany-earmarks-10-billion-for-hydrogen-expansion-idUSKBN23B10L
802 Archibald, David, "The Pure Evil of Hydrogen Hyping," www.WattsUpWithThat.com, September 24, 2020. https://wattsupwiththat.com/2020/09/24/the-pure-evil-of-hydrogen-hyping/?unapproved=3091580&moderation-hash=15fcdf4212c15ec20233e98b2fabbc4f
803 Constable, John, Hydrogen The once and future fuel? www.GWPF.org, (Global Warming Policy Foundation), July 2020. https://www.thegwpf.org/content/uploads/2020/06/Hydrogen-Fuel.pdf
804 Forbes, Viv, "Not So Green," www.SaltBushClub.com, September 11, 2020. https://saltbushclub.com/2020/09/11/not-so-green/
805 West, Andy, via Climate Etc., "Cultural motivations for wind and solar renewables deployment," www.JudithCurry.com, November 19, 2020. https://judithcurry.com/2020/11/19/cultural-motivations-for-wind-and-solar-renewables-deployment/
806 Dr. Lehr, Jay, "Wind and solar add zero value to the grid," www.Cfact.org, April 30, 2020. https://www.cfact.org/2020/04/30/wind-and-solar-add-zero-value-to-the-grid/
807 Gosselin, Pierre, "Analysis Suggests Elon Musk's Vision Of A Battery-Powered Society Remains An Unworkable Fantasy," www.NoTricksZone.com, October 30, 2020. https://notrickszone.com/2020/10/30/analysis-suggests-elon-musks-vision-of-a-battery-powered-society-remains-an-unworkable-fantasy/
808 Institute for Energy Research (IER), The Environmental Impact of Lithium Batteries, www.InstituteforEnergyResearch.com, November 12, 2020. https://www.instituteforenergyresearch.org/renewable/the-environmental-impact-of-lithium-batteries/
809 IER. Ibid. 2020.
810 Katwala, Amit, "The spiraling environmental cost of our lithium battery addiction, www.Wired.co.UK, August 5, 2018. https://www.wired.co.uk/article/lithium-batteries-environment-impact
811 Katwala. Ibid. 2018.

812 Wojick, David, CLINTEL challenges McKinsey's climate alarmism," www.Cfact.org, October 25, 2020. https://www.cfact.org/2020/10/25/clintel-challenges-mckinseys-climate-alarmism/
813 McKinsey & Company, McKinsey on Climate Change, www.McKinsey.com, September 2020. https://www.mckinsey.com/~/media/McKinsey/Business%20Functions/Sustainability/Our%20Insights/McKinsey%20on%20Climate%20Change/McKinsey-on-Climate-Change-Report-v2.pdf
814 McKinsey & Company. Ibid. 2020.
815 Groom, Nichola, Hiller, Jennifer, "Biden administration pauses federal drilling program in climate push," www.Reuters.com, January 21, 2021. https://www.reuters.com/article/us-usa-drilling-interior/biden-administration-suspends-federal-oil-and-gas-permitting-idUSKBN29Q2N1
816 Kennedy, Brian T., "Facing Up to the China Threat," www.Imprimus.Hillsdale.edu, September 2020, Volume 49, Number 9. https://imprimis.hillsdale.edu/facing-china-threat/
817 Ghoshal, Devjyot, for Reuters, "India's top general warns that an ongoing face-off with China could spark a much bigger conflict," www.BusinessInsider.com, November 6, 2020. https://www.businessinsider.com/india-general-warns-china-border-dispute-could-escalate-to-war-2020-11
818 Kennedy. Ibid. 2020.
819 Paraskova, Tsvetana, "China's Oil Imports Are Falling From Record Highs," www.OilPrice.com, November 4, 2020. https://oilprice.com/Energy/Energy-General/Chinas-Oil-Imports-Are-Falling-From-Record-Highs.html
820 Wikipedia, "Socialism with Chinese Characteristics," www.en.Wikipedia.com, Page accessed on November 30, 2020. https://en.wikipedia.org/wiki/Socialism_with_Chinese_characteristics
821 Kennedy. Ibid. 2020.
822 Kennedy. Ibid. 2020.
823 Kennedy. Ibid. 2020.
824 Associated Press (AP) News, "China, top global emitter, aims to go carbon-neutral by 2060," www.APNews.com, September 22, 2020. https://apnews.com/article/climate-climate-change-paris-xi-jinping-emissions-reduction-7a4216ad4026090adb8d600fab210406

825 Institute for Energy Research (IER), "On Carbon Neutrality Pledge, China Says One Thing, but Does Another," www.InstituteforEnergyResearch.org, September 24, 2020. https://www.instituteforenergyresearch.org/international-issues/on-carbon-neutrality-pledge-china-says-one-thing-but-does-another/

826 Centre for Research on Energy and Clean Air (CREA), New Coal Boom in China: Country Accelerates New Coal Plant Permitting And Proposals, Global Energy Monitor, www.GlobalEnergyMonitor.org, Briefing: June 2020. https://globalenergymonitor.org/wp-content/uploads/2020/06/A-New-Coal-Boom-in-China_English.pdf

827 IER. Ibid. 2020.

828 CREA. Ibid. 2020.

829 CREA. Ibid. 2020.

830 British Petroleum (BP) Statistical Review of World Energy 2020 | 69th edition, www.BP.com, June 2020. https://www.bp.com/content/dam/bp/business-sites/en/global/corporate/pdfs/energy-economics/statistical-review/bp-stats-review-2020-full-report.pdf

831 BP Statistical Review of World Energy 2020. Ibid. 2020.

832 Shearer, Christine, "New Report – Boom and Bust 2020: Tracking the Global Coal Plant Pipeline," www.EndCoal.org, March 25, 2020. https://endcoal.org/2020/03/new-report-global-coal-power-under-development-declined-for-fourth-year-in-a-row/

833 Reuters Staff, "CORRECTED-China to cap 2020 coal-fired power capacity at 1,100 GW," www.Reuters.com, June 17, 2020. https://www.reuters.com/article/china-energy-coal/corrected-china-to-cap-2020-coal-fired-power-capacity-at-1100-gw-idUSL4N2DV0ZE

834 The World Bank | Data, "Electricity production from coal sources (% of total), www.Data.WorldBank.org, Page accessed December 1, 2020. https://data.worldbank.org/indicator/EG.ELC.COAL.ZS

835 Multiple Authors, CarbonBrief, "Global coal power," www.CarbonBrief.org, March 26, 2020. https://www.carbonbrief.org/mapped-worlds-coal-power-plants

836 McLaughlin, Timothy "Joe Biden Has a Barack Obama Problem," www.TheAtlantic.com, November 17, 2020. https://www.theatlantic.com/international/archive/2020/11/barack-obama-joe-biden/617116/

837 McLaughlin. Ibid. 2020.

838 Gold, Russell, The Boom: How Fracking Ignited the American Energy Revolution and Changed the World," (Simon & Schuster, New York, NY), Entire book, April 21, 2015. https://www.

amazon.com/Boom-Fracking-Ignited-American-Revolution/dp/1451692293/ref=sr_1_2?s=books&ie=UTF8&qid=1544497960&sr=1-2&keywords=the+boom

839 McCarthy, Tom, "Barack Obama appoints Derek Mitchell as first ambassador to Burma," www.TheGuardian.com, May 17, 2012. https://www.theguardian.com/world/2012/may/17/obama-administration-burma-ambassador

840 BBC News, "Obama Asia tour: US-Japan treaty 'covers disputed islands,'" www.BBC.com, April 24, 2014. https://www.bbc.com/news/world-asia-27137272

841 Sanger, David E., "Commitments on Three Fronts Test Obama's Foreign Policy," www.NYTimes.com, September 3, 2014. https://www.nytimes.com/2014/09/04/world/europe/commitments-on-3-fronts-test-obamas-foreign-policy-doctrine.html

842 Emont, Jon, Gale, Alastair, "Asia-Pacific Countries Sing Major Trade Pact in Test for Biden," www.WSJ.com, November 15, 2020. https://www.wsj.com/articles/asia-pacific-nations-sign-major-china-backed-trade-deal-11605434779

843 McLaughlin. Ibid. 2020.

844 BBC News, "India-China dispute: India hands over soldier who crossed border," www.BBCNews.com, October 21, 2020. https://www.bbc.com/news/world-asia-india-54610763

845 Everington, Keoni, "Xi Jingping (Chinese President & Head of the Chinese Communist Party) orders Chinese soldiers not to fear death," www.TaiwanNews.com.TW, November 27, 2020. https://www.taiwannews.com.tw/en/news/4063776

846 Bengali,Shashank, Uyen Bao Kieu, Vo, "Sunken boats. Stolen gear. Fishermen are prey as China conquers a strategic sea: Beijing's aggressive South China Sea expansion shows its willingness to defy international laws for President Xi Jingping's visions of power," www.LATimes.com, November 12, 2020. https://www.latimes.com/world-nation/story/2020-11-12/china-attacks-fishing-boats-in-conquest-of-south-china-sea

847 Su, Alice, "Dreams of a Red Emperor: The relentless rise of Xi Jingping," www.LATimes.com, October 22, 2020. https://www.latimes.com/world-nation/story/2020-10-22/china-xi-jinping-mao-zedong-communist-party

848 Kerr, Andrew, "Biden's Energy Nominee Divvied Millions In Taxpayer Funds To Alternative Energy Startups That Went Bankrupt," www.

DailyCaller.com, January 27, 2021. https://dailycaller.com/2021/01/27/jennifer-granholm-michigan-energy-companies-bankrupt/
849 Ring, Edward, "John Kerry, Climate Czar in Waiting: A punitively high cost of living is the result of conscious political choices, and the primary force behind these choices is not desire to protect the environment, it is greed," www.AmGreatness.com (American Greatness), November 26, 2020. https://amgreatness.com/2020/11/26/john-kerry-climate-czar-in-waiting/ (Entire paragraph comes from this excellent source. I recommend Mr. Ring highly.)
850 Bradley Jr., Robert, "A Free-Market Energy Vision," www.MasterResource.org, November 4, 2020. https://www.masterresource.org/free-market-energy-overview/free-market-energy-vision-2/
851 Zaremba, Haley, "China Plans To Dominate The Global Nuclear Energy Push," www.OilPrice.com, June 3, 2020. https://oilprice.com/Alternative-Energy/Nuclear-Power/China-Plans-To-Dominate-The-Global-Nuclear-Energy-Push.html
852 Wald, Ellen R., "The U.N. Says America Is Already Cutting So Much Carbon It Doesn't Need The Paris Climate Accords," www.Forbes.com, December 10, 2020. https://www.forbes.com/sites/ellenrwald/2020/12/10/the-un-makes-the-case-for-the-us-to-stay-out-of-the-paris-climate-accord/?sh=34e9008227e5
853 Foster, Peter, "Peter Foster: The IEA's solar spin cycle," www.FinancialPost.com, October 21, 2020. https://financialpost.com/opinion/peter-foster-the-ieas-solar-spin-cycle
854 Reuters Staff, "UK to ban sale of new petrol and diesel cars from 2030: FT," www.Reuters.com, November 14, 2020. https://www.reuters.com/article/britain-autos/uk-to-ban-sale-of-new-petrol-and-diesel-cars-from-2030-ft-idUSKBN27U0DG
855 Kelly, Michael, Until we get a proper roadmap, Net Zero is a goal without a plan, June 8, 2020, https://capx.co/until-we-get-a-proper-roadmap-net-zero-is-a-goal-without-a-plan/
856 Henk Car Sales Statistics, January 16, 2020, https://www.best-selling-cars.com/international/2019-full-year-international-worldwide-car-sales/
857 Green Car Reports, https://www.greencarreports.com/news/1093560_1-2-billion-vehicles-on-worlds-roads-now-2-billion-by-2035-report
858 Bryce. Ibid. 2020.
859 Bryce. Ibid. 2020.

860 Brouillette, Dan, "If EU is serious, it should use more US liquified gas, not less," www.EUObserver.com, December 1, 2020. https://euobserver.com/opinion/150227
861 Slav, Irina, "Natural Gas Demand Will Grow For Decades To Come," www.OilPrice.com, November 3, 2020.
862 Lomberg, Bjorn, Welfare in the 21st century: Increasing development, reducing inequality, the impact of climate change, and the cost of climate policies, www.ScienceDirect.com, Volume 156, July 2020, 119981. https://www.sciencedirect.com/science/article/pii/S0040162520304157
863 Cvrk, Stu, "The Continuing Democrat-Media Narrative That There Is "No Evidence of Election Fraud" Is a Bald-Faced Lie," www.RedState.com, January 28, 2021. https://redstate.com/stu-in-sd/2021/01/28/the-continuing-democrat-media-narrative-that-there-was-no-election-fraud-is-a-bald-faced-lie-n317564
864 Brouillette. Ibid. 2020.
865 Brouillette. Ibid. 2020.

CHAPTER SEVEN: FINANCIAL AND ENVIRONMENTAL RACIAL BIASING

866 Nigel Bruce, Rogelio Perez-Padilla, & Rachel Albalak, Indoor air pollution in developing countries: a major environmental and public health challenge, https://www.who.int/bulletin/archives/78(9)1078.pdf
867 Brad Plumer, Brad, Washington Post, All of the world's power plants, in one handy map, Dec. 8, 2012, https://www.washingtonpost.com/news/wonk/wp/2012/12/08/all-of-the-worlds-power-plants-in-one-handy-map/?noredirect=on
868 Plumer, Bird, Washington Post, All of the world's power plants, in one handy map, December 8, 2012, https://www.washingtonpost.com/news/wonk/wp/2012/12/08/all-of-the-worlds-power-plants-in-one-handy-map/?noredirect=on
869 Coal-fired Power Stations by Region, Global Coal Plant Tracker, January 2021, https://docs.google.com/spreadsheets/d/1ZPbbwBI1cNoS1NqFEnN8PdXGRO0xuYfza1F0AWUDRuY/edit#gid=739846511
870 Coal-fired Power Stations by Country, Global Coal Plant Tracker, January 2021, https://docs.google.com/spreadsheets/

d/1kXtAw6QvhE14_KRn5lnGoVPsHN3fDZHVMlvz_s_ch1w/edit#gid=165011444

871 Global Coal Plant Tracker, July 2020, https://docs.google.com/spreadsheets/d/1I8GeKEfxPpwkQ_t0GQZx1GQm6MASclEtEtrQX3Y1nNc/edit#gid=0

872 Hiroko Tabuchi, New York Times, As Beijing Joins Climate Fight, Chinese Companies Build Coal Plants, July 1, 2017, https://www.nytimes.com/2017/07/01/climate/china-energy-companies-coal-plants-climate-change.html

873 SUNMetrix, What is capacity factor and how do solar and wind energy compare? Data source for the table: Average Capacity Factors by Energy Source, 1998 through 2009, U.S. Energy Information Administration, April 2011, July 25, 2013, https://sunmetrix.com/what-is-capacity-factor-and-how-does-solar-energy-compare/

874 Carmin Chappell, CNBC, Climate change in the US will hurt poor people the most, according to a bombshell federal report, November 26, 2018, https://www.cnbc.com/amp/2018/11/26/climate-change-will-hurt-poor-people-the-most-federal-report.html

875 Shah, Anup, Global Issues, Poverty Facts and Stats, January 7, 2013, https://www.globalissues.org/article/26/poverty-facts-and-stats

876 Worldometer, Countries in the world by population (2021), https://www.worldometers.info/world-population/population-by-country/

877 United States Agency for International Development., Issues in Poverty Reduction and Natural Resource Management, October 2006, https://www.usaid.gov/sites/default/files/documents/1862/issues-in-poverty-reduction-and-natural-resource-management.pdf

878 Do Something.org, 11 FACTS ABOUT GLOBAL POVERTY, 10% of the world's population lives on less than $1.90 a day, https://www.dosomething.org/us/facts/11-facts-about-global-poverty

879 UNICEF, Goal: Reduce child mortality, https://sites.unicef.org/mdg/childmortality.html

880 UNITED NATIONS CONFERENCE ON TRADE AND DEVELOPMENT , COMMODITIES AT A GLANCE Special issue on strategic battery raw materials, https://unctad.org/en/PublicationsLibrary/ditccom2019d5_en.pdf

881 Amnesty International, Industry giants fail to tackle child labour allegations in cobalt battery supply chains, November 15, 2017, https://www.amnesty.org/en/latest/news/2017/11/industry-giants-fail-to-tackle-child-labour-allegations-in-cobalt-battery-supply-chains/

882 700 Refineries Supply Oil Products to the World, https://hrcak.srce.hr/file/65010
883 Wayne B. Gray, Ronald J. Shadbegian, and Ann Wolverton, National Center for Environmental Economics, Environmental Justice: Do Poor and Minority Populations Face More Hazards? https://www.epa.gov/sites/production/files/2014-12/documents/environmental_justice_do_poor_and_minority_populations_face_more_hazards.pdf
884 Hensley, Billy, National Endowment for Financial Education, Racism, Bias and Economic Inequality Impair the Financial Well-Being of Millions, July 2, 2020, https://www.nefe.org/news/the-latest/2020/racism-bias-economic-inequality-impair-financial-well-being-of-millions.aspx
885 Robin Broad and John Cavanag, Science Direct, Poorer Countries and the Environment: Friends or Foes? https://www.sciencedirect.com/science/article/pii/S0305750X15000662
886 Bailey, Ronald, End of Doom: Environmental Renewal in the Twenty-first Century, (Thomas Dunne Books, New York, NY), Entire book for the source, www.Amazon.com, July 21, 2015. https://www.amazon.com/End-Doom-Environmental-Renewal-Twenty-first-ebook/dp/B00O79WFOO/ref=sr_1_1?dchild=1&keywords=ronald+bailey+the+end+of+doom&qid=1607896251&s=digital-text&sr=1-1
887 Zerpa, Fabiola, Millard, Peter, Rosati, Andrew, "Toxic Spills in Venezuela Offer a Bleak Vision of the End of Oil: Maduro's government is squeezing what it can from the collapsing industry – and unleashing an environmental disaster in the process," www.Bloomberg.com, December 15, 2020. https://www.bloomberg.com/news/features/2020-12-15/oil-spills-in-venezuela-offer-bleak-vision-of-what-lies-ahead
888 Bruce, Nigel, Perez-Padilla Rogelio, Albalak, Rachel, Indoor air pollution in developing countries: a major environmental and public health challenge, www.WHO.int, April 2018, entire report for the source. https://www.who.int/indoorair/publications/bulletin/en/
889 Nuclear Energy Agency (NEA), Projected Costs of Generating Electricity – 2020 Edition, www.OECD-NEA, Ongoing from April 2015. https://www.oecd-nea.org/jcms/pl_28612/projected-costs-of-generating-electricity
890 Global Warming Policy Forum, Press Release, "New data casts doubt on Boris Johnson's 2030 offshore wind plans," www.TheGWPF.org,

November 8, 2020. https://www.thegwpf.org/new-data-casts-doub t-on-boris-johnsons-2030-offshore-wind-plans/

891 Homewood, Paul, "Coal Outperforms Wind Power In UK Wind Week!" www.NotaLotofPeopleKnowThat.WordPress.com, November 27, 2020. https://notalotofpeopleknowthat.wordpress. com/2020/11/27/coal-outperforms-wind-power-in-uk-wind-week/

892 Morison, Rachel, "U.K. Power Prices Jump After National Grid Warns of Supply Risks," www.Bloomberg.com, December 5, 2020. https://www.bloomberg.com/news/articles/2020-12-05/u-k-power-pr ices-jump-after-grid-warns-of-narrowing-reserves

893 StopTheseThings.com via Pierre Gosselin's NoTricksZone.com, "RE Reckoning: Germans Suffer Europe's Highest Power Prices – 300,000 Families Can't Afford Electricity," www.StopTheseThings. com, December 17, 2020. https://stopthesethings.com/2020/12/17/ re-reckoning-germans-suffer-europes-highest-power-pr ices-300000-families-cant-afford-electricity/

894 Balch, Oliver, "The curse of 'white oil': electric vehicles dirty secret," www.TheGuardian.com, December 8, 2020. https://www. theguardian.com/news/2020/dec/08/the-curse-of-white-oil-electri c-vehicles-dirty-secret-lithium

895 StopTheseThings.com via The Australian, "Massive & Endless Subsidies Destroy Claim That Wind & Solar Cheapest Of All," www. StopTheseThings.com, December 13, 2020. https://stopthesethings. com/2020/12/13/massive-endless-subsidies-destroy-claim-that-win d-solar-cheapest-of-all/comment-page-1/

896 StopTheseThings.com via The Australian, "Familiar Territory: Chief Minister Busted Lying About Staggering Cost of Renewable Energy Roadmap," www.StopTheseThings.com, December 14, 2020. https://stopthesethings.com/2020/12/14/familiar-territor y-chief-minister-busted-lying-about-staggering-cost-of-renewa ble-energy-roadmap/

897 Armentano, Dominick T., "Are Temperature and Sea Levels Rising Dangerously? Hardly," www.ClimateChangeDispatch.com, December 11, 2020. https://climatechangedispatch.com/are-temperatures-an d-sea-levels-rising-dangerously-hardly/

898 National Oceanic and Atmospheric Administration, National Centers for Environmental Information, National Climate Report – Annual 2019, www.NCDC.NOAA.gov, Page accessed December 14, 2019. https://www.ncdc.noaa.gov/sotc/national/201913

899 National Ocean Service, National Oceanic and Atmospheric Administration, U.S. Department of Commerce, Is sea level rising? Yes, sea level is rising at an increasing rate," www.OceanService.NOAA.gov, Page accessed December 13, 2020. https://oceanservice.noaa.gov/facts/sealevel.html

900 Armentano. Ibid. 2020.

901 Armentano. Ibid. 2020.

902 Institute for Energy Research (IER), Commentary, "China's New Export Control Law," www.InstituteforEnergyResearch.org, December 8, 2020. https://www.instituteforenergyresearch.org/international-issues/chinas-new-export-control-law/

903 Faculty of Science – University of Copenhagen, "Satellite images confirm uneven impact of climate change," www.ScienceDaily.com, November 26, 2020. https://www.sciencedaily.com/releases/2020/11/201126085919.htm

904 Turner, Daniel, "TURNER: Biden's Green Energy Plan Will Leave The US Dependent On China," www.DailyCaller.com, December 3, 2020. https://dailycaller.com/2020/12/03/turner-bidens-green-energy-plan-will-leave-the-us-dependent-on-china/

905 JunkScience.com, "Communist China to run Biden climate policy?" www.JunkScience.com, December 16, 2020. https://junkscience.com/2020/12/communist-china-to-run-biden-climate-policy/

906 JunkScience.com. Ibid. 2020.

907 Adams, Patricia, The Red and the Green China's Useful Idiots, The Global Warming Policy Foundation, Briefing 51, www.TheGWPF.org, December 2020. https://www.thegwpf.org/content/uploads/2020/12/Green-reds.pdf

908 Adams. Ibid. 2020.

909 Adams. Ibid. 2020.

910 Devine, Miranda, "US companies riddled with members of Chinese Communist Party: Devine," www.NYPost.com, December 13, 2020. https://nypost.com/2020/12/13/us-companies-riddled-with-members-of-chinese-communist-party/

911 Jacobs, Emily, "Top Chinese professor boasts of operatives in top of US 'core inner circle,'" www.NYPost.com, December 8, 2020. https://nypost.com/2020/12/08/professor-claims-china-has-people-in-americas-core-inner-circle/

912 Dr. Peiser, Benny, "Green Groups Are China's 'Useful Idiots'" www.ClimateChangeDispatch.com, December 11, 2020. https://climatechangedispatch.com/green-groups-are-chinas-useful-idiots/
913 Wood, Graeme, Glacier Media, "B.C. politicians should lift mute button on China's human rights: top critic," www.PRPeak.com, October 22, 2020. https://www.prpeak.com/b-c-politicians-shoul d-lift-mute-button-on-china-s-human-rights-top-critic-1.24225580
914 Canada House of Commons, Special Committee on Canada-China Relations, Number 007, 1st Session, 43rd Parliament, Evidence, Monday, 24 February 2020 at https://www.ourcommons.ca/DocumentViewer/en/43-1/ CACN/meeting-7/evidence.
915 Canada House of Commons, Special Committee on Canada-China Relations, Number 008, 1st Session, 43rd Parliament, Evidence, Monday, 9 March 2020 at https://www.ourcommons.ca/DocumentViewer/en/43-1/ CACN/meeting-8/evidence.
916 Energy via FT.com (Financial Times. There is a paywall which prevented me from accessing the original source. Link at the end of this source will give you the opportunity to access www.FT.com and pay for the original material if desired.), "Green power needs to account for all its costs," www.NewsCabal.co.UK, September 7, 2020. https://www.newscabal.co.uk/green-power-needs-to-accoun t-for-all-its-costs/
917 Meyer, Gregory, Waters, Richard, "Californians face dark, hot summer as green energy is sapped: Warning of rolling blackouts for millions as heatwave overwhelms grid," www.FT.com, August 18, 2020. https://www.ft.com/content/27f4c697-8698-42be-b91d-9d7c8def3031
918 Preston, Bryan, "Toyota CEO Agrees With Elon Musk: We Don't Have Enough Electricity to Electrify All the Cars," www.PJMedia.com, December 21, 2020. https://pjmedia.com/news-and-politics/bryan-preston/2020/12/21/toyota-ceo-agrees-with-elon-musk-we-dont-have-enough-electricity-to-electrify-all-the-cars-n1222999
919 Volcovici, Valerie, "U.S. green groups say honeymoon is over, turn up heat on Biden," www.in.Reuters.com, December 7, 2020. https://in.reuters.com/article/usa-biden-green/u-s-green-groups-say-honeym oon-is-over-turn-up-heat-on-biden-idUSKBN28H1CC
920 Wall Street Journal Editorial Board, "Big Oil to the Coronavirus Rescue: Look whose products are crucial for fighting off COVID-19," www.WSJ.com, April 23, 2020. https://www.wsj.com/articles/big-oi l-to-the-coronavirus-rescue-11587683239

921 Volcovici. Ibid. 2020.
922 Ring, Edward, "California's Cruel Green Cramdown," www.AmGreatness.com, (American Greatness), December 11, 2020. https://amgreatness.com/2020/12/11/californias-cruel-green-cramdown/
923 Gosselin, Pierre, "1.35 Million Tonnes of "Hazardous Materials", Germany Admits No Plan To Recycle Used Wind Turbine Blades," www.NoTricksZone.com, November 21, 2020. https://notrickszone.com/2020/11/21/1-35-million-tonnes-of-hazardous-material-germany-admits-no-plan-to-recycle-used-wind-turbine-blades/
924 Gosselin, Pierre, "Environment of Dystopia: Germany Plans To Wipe Out 20 Million Sq M of 1000-Year Old Forest, For Wind Parks!" www.NoTricksZone.com, December 8, 2020. https://notrickszone.com/2020/12/08/environment-of-dystopia-germany-plans-to-wipe-out-20-million-sq-m-of-1000-year-old-forest-for-wind-parks/
925 Diouhy, Jennifer A., Natter, Ari, "Biden Gets Earful From Progressives on Environment Jobs," www.Bloomberg.com, December 10, 2020. https://www.bloomberg.com/news/articles/2020-12-10/progressives-push-biden-to-pick-activists-for-environmental-jobs
926 Bryce, Robert, Narrator, Writer, Producer, Culver, Tyson, Director, Writer, Producer, Documentary Film: Juice: How Electricity Explains the World," www.JuicetheMovie.com, available on iTunes and Amazon Prime Video, Released 2019. http://juicethemovie.com
927 Murphy, Peter, "Biden-Harris administration taking shape – beware," www.Cfact.org, December 17, 2020. https://www.cfact.org/2020/12/17/biden-harris-administration-taking-shape-beware/
928 Johnson, Collister, "The folly of "climate emergency,"" www.Cfact.org, July 27, 2019. https://www.cfact.org/2019/07/27/the-folly-of-climate-emergency/
929 Bell, Larry, "EPA's McCarthy admits regs (regulations) are for show, not results!" www.Cfact.org, September 23, 2013. https://www.cfact.org/2013/09/23/epas-mccarthy-admits-regs-are-for-show-not-results/
930 Murphy. Ibid. 2020.
931 Russell Mead, Walter, "How American Fracking Changes the World," www.WSJ.com, November 26, 2018. https://www.wsj.com/articles/how-american-fracking-changes-the-world-1543276935
932 Kandrach, Matthew, "Fracking in NM helps create jobs, lower carbon emissions," www.LCSUN-News.com, (Las Cruces Sun News), January 8, 2020. https://www.lcsun-news.com/story/opinion/2020/01/08/fracking-nm-helps-create-jobs-lower-carbon-emissions/2833434001/

933 Dr. Lehr, Jay, Harris, Tom, "Brace for nationwide blackouts under Biden," www.Cfact.org, December 8, 2020. https://www.cfact.org/2020/12/08/brace-for-nationwide-blackouts-under-biden/
934 Kluger, Jeffrey/Time Books, "130 Years After Hitler's Birth, He Continues to Live as a Symbol of Evil," www.Time.com, Originally published April 19, 2019, Updated December 12, 2019. https://time.com/5573720/hitler-world-influence/
935 Zycher, Benjamin, "The climate empire strikes out: The perils of policy analysis in an echo chamber," www.AEI.org, September 26, 2018. https://www.aei.org/research-products/report/the-climate-empire-strikes-out-the-perils-of-policy-analysis-in-an-echo-chamber/
936 Darwall, Rupert, "China's Green NGO Climate Propaganda Enablers," www.RealClearEnergy.org, December 21, 2020. https://www.realclearenergy.org/articles/2020/12/21/chinas_green_ngo_climate_propaganda_enablers_654042.html
937 United Nations Environment Programme and UNEP DTU Partnership, Emissions Gap Report 2020, www.UNEP.org, December 9, 2020. https://www.unep.org/emissions-gap-report-2020
938 Wald, Ellen R., "The U.N. Says America Is Already Cutting So Much Carbon It Doesn't Need The Paris Climate Accord," www.Forbes.com, December 10, 2020. https://www.forbes.com/sites/ellenrwald/2020/12/10/the-un-makes-the-case-for-the-us-to-stay-out-of-the-paris-climate-accord/?sh=34e9008227e5
939 Ring, Edward, "Abundance, Not Scarcity, Can Be the Immediate Future for Humanity," www.AmGreatness.com, December 22, 2020. https://amgreatness.com/2020/12/22/abundance-not-scarcity-can-be-the-immediate-future-for-humanity/
940 Squires, Delano, "Dear Democrats: Stop Treating Black Men Like We're Stupid Or Lose More Votes," www.TheFederalist.com, December 22, 2020. https://thefederalist.com/2020/12/22/dear-democrats-stop-treating-black-men-like-were-stupid-or-lose-more-votes/
941 U.S. Senator Merkley, Jeff, "How Joe Biden can act boldly to address the climate crisis," www.WashingtonPost.com, December 21, 2020. https://www.washingtonpost.com/opinions/2020/12/21/jeff-merkley-biden-climate-crisis-executive-action/
942 Gingrich, Newt, "Why I will not accept Joe Biden as president," www.WashingtonTimes.com, December 21, 2020. https://www.washingtontimes.com/news/2020/dec/21/why-i-will-not-accept-joe-biden-as-president/

943 Marks, Allan, "Big Stimulus For Clean Energy: Covid Relief Bill To Include Bipartisan Support For Green Tax Credits," www.Forbes.com, December 21, 2020. https://www.forbes.com/sites/allanmarks/2020/12/21/clean-energy-investments-get-a-bipartisan-boost-from-congress-in-relief-bill/?sh=235bc2d536f2
944 Landers, Peter, "Toyota Chief Says Electric Vehicles Are Overhyped," www.WSJ.com, December 17, 2020. https://www.wsj.com/articles/toyotas-chief-says-electric-vehicles-are-overhyped-11608196665
945 Landers. Ibid. 2020.
946 Clear Energy Alliance, "EV Dream," www.ClearEnergyAlliance.com, via www.YouTube.com, February 28, 2020. https://www.youtube.com/watch?v=GaesOIY3I4U&feature=youtu.be
947 Richardson, Valerie, "'Devastating': Study warns Western states would pay for Biden drilling ban," www.WashingtonTimes.com, December 15, 2020. https://www.washingtontimes.com/news/2020/dec/15/joe-biden-drilling-ban-devastating-western-states-/
948 Dr. Considine, Timothy J., Professor of Economics, School of Energy Resources, University of Wyoming, The Fiscal and Economic Impacts of Federal Onshore Oil and Gas Lease Moratorium and Drilling Ban Policies, www.Wyoenergy.org, December 14, 2020. https://www.wyoenergy.org/wp-content/uploads/2020/12/Final-Report-Federal-Leasing-Drilling-Ban-Policies-121420.pdf
949 Richardson. Ibid. 2020.
950 Finley, Allysia, "Now We're Cooking With Gas. But Tomorrow?" www.WSJ.com, December 16, 2020. https://www.wsj.com/articles/now-were-cooking-with-gas-but-tomorrow-11608160284?mod=opinion_lead_pos9
951 Hayes, Jason, "Biden's energy and climate appointments: A return to 'the swamp,'" www.TheHill.com, December 17, 2020. https://thehill.com/opinion/energy-environment/530586-bidens-energy-and-climate-appointments-a-return-to-the-swamp#.X9zQyryaJE0.twitter
952 Rotter, Charles, from the New York Times and used by WattsUpWithThat.com, "1,600 new coal-fired power plants are planned or under construction in 62 countries," www.WattsUpWithThat.com, July 3, 2017. https://wattsupwiththat.com/2017/07/03/forget-paris-1600-new-coal-power-plants-built-around-the-world/
953 Hartnett White, Kathleen, Dr. Rossiter, Caleb Stewart, New-tech American Coal-fired Electricity for Africa: Clean Air, Indoors and Out, www.CO2Coalition.org, November 2020. https://co2coalition.

org/wp-content/uploads/2020/11/American-Coal-fired-Electricit y-for-Africa.pdf

954 Inskeep, Steve, Westerman, Ashley, "Why is China Placing A Global Bet On Coal?" www.NPR.org, (U.S. National Public Radio), April 29, 2019. https://www.npr.org/2019/04/29/716347646/why-is-chin a-placing-a-global-bet-on-coal

955 Flanakin, Duggan, "China Recolonizes Africa," www.Townhall.com, December 23, 2020. https://townhall.com/columnists/dugganflanakin/2020/12/23/china-recolonizes-africa-n2582000

956 Driessen, Paul, Mamula, Ned, "Democrats' Green New Deal would make US reliance on China much worse," www.WattsUpWithThat.com, July 24, 2020. https://wattsupwiththat.com/2020/07/24/democrat s-green-new-deal-would-make-us-reliance-on-china-much-worse/

957 Dr. Lehr, Jay, "The potential power of politicians is astounding," www.Cfact.org, October 13, 2020. https://www.cfact.org/2020/10/13/th e-potential-power-of-politicians-is-astounding/

958 Johnson, Drew, "The North Face is the new king of environmental hypocrisy," www.WashingtonExaminer.com, December 22, 2020. https://www.washingtonexaminer.com/opinion/op-eds/the-north-fac e-is-the-new-king-of-environmental-hypocrisy

959 Slav, Irina, "Is The Energy Transition Creating A Major Investment Bubble?" www.OilPrice.com, December 28, 2020. https://oilprice.com/Energy/Energy-General/Is-The-Energy-Transitio n-Creating-A-Major-Investment-Bubble.html

960 Wojick, David, "CLINTEL study finds most of the CO2 increase is natural," www.Cfact.org, December 19, 2020. https://www.cfact.org/2020/12/19/clintel-study-finds-most-of-the-co2-increase -is-natural/

961 Bohleber, P., Schwikowski, M., Stocker-Waldhuber, M. et al. New glacier evidence for ice-free summits during the life of the Tyrolean Iceman. Sci Rep 10, 20513 (2020). https://doi.org/10.1038/s41598-020-77518-9. Original http found at: https://www.nature.com/articles/s41598-020-77518-9#citeas (Authors requested this source be cited this way)

962 Wojick, David, "New York can't buy its way out of blackouts," www.Cfact.org, December 24, 2020. https://www.cfact.org/2020/12/24/new-york-cant-buy-its-way-out-of-blackouts/

963 Wojick. Ibid. 2020.

964 Wojick. Ibid. 2020. Entire paragraph for the source.

965 Wojick. Ibid. 2020.

966 Taraldsen, Lars Erik, Paulsson, Lars, Starn, Jesper, “Wind Farm Backlash Grows in Oil-Rich Norway Ahead of Election,” www.Bloomberg.com, December 9, 2020. https://www.bloomberg.com/news/articles/2020-12-09/wind-farm-backlash-grows-in-oil-rich-norway-ahead-of-election

967 Ring. Ibid. 2020.

968 Plumer, Bird, Washington Post, All of the world’s power plants, in one handy map, December 8, 2012, https://www.washingtonpost.com/news/wonk/wp/2012/12/08/all-of-the-worlds-power-plants-in-one-handy-map/?noredirect=on

969 Coal-fired Power Stations by Region, Global Coal Plant Tracker, January 2021, https://docs.google.com/spreadsheets/d/1ZPbbwBI1cNoS1NqFEnN8PdXGRO0xuYfza1F0AWUDRuY/edit#gid=739846511

970 Coal-fired Power Stations by Country, Global Coal Plant Tracker, January 2021, https://docs.google.com/spreadsheets/d/1kXtAw6QvhE14_KRn5lnGoVPsHN3fDZHVMlvz_s_ch1w/edit#gid=165011444

Printed in the United States
by Baker & Taylor Publisher Services